HIGH-ENERGY SOLAR PHENOMENA—A NEW ERA OF SPACECRAFT MEASUREMENTS

AIP CONFERENCE PROCEEDINGS 294

HIGH-ENERGY SOLAR PHENOMENA—A NEW ERA OF SPACECRAFT MEASUREMENTS

WATERVILLE VALLEY, NH 1993

EDITORS: JAMES M. RYAN
W. THOMAS VESTRAND
INSTITUTE FOR THE STUDY OF EARTH, OCEANS, AND SPACE
UNIVERSTIY OF NEW HAMPSHIRE

American Institute of Physics New York

Authorization to photocopy items for internal or personal use, beyond the free copying permitted under the 1978 U.S. Copyright Law (see statement below), is granted by the American Institute of Physics for users registered with the Copyright Clearance Center (CCC) Transactional Reporting Service, provided that the base fee of $2.00 per copy is paid directly to CCC, 27 Congress St., Salem, MA 01970. For those organizations that have been granted a photocopy license by CCC, a separate system of payment has been arranged. The fee code for users of the Transactional Reporting Service is: 0094-243X/87 $2.00.

© 1994 American Institute of Physics.

Individual readers of this volume and nonprofit libraries, acting for them, are permitted to make fair use of the material in it, such as copying an article for use in teaching or research. Permission is granted to quote from this volume in scientific work with the customary acknowledgment of the source. To reprint a figure, table, or other excerpt requires the consent of one of the original authors and notification to AIP. Republication or systematic or multiple reproduction of any material in this volume is permitted only under license from AIP. Address inquiries to Series Editor, AIP Conference Proceedings, AIP, 500 Sunnyside Boulevard, Woodbury, NY 11797-2999

L.C. Catalog Card No. 93-074147
ISBN 1-56396-291-8
DOE CONF-9303219

Printed in the United States of America.

CONTENTS

Preface ix

γ-RAY OBSERVATIONS AND THE ROLE OF PROTONS IN FLARES

X-Ray and Gamma-Ray Observations of Solar Flares by GRANAT 3
    G. Trottet
OSSE Observations of the 4 June 1991 Solar Flare 15
    R. J. Murphy, G. H. Share, J. E. Grove, W. N. Johnson,
    R. L. Kinzer, R. A. Kroeger, J. D. Kurfess, M. S. Strickman,
    S. M. Matz, W. R. Purcell, M. P. Ulmer, D. A. Grabelsky,
    G. V. Jung, C. M. Jensen, D. J. Forrest, and W. T. Vestrand
An Overview of Solar Flare Results from COMPTEL 21
    M. McConnell
Theoretical Models for High-Energy Solar Flare Emissions 26
    R. Ramaty and N. Mandzhavidze
High-Energy Gamma-Ray Signature of Proton Acceleration During
1991 June 15 Solar Flare... 45
    G. E. Kocharov, E. I. Chuikin, G. A. Kovaltsov,
    I. G. Usoskin, and L. G. Kocharov
COMPTEL's Solar Flare Catalog 51
    R. Suleiman, D. Forrest, M. McConnell, J. Ryan, R. Diehl,
    G. Lichti, G. Rank, V. Schönfelder, A. Strong, M. Varendorff,
    K. Bennett, L. Hanlon, C. Winkler, H. Bloeman, W. Hermsen,
    and B. Swanenburg
COMPTEL Observations of Gamma-Ray Flares in October 1991 55
    M. Varendorff, D. Forrest, M. McConnell, J. Ryan,
    R. Suleiman, R. Diehl, G. Lichti, G. Rank, V. Schönfelder,
    K. Bennett, L. Hanlon, C. Winkler, and B. N. Swanenburg
A Search for Low-Energy Protons in a Solar Flare from October 1992:
Preliminary Results 59
    T. Metcalf, D. Mickey, R. Canfield, and J.-P. Wülser
A Correlation Between 4–8 MeV Gamma-Ray-Line Fluence and >50 keV
X-Ray Fluence in Large Solar Flares 65
    E. W. Cliver, N. B. Crosby, and B. R. Dennis
Trapping of Protons in Twisted Magnetic Loops 71
    Y.-T. Lau and R. Ramaty

LONG DURATION EVENTS AND EXTENDED ACCELERATION

Solar Flare Neutrons and Gamma-Rays 77
    R. E. Lingenfelter
**Neutron and Gamma-Ray Measurements of the Solar Flare
of 1991 June 9.** 89
    J. Ryan, D. Forrest, J. Lockwood, M. Loomis, M. McConnell,
    D. Morris, W. Webber, K. Bennett, L. Hanlon, C. Winkler,
    H. Debrunner, G. Rank, V. Schönfelder, and B. N. Swanenburg
**EGRET Observations of Extended High-Energy Emissions
from the Nuclear Line Flares of June 1991** 94
    E. J. Schneid, K. T. S. Brazier, G. Kanbach, C. von Montigny,
    H. A. Mayer-Hasselwander, D. L. Bertsch, C. E. Fichtel,
    R. C. Hartman, S. D. Hunter, D. J. Thompson, B. L. Dingus,
    P. Sreekumar, Y. C. Lin, P. F. Michelson, P. L. Nolan,
    D. A. Kniffen, and J. R. Mattox
Observations of the 1991 June 11 Solar Flare with COMPTEL 100
    G. Rank, R. Diehl, G. G. Lichti, V. Schönfelder, M. Varendorff,
    B. N. Swanenburg, R. van Dijk, D. Forrest, J. Macri, M. McConnell,
    M. Loomis, J. Ryan, K. Bennett, and C. Winkler
**Some Evidences of Prolonged Particle Acceleration in the High-Energy
Gamma-Ray Flare of June 15, 1991.** 106
    V. V. Akimov, N. G. Leikov, A. V. Belov, I. M. Chertok, V. G. Kurt,
    A. Magun, and V. F. Melnikov
"Extended Phase" of Solar Flares Observed by SMM 112
    P. P. Dunphy and E. L. Chupp
**Numerical Modeling of Particle Transport and Acceleration
of Solar Flare Particles in a Coronal Loop** 118
    E. Bennett, M. A. Lee, and J. M. Ryan
Directivity of Gamma-Rays from Neutral Pi-Mesons Decay 124
    D. Heristchi and R. Boyer
The GAMMA-1 Data on the March 26, 1991 Solar Flare 130
    V. V. Akimov, N. G. Leikov, V. G. Kurt, and I. M. Chertok
Stochastic Fermi Acceleration and Solar Cosmic Rays 134
    M. A. Lee
Gamma-Rays from an "Over-the-Limb" Flare 143
    W. T. Vestrand and D. J. Forrest

YOHKOH OBSERVATIONS/HIGH-ENERGY ELECTRON PHENOMENA

The Yohkoh Context for High-Energy Particles in Solar Flares 151
    H. S. Hudson
Acceleration of Electrons in Solar Flares 162
    V. Petrosian

Temporal Evolution of Bremsstrahlung-Dominated Gamma-Ray
Spectra of Solar Flares 171
    H. Marschhäuser, E. Rieger, and G. Kanbach
EGRET Observation of the June 30 and July 2, 1991 Energetic
Solar Flares... 177
    B. L. Dingus, P. Sreekumar, D. L. Bertsch, C. E. Fichtel,
    R. C. Hartman, S. D. Hunter, D. J. Thompson, E. J. Schneid,
    K. T. S. Brazier, G. Kanbach, C. von Montigny, H. A. Mayer-Hasselwander,
    Y. C. Lin, P. F. Michelson, P. L. Nolan, D. A. Kniffen, and J. R. Mattox
Observations of Small Solar Flares with BATSE 183
    D. A. Biesecker, J. M. Ryan, and G. J. Fishman
Energetic Electron Injection into the High Corona During
the Gradual Phase of Flares: Evidence Against Acceleration
by a Large Scale Shock 187
    K.-L. Klein and G. Trottet
New Technique for Dynamic Burst Position Determination with
High Spatial Definition/Time Resolution at a MM-Wavelength 193
    J. E. R. Costa, E. Correia, P. Kaufmann, R. Herrmann,
    and A. Magun
Energetic Electron Populations in Solar Flares 199
    S. M. White

NEUTRON MONITOR AND GLE EVENTS

Neutron Monitor Measurements as a Complement to Space
Measurements of Energetic Solar Particle Fluxes 207
    H. Debrunner
The Relativistic Solar Proton Ground-Level Enhancements
Associated with the Solar Neutron Events of 11 June
and 15 June 1991 222
    D. F. Smart, M. A. Shea, and L. C. Gentile
The NASA High-Energy Solar Physics Mission (HESP).................... 230
    B. R. Dennis, A. G. Emslie, R. Canfield, G. Doschek, R. P. Lin,
    and R. Ramaty
Author Index 245

PREFACE

There has been an explosion in the number of gamma-ray flare observations since the Solar Maximum Mission satellite started monitoring flares in 1980. During the current Solar Cycle exciting measurements have been made by all four instruments on the Compton Gamma-Ray Observatory as well as by GRANAT, Yohkoh, SMM, and other spacecraft. There have also been major advances in ground-based measurements which support and provide context information for the high-energy measurements. As a result, the study of high-energy solar phenomena has reached a level of maturity and complexity that has outgrown single sessions at general meetings like the Compton Symposia. The realization that a forum was needed for the detailed presentation and examination of these new measurements provided the motivation for organizing a narrowly focused workshop: "High-Energy Solar Phenomena—A New Era of Spacecraft Measurements."

During the first week of March in 1993 scientists from more than ten countries met in Waterville Valley, New Hampshire to attend this workshop which was hosted by the University of New Hampshire. The generous support provided by the Compton Gamma-Ray Observatory project at Goddard Space Flight Center allowed us to gather together a group of 50 experts which included a representative of nearly every instrument that measured high-energy flare emission during Cycle 22. Review and contributed talks were presented on the new flare measurements as well as talks on the theory of high-energy processes in flares. As the papers published in this conference proceedings demonstrate, the detailed discussions that took place were some of the most productive in recent years. They also made it clear that regular workshops dealing with these high-energy solar measurements would be very productive. The high-energy processes that proximity allows us to study in detail for this nearby star are certainly at play elsewhere in the Universe. We are confident that what one learns from the study of high-energy solar phenomena enhances our understanding of all disciplines in high-energy astrophysics.

James M. Ryan and W. Thomas Vestrand
Durham, New Hampshire

γ-RAY OBSERVATIONS AND THE ROLE OF PROTONS IN FLARES

X-RAY AND GAMMA-RAY OBSERVATIONS OF SOLAR FLARES BY GRANAT

G. Trottet
Observatoire de Paris-Meudon DASOP CNRS-ERS76 F92190 Meudon

ABSTRACT

The soviet mission GRANAT was launched on 1 December 1989. This review presents observations of hard X-ray/gamma-ray solar flares obtained with the french experiments PHEBUS and SIGMA which provide spectral data in the (0.075-124) MeV energy range. After a brief description of the two instruments, the main temporal and spectral characteristics of the detected bursts are illustrated and discussed. Some specific events are described in more details and compare, when possible, with CGRO observations.

INTRODUCTION

The soviet satellite GRANAT has been launch on 1 December 1989 on an excentric orbit (2000-200000 km). It will remain in operation till at least the end of 1993. Solar flare observations are obtained in a wide photon energy domain (6 keV-124) MeV by the KONUS-B (0.01-8 MeV)[1], the WATCH (6-180 keV)[2], the SIGMA (0.25-15 MeV)[3] and the PHEBUS (0.075-124 MeV)[4] experiments. The GRANAT observations of XR/GR solar events are the only one available from the end of the SMM mission to the launch of the CGRO satellite and later on of YOHKOH. The GRANAT observations are thus essential because they allow: (i) to get a continuous recording of high energy solar flares since 1980; (ii) to compare observations of the same flares obtained with detectors using different technologies. In this paper we give a brief overview of the hard X-ray/gamma-ray (HXR/GR) flares detected by PHEBUS and SIGMA. The capabiblities of both instruments are summarized in the next section. The three following sections illustrate and discuss the temporal charcteristics, the GR line emission and the high energy continuum emission (above 10 MeV) of HXR/GR solar flares observed with GRANAT. Concluding remarks are made in the last section.

INSTRUMENTS

The orbital period of GRANAT is 4.05 days. The passage of the satellite through the Earth's radiation belts, during which data are not recorded, represents about 30% of the orbital period. Both PHEBUS and SIGMA switch automatically to a burst mode when the count rate exceed the background level by a given number of σ's.

1 The PHEBUS experiment

The PHEBUS experiment has been described in details in[4,5]. It consists of 6 BGO cylindrical cristals (78 mm in diameter by 120 mm in height) surrounded by a plastic anti coincidence shield. Depending on the pointing direction of GRANAT, the sensitive area viewing the Sun varies from about 187 cm^2 to 280 cm^2. The nominal energy range covered by PHEBUS is (0.075-124) MeV. The instrument works in two modes:

a) Waiting mode: For each detector the records consist in: (i) a continuous time history of the (0.1-1.6) MeV count rate with a time resolution of 64 s; (ii) 2 background spectra, in 225 energy channels, recorded every 4 hours. These data provide in flight calibration of the detectors and may be used to determine the background level for solar flares[5].

b) Burst mode: For each detector the following data are recorded:

- integral count rate in the (0.075-1.6) MeV range with a time resolution varying from 1/128 s to 1/32 s during the first 86.6 s of the burst.

- time intervals for the accumulation of 24 photons in the (0.075-1.6) MeV band (time to spill mode).

- 176 spectra in 116 channels in the time to spill mode.

- 640 spectra in 40 channels with a time resolution of 1 s below 10 MeV (26 channels) and of 4 s above 10 MeV (14 channels).

Since the launch of GRANAT more than 80 solar bursts with significant flux above 100 keV have been detected by PHEBUS.

2 The SIGMA burst detector

The anti-coincidence shield of the SIMA telescope, which is used as a burst detector, has been descibed in[6]. It is made of CsI cristals of 3 and 4 cm thickness with a total area of 3000 cm^2 viewing the Sun. This area is divided into two 1500 cm^2 sectors which are analysed independently. The covered energy range is roughly (0.25-15) MeV. The following data are recorded in the burst mode:

- integral (0.25-2) MeV count rate with a time resolution of 8 ms during 30 s.

- 12288 time intervals for the acquisition of 32 photons in the (0.25-2) MeV band.

- one spectrum in 80 channels integrated over the 128 s preceeding the time of the burst mode trigger.

- spectra in 80 channels with accumulation times ranging from 0.125 s to 1 s for the first minute of the burst.

- for the following 10 minutes count rates are recorded in 80 channels every 20 s and in 25 channels every second.

Complementary data are obtained with the Gamma camera which continuously records the count rates in 4 energy channels every 4 s. These data cannot be used for spectral analysis because the photons reach the camera after propagation through the CsI cristals so that their energies are not accuratly known. Due to the high energy threshold of the SIGMA anti-coincidence shield, only 20 solar flares have been triggered since the launch of GRANAT.

TEMPORAL CHARACTERISTICS

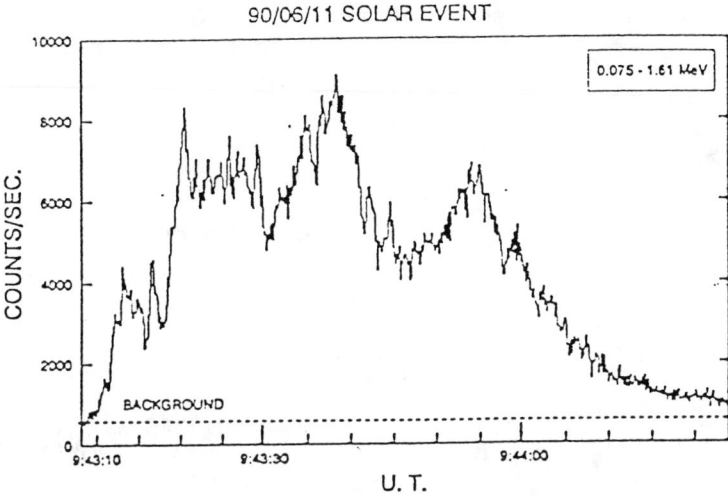

Figure 1: Time profile of the 11 June 1990 observed with PHEBUS (from[5]). Each bin is 125 ms.

Figure 1 shows the time profile of the 11 June 1990 flare in the (0.075-1.6) MeV range with a time accumulation of 0.125 ms. Subsecond time variations are evident. 20 peaks with durations ranging from 0.125 to 0.625 ms were found to be significant[7]. The existence of subsecond time structures seems to be a general property of impulsive (duration <5 min). HXR bursts detected by PHEBUS around 100 keV[5,7]. Figure 2

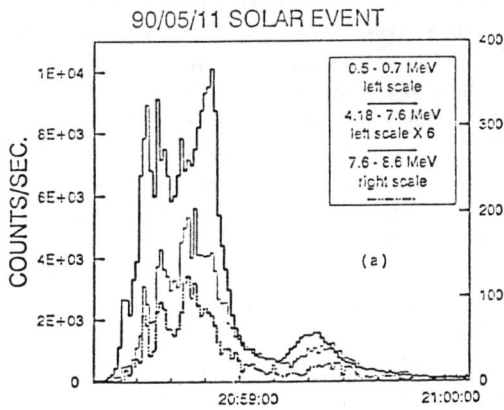

Figure 2: Time profiles (1 s time resolution) of the 11 May 1990 flare observed with PHEBUS (from[13]).

displays the time profile of the 11 May 1990 event. This event shows one second duration peaks which are significant up to the (7.6-8.6) MeV channel. As one second corresponds to the highest time resolution available these peaks are not resolved in time. This suggests that subsecond time variations may also exist in the MeV domain. These findings are in agreement with the more sensitive observations of BATSE and OSSE

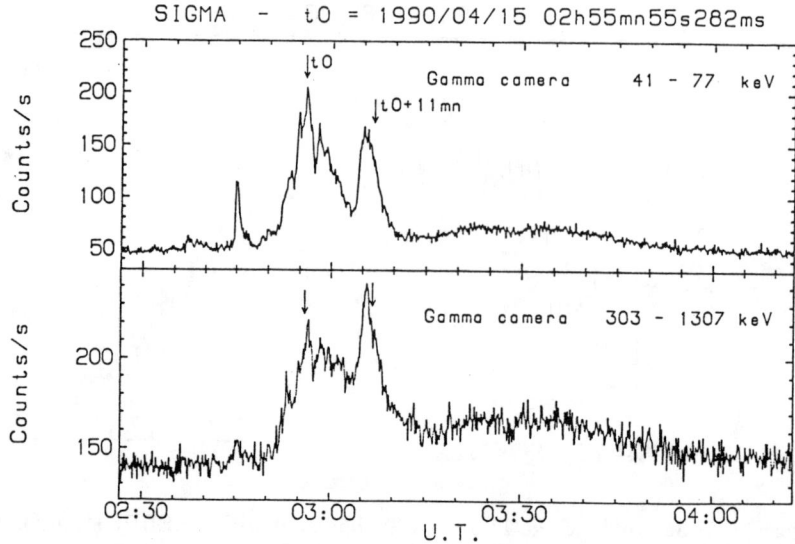

Figure 3: Time profiles of the 15 April 1990 flare observed with the SIGMA Gamma-camera (from[6]). The arrows indicate where burst mode data were recorded.

which detect subsecond spikes in many events[8,9,10]. On the other hand they contrast with HXRBS/SMM results which indicate that less than 10% of the HXR events show subsecond time structures around 30 keV[11]. This apparent discrepancy may be due to: (i) the fact that the analysis of the SMM data was performed at an energy (about 30 keV) where the contribution of the hot thermal plasma may be large while it is negligible at higher energies; (ii) a selection effect since the burst mode is generally triggered for events with fast rise times.

Contrary to impulsive bursts, longer duration events (> 5 min.) do not, in general, exhibit fast temporal variations[5,6,7]. Figure 3 shows the time profile of the 15 April 1990 event recorded by the SIGMA gamma-camera. The energy range indicated in the figure is only indicative. This event is the longest one detected by GRANAT. It consists of a succession of peaks of one to several minutes typical duration for a period of about 30 min.. These peaks are superposed on a hard gradual component which is detected from the very beginning of the event to its end. The existence of two timescales in long duration events have been observed for several flares and it was found that the slowly varying component is sometimes the hardest one[5,6,7]. During this event the PHEBUS was triggered twice at 02:44:43 and 03:04:21 UT. The Sigma burst mode was also triggered at 02:55:55 UT. This illustrates the complementarity of PHEBUS and SIGMA observations. Indeed the combination of both sets of data provide spectral observations over a time interval which is longer than that covered by each instrument. These spectral data indicate that the HXR spectrum hardens between peaks at 02:56 UT and at 03:06 UT and that the time of the maximum of this latter peak increases with energy. There is about a 20 s delay betweem the maximum in the (0.2-8) MeV band and that in the (0.8-15) MeV band[5,6]. The existence of such delays, which has been reported in the past[12], is observed in several GRANAT events.

The systematic detection of subsecond time variations up to MeV energies during impulsive events confirms that HXR producing particles must interact in dense regions ($>10^{11} cm^{-3}$ for electrons) where the time to loose their energy by collisions is smaller than the typical duration of the HXR spikes. Multi wavelength studies of flares combining radio and HXR/GR observations indicate that the magnetic structures in which particles are released and propagate suffer changes on timescales of 10 s or less[13,14]. This supports the idea that subsecond variations more likely reflect the characteristic timescale of the acceleration rather than some leakage time from a single region where particles are accelerated during the whole flare. This is further supported by the observed similarity between the time histories of the production rate of decimeter radio spikes and of the HXR emission[15]. The absence of fast time variations in long duration HXR/GR events does not necessarily imply long timescales for the acceleration. This may simply indicate that particles interact in less dense regions where the collisional lifetime is longer than the acceleration time. This idea is supported by models of long duration flares[16,17] and by observations in the HXR and radio domains, see e.g.[18]. Such a picture is consistent with the existence of two componenets with different temporal and spectral characteristics: the low energy and more rapidly varying one may reflect partial precipitation from different regions, while the high energy and slowly varying one reflects more efficient trapping of higher energy particles in the same regions.

OBSERVATIONS OF GAMMA-RAY LINE EMISSION

Several events with canonical GR line spectra have been observed by PHEBUS and SIGMA. In this section the PHEBUS observations of the 11 June 1990 flare are taken as an example of such GR line events and discussed in some details. Figure 4 shows the count spectrum measured around the maximum of the event (C. Barat private coomunication). It clearly exhibits strong emission of the largest prompt GR lines and of the delayed neutron capture line at 2.23 MeV.

Figure 5 displays the time histories of the HXR emission in the (180-310) keV and (310-540) keV energy ranges. It also shows the (4.2-7.3) MeV excess count rates, which is entirely due to prompt GR line emission, and the (2.13-2.38) MeV excess count rate which is almost entirely due to the emission of the 2.23 MeV line[19]. These excess count rates have been estimated by substracting the bremssstrahlung continuum spectrum fitted to a power law in the (0.6-1) MeV region. Figure 5 shows that:

> The HXR emission lasts for more than 10 min. and presents peaks of 1 min. typical duration. No energy dependent delays are observed in the peak times even

8   Observations of Solar Flares by GRANAT

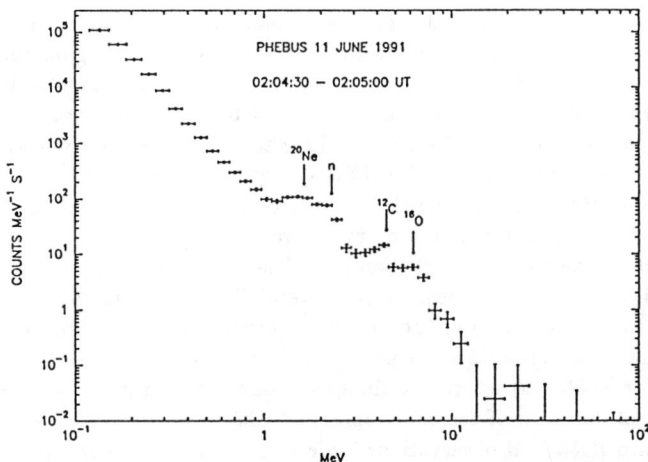

Figure 4: Spectrum of the 11 June 1991 burst recorded by PHEBUS (C. Barat, private communication).

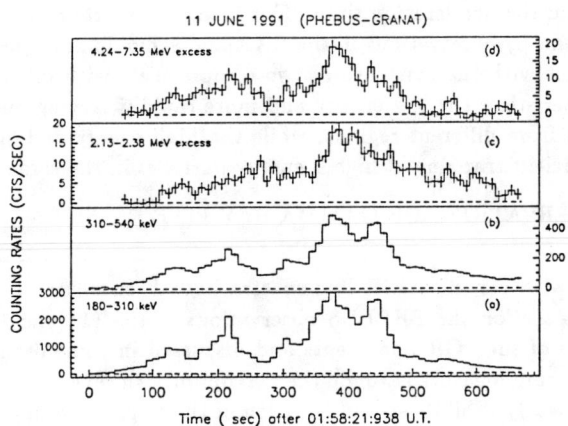

Figure 5: Time profiles (10 s averaged) of the HXR emission and of the excess count rates due to (4.2-7.3) MeV prompt GR lines and to the 2.2 MeV line recorded by PHEBUS during the 11 June 1991 event (from[19]).

with a time resolution of 1 s. Thus efficient coronal trapping of energetic electrons does not play a major role during the main phase of this event (see e.g.[16]).

- The time history of the promt GR line emission, i.e. of the (10-100) MeV/nucl. ions, is similar to that of the HXR emission produced by electrons. This is in agreement with SMM results (see e.g.[20]) and indicates that electrons and ions are produced simultaneously. Consequently most of the (10-100) MeV/nucl. ions also precipitate during the main burst.

- The (2.13-2.38) MeV time profile shows smoother peaks and a slower decay than the (4.2-7.3) MeV one.

The proton spectrum has been estimated in[19] by using calculations performed in[21]. It was found that this spectrum can be represented either by a Bessel function with $\alpha T=0.02$ or by a power of index 3.5. Using the procedure descibed in[22], the 2.23 MeV line emission was found to fall off with a decay time of about 70 s[19]. This value led, in agreement with previous determinations[23,24], to a ^{3}He/H ratio of about 3×10^{-5}. It should be noted that the decay time of the 2.23 MeV line, measured from COMPTEL obervations after 02:13 UT, is of the order of 10 min.[25]. This indicates that the 2.2 MeV line decays with highly different time constants during the main and late phases of this event, the latter being not observed by PHEBUS.

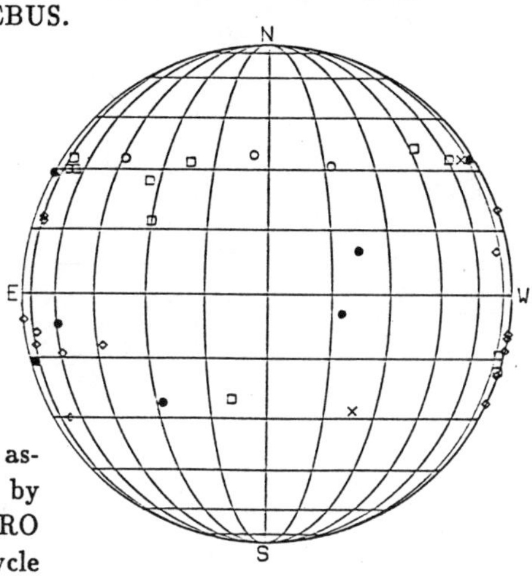

Figure 6: Heliocentic locations of Hα flares associated with >10 MeV emission observed by SMM during Cycle 21 ◊ and by SMM □ , CGRO o , GAMMA-1 X and GRANAT • during Cycle 22.

PHEBUS did not detect any significant flux at energies above 10 MeV (see figure 4). This is probably because of a high orbital background due to enhanced particle fluxes. However EGRET has observed significant photon emission up to GeV energies during more than 8 hours after the main phase[26]. This long lasting high energy photon emission has been successfully interpreted in terms of high energy protons and, in a lesser extent, in terms of relativistic electrons trapped in low density loops[27]. It is remarkable to note the the proton spectrum index, deduced from this model of the late phase, is similar to that obtained from PHEBUS observations during the main phase. However alternative interpretations of the late phase in terms of continuous acceleration of high energy ions cannot be definitely ruled out[28,29,30]. A further investigation of this flare, combining GRANAT and CGRO observations, is necessary to understand the link between the main and late phases.

EVENTS WITH ABOVE 10 MEV CONTINUUM EMISSION

Figure 6 shows the heliocentric distribution of the Hα flares giving rise to GR bursts with above 10 MeV emission. This figure combines observations obtained with SMM[31,32], GAMMA-1[33], CGRO[26] and GRANAT[34,35] during Cycles 21 and 22. It shows that the clustering of above 10 MeV events towards the solar limbs, which is evident for Cycle 21, is not obviously observed for Cycle 22. The question wether Cycle 22 behaves differently or not from Cycle 21 is still open[32,36]. In order to approach this question, we discussed, in the following, two above 10 MeV impulsive flares, with highly different spectral characteristics, that occured respectively close to the disk center and close to the limb.

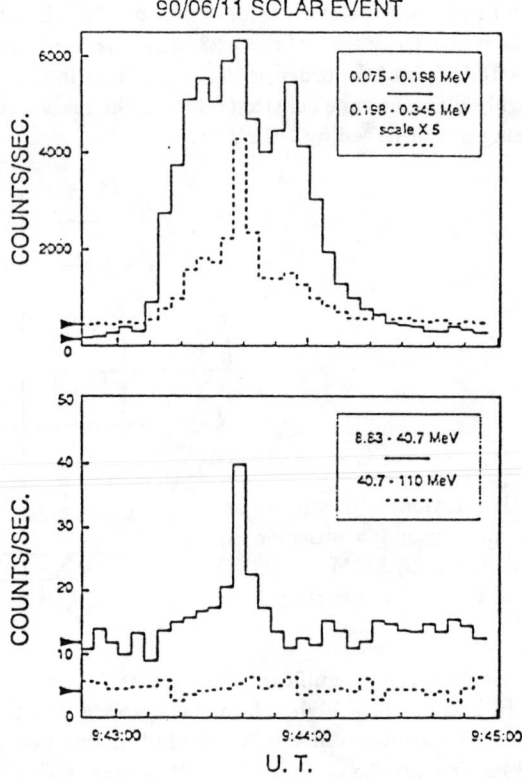

Figure 7: Time profiles (4 s averaged) of the 11 June 1990 event detected by PHEBUS (from[5]).

Figure 7 (from[5]) shows the time profile of the 11 June 1990 event in 4 energy bands. This event was associated with a Hα flare close to the disk center at N10 W22. The emission from the hardest peak extends up to about 40 MeV but no significant emission is detected above this energy. GR line emission is weak, if any. This burst has thus similar characteristics as "electron dominated events"[37]. Other events of this type have been observed by PHEBUS close to the limb as e.g. the 25 January 1991 flare[35].

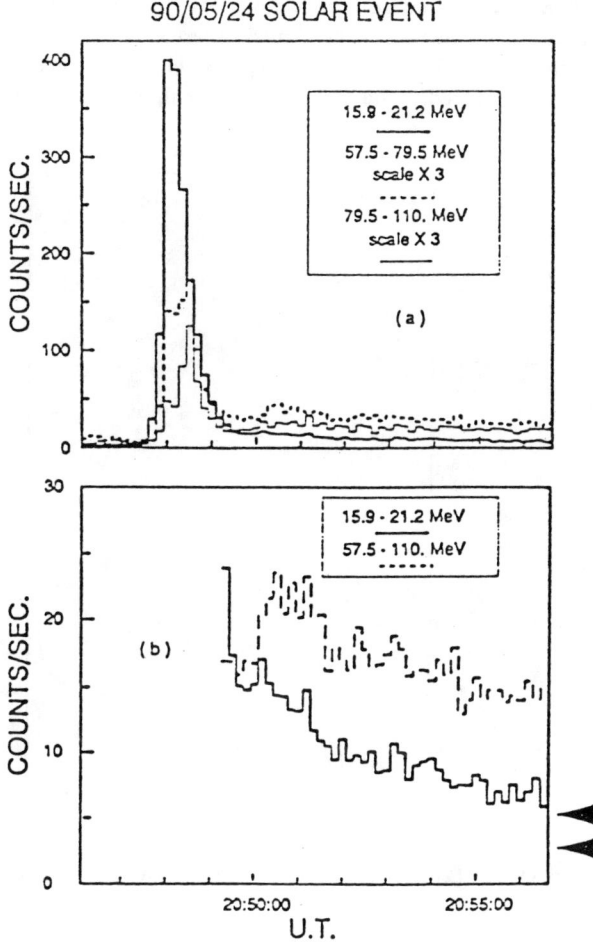

Figure 8: Time profiles (12 s averaged) of the >10 MeV emission detected by PHEBUS during the 24 May 1990 event (from[5]).

Contrary to the above event, the 24 May 1990 flare, shown in figure 8 (from[5]), exhibits a strong emission up to the (80-110) MeV energy range which corresponds to the highest energy band available for this event. The high energy count rate comprises essentially two peaks. While the intensity of the first peak is higher in the (15.9-21.2) MeV range, the two peaks have comparable amplitudes in the (57.5-79.5) MeV domain and the second peak is larger in the (79.5-110) MeV region. This hardening of the spectrum above 60 MeV is consistant with emission from neutral pion decay photons. Such a process may be the dominant one during the second peak[5]. Figure 8 also shows the presence of a long lasting high energy tail which is still present at the end of PHEBUS observations. The enhanced tail emission between 20:50 and 20:52 UT, which is in rough time coincidence with the detection of neutrons at the Earth's ground level[38,39], is certainly the signature of neutrons reaching GRANAT[5]. This ensemble of observations indicates that numerous above 300 MeV/nucl. ions have been produced during the main phase of this flare. Thus, unless there exists some energy cut off in the (10-300) MeV region, which is unlikely, one expects that numerous (10-100) MeV/nucl. ions have also been produced. The striking feature shown in figure 9 (from[6]) is that

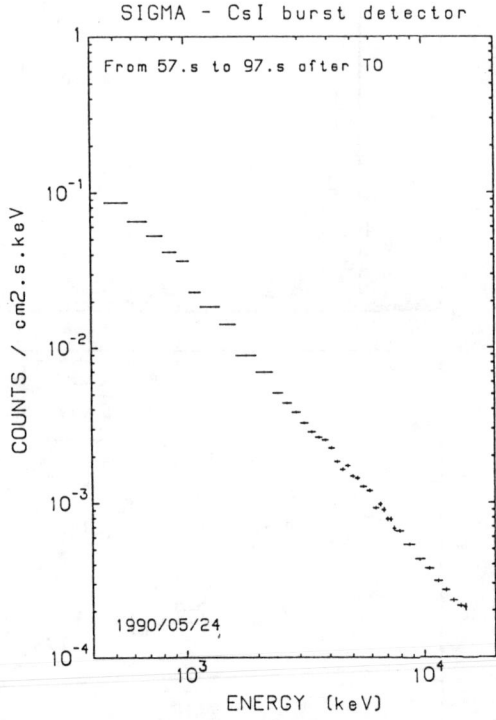

Figure 9: Raw spectrum (20:47:52 - 20:48:32 UT) of the 24 May 1990 event detected with SIGMA (from[6]).

there is no clear excess in the (4-7) MeV prompt GR line domain during the most intense part of this event. However strong 2.23 MeV line emission has been observed during the decay and late phases[40], indicating that (10-100) MeV/nucl. ions have been effectively produced during the main phase of this flare. One event (22 March 1991) with similar spectral characteristics and neutron detection at Earth has been observed close to the center of the disk (S26 E28)[35]. A few other events with intermediate characteristics (emission above 60 MeV, no clear hardening above this energy and no clear prompt line emission during the main phase) have been detected by GRANAT both close to the limbs and on the disk[5,34,35].

In summary GRANAT observations tend to indicate that the spectral properties of HXR/GR bursts with above 10 MeV emission are not strongly dependent on the location of the associated Hα flare. Moreover only weak, if any, prompt GR line emission is detected during the main phase of these events. These findings, which are based on a small sample of events, are not sufficient to conclude that above 10 MeV events are isotropically distributed in longitude. One way to further investigate this problem would be to intercalibrate the different experiments in order to compare the relative fluxes of events with similar spectral characteristics which occur at different longitudes.

CONCLUSION

The PHEBUS and SIGMA observations of solar flares complement those obtained from other satellites and provide unique information on extremely large flares that occured in May-June 1990 and in March 1991. The preliminary conclusions that can be drawn from the previous sections are:

- Particles up to MeV energies are accelerated on subsecond timescales. A HXR/GR burst appears then as the result of many short accelerations. Among other possibilities, this may be taken as an argument in favor of a fragmented accelerator as discussed in[41].

- It seems that events with strong emission above 10 MeV do not show significant GR line emission during their maximum phase. Though a case by case modelling and the detailed knowledge of the instrumental responses would be necessary to reach any firm conclusion, GRANAT observations suggest that this lack of GR line emission does not necessarly imply that these events are electron dominated. Indeed bursts with strong 2.2 MeV line emission in the decay and late phases and with large emission from pion and large neutron emissivity may be, on the contrary, proton dominated. In this latter case prompt GR line emission may be overwhelmed by continuum emission from e.g.: (i) neutral pion decay photons which under certain circumstances may be significant in the GR line domain[42]; (ii) secondary electrons and positrons from charged pion decay; (iii) secondary electrons produced by knock on acceleration on high energy ions.

- The spectral characteristics of HXR/GR bursts with above 10 MeV emission do not strongly depend on the flare longitude. This may simply reflect, as suggested by joint HXR/GR-radio studies of some events (see e.g.[13,14]), that particle acceleration and transport takes place in complex magnetic configurations which are not static during a flare and which vary from flare to flare.

ACKNOWLEDGMENTS

The author thanks C. Barat, P. Mandrou, J. Paul and R. Sunyaev, principal investigators of the PHEBUS and SIGMA experiments, who made the data available. This work was supported by Centre National d'Etudes Spatiales.

REFERENCES

1. S. V.Golenetskii et al., Sov. Astron. Lett., 17,83 (1991).
2. N. Lund, Adv in Space Res., 8, 17 (1991).
3. J. Paul et al., Adv. Space Res., 11, No 8, 289 (1991).
4. C. Barat et al., A.I.P., 170,395 (1988).
5. R. Talon et al., Solar Phys., in press (1993).
6. F. Pelaez et al., Solar Phys., 140, 121 (1992).
7. R. Talon et al., Adv. Space Res., in press (1993).
8. M. E. Machado et al., Adv. in Space Res., in press (1993).
9. M. E. Machado et al., A.I.P., this volume (1993).
10. R. J. Murphy et al., Adv. Space Res., in press (1993).
11. A. L. Kiplinger et al., Astrophys. J., 265, 199 (1983).
12. T. Bai and B. R. Dennis, Astrophys. J., 292, 699 (1985).
13. G. Trottet et al., Adv. Space Res., in press (1993).
14. E. L. Chupp et al., Astron.Astrophys., in press (1993).
15. M. Güdel, M. J. Aschwanden and A. O. Benz, Astron.Astrophys., 251, 285 (1991).
16. E. Hulot et al., Astron.Astrophys., 256, 273 (1992).
17. G. Bruggmann et al., Solar Phys., submitted (1993).
18. K. Kai et al., Solar Phys., 105, 393 (1986).
19. G. Trottet et al., Astron.Astrophys suppl., 97, 337 (1993).
20. E. L. Chupp, Science, 250, 229 (1990).
21. X. M. Hua and R. E. Lingenfelter, Solar Phys. 107, 351 (1987).
22. T. AT. Prince et al., Proc. 18th ICRC, 4, 79 (1983).
23. E. L. Chupp et al., Astrophys. J.,244, L171 (1981).
24. X. M. Hua and R. E. Lingenfelter, Astrophys. J., 319, 555 (1997).
25. J. Ryan et al., Astrophys. J. suppl. submitted (1993).
26. G. Kanbach et al., Astron.Astrophys. suppl., 97, 349 (1993).
27. N. Mandzhavidze and R. Ramaty, Astrophys. J., 396, L111 (1992).
28. D. Forrest, A.I.P., this volume (1993).
29. V. V. Akimov et al., A.I.P., this volume (1993).
30. G. Kocharov, A.I.P. this volume (1993).
31. W. T. Vestrandet al., Astrophys. J., 322, 1010 (1987).
32. W. T. Vestrand, D. J. Forrest and E. Rieger, Proc. 22nd ICRC, SH2.3-5, p. 69 (1991).
33. N. G. Leikov et al., Astron.Astrophys. suppl., 97, 345 (1993).
34. F. Pelaez et al., Proc. 22nd ICRC, SH2.3-11, 89 (1991).
35. N. Vilmer et al. Proc 7th European Meeting on Solar Physics, submitted (1993).
36. R. Ramaty and N. Mandzhavidze, Proc. Compton Symp., in press (1993).
37. E. Rieger and H. Marschhäuser, Proc. 3rd MAX'91/SMM Workshop on Solar Flares, eds; R. M. Winglee and A. L. Kiplinger, p. 68, (1990).
38. K. R. Pyle, M. A. Shea and D. F. Smart, Proc. 22nd ICRC, SH2.2-5, p. 7 (1991).
39. M. A. Shea, D. F. Smart and K. R. Pyle, Geophys; Res. Lett., 18, 1655 (1992).
40. O. Terekhov et al., Adv. Space Res., in press (1993).
41. L. Vlahos, Adv. Space Res., in press (1993).
42. Dj. Heristchi, A.I.P., this volume (1993).

OSSE OBSERVATIONS OF THE 4 JUNE 1991 SOLAR FLARE

R. J. Murphy, G. H. Share, J. E. Grove, W. N. Johnson, R. L. Kinzer,
R. A. Kroeger, J. D. Kurfess, M. S. Strickman
Naval Research Laboratory, Washington, DC 20375

S. M. Matz, W. R. Purcell, M. P. Ulmer, D. A. Grabelsky
Northwestern University, Evanston, IL 60201

G. V. Jung
Universities Space Research Association, Washington, DC 20024

C. M. Jensen
George Mason University, Fairfax, VA 22030

D. J. Forrest, W. T. Vestrand
University of New Hampshire, Durham, NH 03824

ABSTRACT

We present time profiles of the 2.223 MeV neutron-capture line and the 4.44 MeV ^{12}C nuclear-deexcitation line derived from observations of the 4 June 1991 X12+ solar flare obtained by the Oriented Scintillation Spectrometer Experiment (OSSE) on board the Compton Gamma-Ray Observatory *(CGRO)*. We discuss the OSSE instrument, the solar observation mode used during the June period, the data analysis technique employed, and derive an estimate of the accelerated-particle spectrum and a lower limit to the number of interacting particles.

INTRODUCTION

In this paper, we present time profiles of the 2.223 MeV neutron-capture and the 4.44 MeV ^{12}C nuclear-deexcitation lines derived from OSSE observations of the 4 June 1991 X12+ solar flare. Using a combination of data from on- and off-pointed detectors, we have determined these fluxes throughout most of the observable emission period, excluding only three short periods when the detectors were saturated due to intense emission. Measurement of these 2 line fluxes can provide information on conditions at the Sun during the flare. Since the ^{12}C line is prompt, its time profile represents the time profile of the nuclear reactions themselves. Also, the total, time-integrated fluence in either of these two lines is a measure of the number of interacting accelerated particles. Finally, the ratio of the total neutron-capture-line fluence to the total ^{12}C-line fluence is a measure of the flare-averaged kinetic-energy spectrum of the interacting particles.

In June of 1991, solar active region 6659 produced some of the largest *GOES* flare events ever recorded by the satellites. Fortunately, the *CGRO* Phase I Viewing Period 2 began on 30 May with the Sun accessible to the OSSE field of view (FoV). On 1 June, AR 6659 appeared at the East limb and produced an X12+ flare. At the time of flare onset, OSSE was shut down due to a South Atlantic Anomaly transit. However, as a result of the high probability for intense flare production, the Sun was declared an OSSE Target of Opportunity and replaced the existing secondary celestial target. On 4 June, AR 6659 produced a second X12+ flare while OSSE was viewing the Sun. Excellent observations were obtained of the rise, peak and decay of the event. The decay was interrupted by spacecraft night, but observations were resumed at sunrise of the next two orbits and additional observations were obtained.

In Section 2 we discuss the OSSE instrument and the solar observation mode used during the June period. In Section 3 we discuss the analysis technique and in Section 4 we present the data and discuss the results.

OSSE INSTRUMENT AND DATA COLLECTION MODES

OSSE consists of four, independently-oriented phoswich scintillation detectors with both passive and active shielding for reducing background and defining its aperture. The principal detector element is a 33-cm-diameter phoswich (effective area 480 cm^2 at 511 keV), consisting of 10.2-cm-thick NaI and 7.6-cm-thick CsI crystals, optically coupled to each other and viewed by seven photomultiplier tubes. The CsI and NaI pulses are electronically separated by pulse-shape discrimination, thus providing a compact anticoincidence system for charged particles and background γ-rays. The energy resolution is 3.8% full width at half maximum (FWHM) at 6.1 MeV, increasing to 8.2% at 662 keV. The combined sensitivity for narrow-line detection from a 1000-sec observation is 1×10^{-3} photons cm^{-2} sec^{-1}. A tungsten collimator above the phoswich defines a 3.8° × 11.4° FWHM FoV. Each detector is mounted in an independent elevation-angle gimbal which provides 192° of rotation about the spacecraft Y-axis. Surrounding the phoswich and collimator is a NaI annular shield, made up of four segments and having the capability of providing 0.1- to 8-MeV count spectra in 256 channels at ~10% FWHM energy resolution.

When viewing the Sun during the June 1991 period, the instrument was configured to obtain a range of solar data: (1) Count spectra covering the photon-energy range from 0.05 to >200 MeV in 528 channels along with 16 channels of neutron count spectra (>10 MeV) at a spectral accumulation time of ~8.2 sec; (2) Count rates at 16-msec accumulation times in four broad energy windows (200–450 keV, 570–750 keV, 4–7 MeV and >10 MeV); (3) If a BATSE burst trigger was received, 4096 16-msec shield count rates above threshold (~93 keV), with 256 of these accumulated prior to the trigger; and (4) If the BATSE trigger identified the burst as solar, 1000 seconds of shield count spectra at ~32-sec temporal resolution. Two of the detectors maintained a fixed orientation with the Sun in the FoV while the other two "chopped" (i.e., alternately pointed on and off the Sun at ~2-minute intervals) to facilitate background determination and to provide non-saturated spectra for intense flares. The Sun was therefore viewed by 3 of the detectors at all times, while one of the detectors was off-pointed.

DATA ANALYSIS TECHNIQUE

During the three orbits associated with the OSSE observations of the 4 June X12+ flare, the instrument was viewing the Sun during 3:01:56–4:04:52, 4:36:00–5:37:27, and 6:09:32–7:09:53 UT. The corresponding times of satellite day were 3:06:04–4:08:57, 4:39:32–5:42:28 and 6:12:59–7:15:59 UT. OSSE was therefore already viewing the Sun at satellite sunrise for each of these orbits. The onset, peak and end of the *GOES* soft X-ray event was reported to be 3:37, 3:52 and 8:00 UT, respectively. The flare was located at N30E65.

The flare saturated the *GOES* detectors for more than 30 minutes during the peak of the emission. Similarly, the OSSE detectors suffered from several saturation effects. The effects addressed in this analysis are:

(1) Extremely low livetimes were experienced due to the large event rates suffered by both the shields and the central detectors. In this analysis, we have eliminated data for which the fractional livetime fell below 25%. This occurred during three intervals at the peak of the emission: 3:38:23–3:40:35, 3:40:59–3:41:40 and 3:43:27–3:44:08UT.

(2) Pulse-pile-up distortion of the central-detector pulse shapes due to low-energy flare photons interacting simultaneously with an event of interest can cause the event to be rejected by pulse-shape discrimination. The result is a reduction of detector efficiency and a distortion (hardening) of the count spectrum. Pulse pile-up occurred during the intense portions of the flare in those detectors directly viewing the Sun. The off-pointed detector was adequately protected by absorption of the low-energy photons in the shields and collimator. We therefore use data only from the one off-pointed detector during these times, with the count rate adjusted for the off-axis response. We determine the time after which data from the on-pointed detectors can be used by comparing the on-pointed rates of a "chopping" detector to its off-pointed rates (adjusted for off-axis response). When the two rates are in agreement, we use data from the three on-pointed detectors to improve the sensitivity. This occurred for times after 3:53:20 UT.

Background spectra were obtained by propagating an initial background spectrum forward through time using a technique based on the measured variations of background spectra observed in orbit. The initial background spectrum was obtained by summing 8-sec spectra accumulated during the Sun-viewing portion of the orbit prior to the flare-onset orbit. This previous orbit was free of significant solar activity, as determined by inspection of the *GOES* flare reports. Background-subtracted count spectra were then used in the subsequent analyses.

Summed count spectra for each of the 4 detectors were constructed, derived from 8-sec spectra obtained during 3:53:20–4:04:52 UT when the data problems mentioned above were minimal. Each of the 4 summed spectra were fit for the

2.223 and 4.44 MeV lines to provide a fitting template for the 8-sec spectral fits. To improve statistics during the second and third orbits, the 8-sec spectra for each detector were summed into ~2 minute spectra (16 8-sec spectra each). For each detector, the 8-sec spectra from the first orbit and the 2-minute spectra from the last two orbits were then fit for the two line amplitudes, with each line's central energy and width held fixed at the best-fit values obtained from the fit to the summed-spectra. To obtain the corresponding photon fluxes, the line amplitudes were divided by the photopeak effective area (derived from calibration measurements and Monte Carlo calculations) appropriate for the line energy and source position in the FoV. For times when data from the 3 on-pointed detectors could be used, a weighted mean of the 3 fluxes was calculated.

RESULTS AND DISCUSSION

The results are shown in Figures 1 and 2 for the 2.223 and 4.44 MeV lines, respectively, where the derived line fluxes are plotted versus time. The error bars represent 1-σ statistical uncertainties; systematic uncertainties associated with background subtraction are not included. A study of the level of this uncertainty is in progress.

Figures 1 and 2 show that nuclear interactions and the resultant γ-ray emission continue after the peak of the emission (at about 3:42 UT) for more than 110 minutes (until at least 5:37 UT, corresponding to the end of the second daylight period). Measured flux during the third orbit may only represent the level of systematic uncertainty associated with background subtraction rather than actual flux from the flare.

The total, time-integrated fluences of the 2.223 and 4.44 MeV lines are 692 ± 9 and 68 ± 4 photons cm^{-2}, respectively. This 2.223 MeV line fluence is a factor of two larger than the lower limit reported[1] at the January 1992 American Astronomical Society meeting in Atlanta. That estimate was based on fits to data obtained from the on-pointed detectors whose sensitivities had been significantly reduced by pulse-pile-up effects as discussed above. The fluence values reported here, of course, must be still be considered lower limits since we have not accounted for any emission occurring during the data gaps due to satellite night and detector saturation. If the emissions are interpolated linearly between the measured values bracketing each data gap, an additional 230 and 32 photons cm^{-2} would be added to the 2.223 and 4.44 MeV line fluences, respectively. The 2.223 MeV fluence of 692 photons cm^{-2} can be compared to the 314 photons cm^{-2} observed[2] by the Solar Maximum Mission *(SMM)* Gamma-Ray Spectrometer (GRS) from the 3 June 1983 flare.

We use recent theoretical calculations[3,4] to derive estimates for the flare-averaged kinetic-energy spectrum of the interacting, accelerated particles (using the 2.223-to-4.44 MeV line fluence ratio) and the total number of interacting protons (using the total 2.223 MeV line fluence). Since the data coverage is incomplete, the total-number estimate will be a lower limit. The ratio, however, should be reasonably independent of the unknown emission profile during the data gaps.

Assuming a power-law form for the particle spectrum, we find that the measured ratio (10.2 ± 0.6) implies an index of ~2.8, independent of either the assumed angular distribution of the interacting particles or the assumed abundances. This can be compared to indices calculated[3] for a number of flares using a similar technique. The values ranged from 2.7 (for the 16 December 1988 flare) to 4.5. (Note: The derived index for the large flare of 4 August 1972 was 3.4–3.7.)

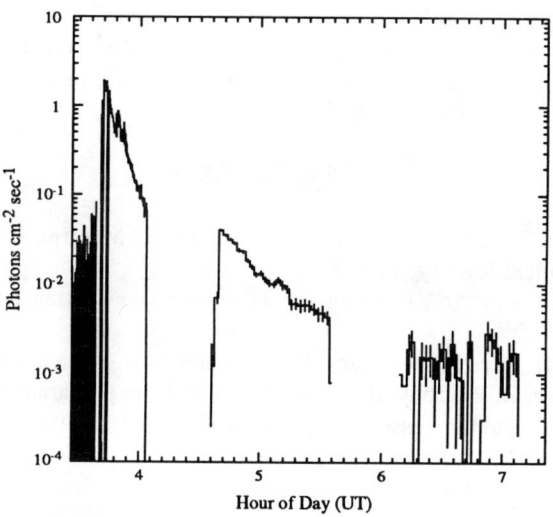

Figure 1. 2.223 MeV neutron-capture line time profile.

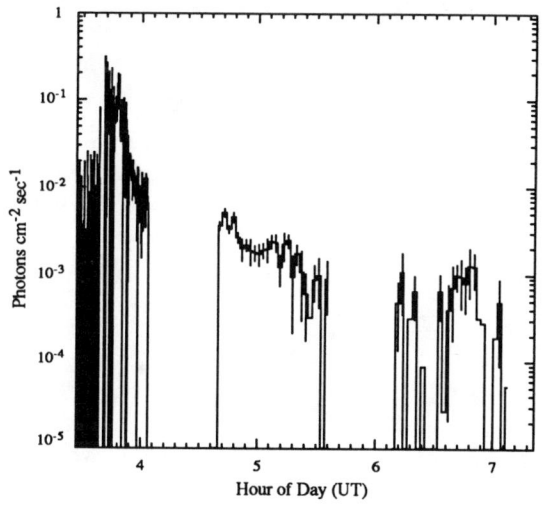

Figure 2. 4.44 MeV ^{12}C deexcitation line time profile.

The implied number of interacting protons with energy greater than 30 MeV [N_p(>30 MeV)] depends somewhat on the assumed abundances at the interaction site. Using photospheric abundances[5] for both the ambient and accelerated particles, the measured 2.223 MeV line fluence (692 photons cm^{-2}) implies N_p(>30 MeV) $\cong$ 1.5×10^{33}. Using the enhanced-heavy-element abundances derived[6] for the 27 April 1981 flare observed by *SMM*/GRS, we find N_p(>30 MeV) $\cong 4 \times 10^{32}$. We emphasize again that these are lower limits since we have not accounted for emission during data gaps. These values can be compared to those derived[3] for a number of flares which ranged from 1.6×10^{30} to 1×10^{33} (for the 4 August 1972 flare). We see that the 4 June 1991 flare was comparable in size to the 4 August 1972 flare but had a significantly harder spectrum.

REFERENCES

1. Share. G. H., *et al.*, Abstracts of the 179th Meeting of the American Astronomical Society (1992.)
2. Prince, T. A., *et al.*, Proc. of the 18th Internat. Cosmic Ray Conf., **4**, 79 (1983).
3. Ramaty, R., *et al.*, Adv. Space. Res., accepted for publication (1993).
4. Kozlovsky, B., Murphy, R. J., and Ramaty, R., in preparation (1994).
5. Anders, E., and Grevesse, N., Geochim. Cosmochim. Acta, **53**, 197 (1989).
6. Murphy, R. J., *et al.*, ApJ, **371**, 793 (1991).

AN OVERVIEW OF SOLAR FLARE RESULTS FROM COMPTEL

Mark McConnell
Space Science Center, University of New Hampshire,
Durham, NH 03824

ABSTRACT

The COMPTEL experiment on the Compton Gamma Ray Observatory (CGRO) has been operating in orbit since April of 1991. During that time, COMPTEL has observed several large flares, the most notable of which were several X-class flares which took place in June of 1991. As a solar instrument, COMPTEL has the capability to measure solar flare radiation in two parallel observing modes. In its telescope mode, COMPTEL is capable of measuring both solar flare photons (in the 0.75-30 MeV range) and solar flare neutrons (in the 20-150 MeV range) using the double scatter technique with a field-of-view of ~1 steradian. This approach also permits the imaging of the incident solar radiations (both photons and neutrons). The burst mode of COMPTEL utilizes two of the lower D2 detectors as large-area spectroscopy detectors to provide additional data in the 0.6-10 MeV range. Here we shall review both modes of COMPTEL operation and provide an overview of solar flare results which are presently available.

INSTRUMENT DESCRIPTION

The design of COMPTEL is illustrated schematically in Figure 1. COMPTEL has two modes of measuring γ-rays[1]. These two modes operate in parallel. The primary mode of operation is the double-scatter (or telescope) mode which is used for imaging incident radiations within a field-of-view of ~1 steradian. The burst mode utilizes two large area NaI detectors to provide additional spectroscopic data.

In the telescope mode an incoming γ-ray scatters off of an electron in one of seven D1 detectors and proceeds down to interact in one of fourteen D2 detectors. Each D1 detector is 28 cm in diameter by 8.5 cm thick and filled with liquid organic scintillator (NE213A). The low density and low Z of Ne213A maximizes the probability of a single photon scatter followed by the subsequent escape of the scattered photon. Each D2 detector is composed of NaI (Tl) 28 cm in diameter by 7.5 cm thick. Events which scatter only once in D1 and are fully absorbed in D2 constitute an ideal telescope events. The photon scattering events in D1 take place according to the Compton kinematic formula

$$\phi = \cos^{-1}\{(1 - \varepsilon/E_2 + \varepsilon/(E_1 + E_2))\}, \qquad (1)$$

where ε is the electron rest mass energy, E_1 is the energy deposit in D1, E_2 is the energy deposit in D2 and ϕ is the Compton scatter angle provided $E_1 + E_2$ is the full incident γ-ray energy.

As an imaging telescope, COMPTEL relies on the full energy deposit of the scattered γ-ray to correctly estimate the scattering angle φ. For solar flares, we can select only those events where the inferred scatter angle φ about the vector of the scattered γ-ray is consistent with the solar direction. These events correspond to those which deposit the full energy of their scattered photon in D2. This type of event selection has the advantage that the the response of the telescope becomes Gaussian in shape with a heavily suppressed Compton tail at low energies. Since the solar γ-ray spectra are rich in nuclear lines, a simple response function facilitates the de-convolution of the energy-loss spectra.

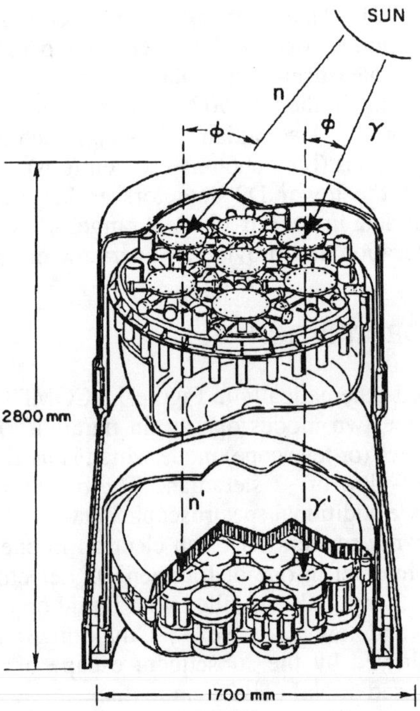

The D1 and D2 subsystems of the telescope are each completely surrounded by charged particle detectors. Four domes of plastic scintillator (NE110) are 1.5 cm thick and do not significantly attenuate the incident energetic γ-ray or neutron fluxes, yet they are virtually 100% efficient in identifying charged cosmic rays. The charged particle shields and other intervening material heavily attenuate the solar flare hard X-ray flux, minimizing pulse pile-up effects in the D1 and D2 detectors. They do, however, saturate under the intense soft thermal X-ray flux associated with flares and, for intense flares, may become a significant source of deadtime.

Since neutrons (like photons) are not detected by the charged particle veto domes, neutrons can produce photon-like events via elastic proton scattering in D1. COMPTEL can distinguish between photons and neutrons in two ways. First, the liquid scintillator in D1 possesses pulse shape discrimination (PSD) properties. This capability allows for efficient identification of signals from recoil protons produced by the elastic neutron scatters. Secondly, time-of-flight (TOF) measurements of the scattered particle traveling between D1 and D2 permit the identification of scattered neutrons.

The ideal type of neutron interaction in COMPTEL occurs when the incoming neutron scatters elastically off a hydrogen nucleus in the D1 detector. The scattered neutron then proceeds to the D2 detector where it interacts, depositing some of its

COMPTEL
IMAGING COMPTON TELESCOPE

Fig. 1. Schematic of COMPTEL with typical g-ray and neutron interactions.

energy to produce a trigger signal. The energy of the incident neutron is computed by summing the proton recoil energy E_1 in the D1 detector with the energy of the scattered neutron E_s deduced from the TOF measurement. The scatter angle for non-relativistic neutrons (< 150 MeV) can be computed by the formula:

$$\tan^2 \phi = E_1 / E_s. \qquad (2)$$

As with γ-rays, neutrons can be traced backwards from D2 to D1 through the angle φ to a cone mantle restricting the incident direction to include the Sun. The pulse shape from recoil protons is sufficiently different from that of electrons to reject more than 95% of electron-recoil events greater than about 1 MeV, the energy threshold in D1 for neutron detection. The PSD and TOF criteria are normally set such that solar neutrons incident on D1 in the energy range from about 10 MeV to 150 MeV are recorded. In this energy interval COMPTEL can observe neutrons from about 14.5 to 55 minutes after release from the Sun.

The burst mode of the COMPTEL instrument can also be used to detect solar γ-rays[2]. One D2 detector module (the low-range module) covers the energy interval from 0.1 to 1 MeV and another (the high-range module) covers the interval from ~ 1 to 10 MeV. Each detector module has an unobscured field-of-view of about 2.5 sr and a physical area of ~ 600 cm[2]. Outside this field-of-view varying amounts of intervening material exist which attenuate the solar γ-ray flux. These detectors operate in parallel with the telescope data, continuously accumulating spectra with 100 second integration time. Upon receipt of a BATSE trigger, these detectors begin a series of shorter accumulations (from 0.1 to 6 seconds in duration) which permit a more careful study of the burst (or flare) event. During the time period from May of 1991 to May of 1992, the low range detector of COMPTEL was turned off due to an intermittent PMT (one of seven PMTs on each D2 module). This PMT has now failed completely, allowing us to make full use of this module for burst and solar flare studies.

OBSERVATIONS

Since the time of the GRO launch (on 5 April 1991), the most notable period of solar activity was in June of 1991 with the appearance of active region 6659. From the emergence of this region on the east limb on 1 June until 8 June the Sun was outside the field-of-view of COMPTEL. The occurence of several major flares during this time prompted the declaration of a CGRO Target of Opportunity. From 8 June until passage of the region around the west limb on 15 June, CGRO was reoriented so that the Sun remained within the COMPTEL field-of-view (at a zenith angle of ~ 10° to 15°). Major (X-class) flares were recorded on 9, 11 and 15 June 1991. Intial results from these flares have already been reported[3-6]. A more detailed discussion of the June 9 flare is given elsewhere in these proceedings[7]. Here we shall provide a few typical examples of COMPTEL solar flare data.

Consider first, the flare of June 11. This flare took place within the COMPTEL FoV shortly after satellite sunrise. The impulsive phase (according to the GOES X-ray data) extended from 0158 UT until approximately 0210 UT, after which there was emission lasting until at least 0225 UT. Figure 2 shows a time history of this flare as observed by the COMPTEL high-range burst module (0.6-10 MeV). A telescope-mode energy-loss spectrum from the impulsive phase of the flare is shown in Figure

3. Here we see evidence for the strong 2.2 MeV line emission which is also measured by COMPTEL in each of the two orbits following the impulsive phase[6]. This observation of extended emission is especially interesting given the observation by EGRET of emission at energies > 50 MeV lasting for up to 10 hours after the impulsive phase[8].

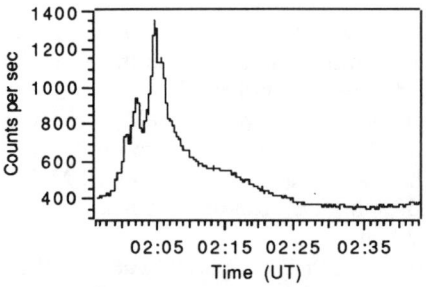

Fig. 2. Time history of the June 11 flare as measured by the high-range burst module (0.6-10 MeV)

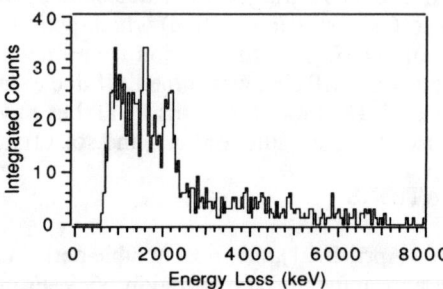

Fig. 3. The energy loss spectrum from the impulsive phase of the June 11 flare.

Extended emission (lasting on the order of hours after the impulsive phase) appears to have been a rather common feature of those flares associated with AR6659. Another example is the flare of June 15. The impulsive phase of this X12+ flare, which peaked at 0820 UT, was unobserved by CGRO due to orbital darkness. It was not until some 40 minutes after the flare onset that COMPTEL was able to collect its first data from this event. This flare was also measured by the GAMMA-1 instrument[9] and was reported to have emission > 30 MeV after ~ 0837 UT lasting for at least 20 minutes after the X-ray maximum. The COMPTEL measurements began at ~ 0859 UT, while the GAMMA-1 measurements ended at 0902 UT. The COMPTEL instrument continued making measurements until 0930 UT. There is no apparent signature of this event in the burst mode data, a testament to the sensitivity of the telescope mode. The light curve as measured with the telescope mode data is shown in Figure 4. The e-folding decay time of this emission is measured to be 910±135 seconds.

Figure 5 demonstrates the neutron capability of COMPTEL using data from the June 15 flare. The plot shows the neutron production time history. The data points are plotted at the arrival time of photons created simultaneously with the neutrons. The curve suffers from velocity dispersion effects, in that the counts at the earliest times in the plot are the slowest neutrons, while the counts at the later times come from higher energy neutrons. The time intervals do not sample uniformly the available neutron energy spectrum. The plot does show, however, that neutron emission occurred at times near the X-ray maximum and extended for over an hour. This, coupled with the detection of γ-rays up to 0930 UT, means that proton acceleration or precipitation persisted for at least 70 minutes.

Several other flares have been observed in the COMPTEL telescope mode. Many of these flares have not yet been studied in the same detail as the large flares of June 1991. An effort to catalog all of the telescope mode flare events is underway. A first report of this effort, concentrating on the flares of October 1991, can be found elsewhere in these proceedings[10].

Similarly, a catalog is being compiled of all of those events detected by the COMPTEL burst mode detectors. This list will be considerably larger than the telescope mode catalog because the burst modules are sensitive to a much larger fraction of the sky. A status report on this effort can also be found in these proceedings[11].

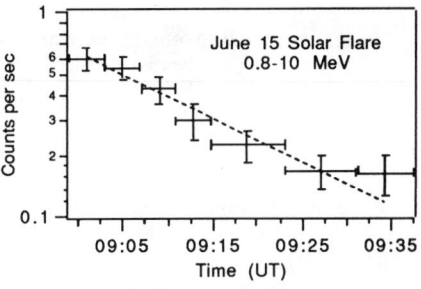

Fig. 4. Decay of the June 15th flare as seen in telescope mode data.

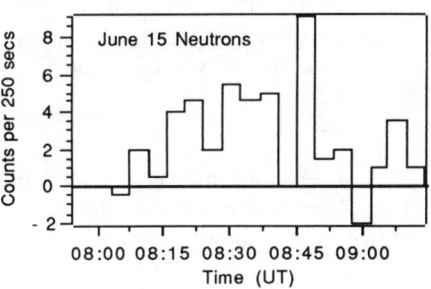

Fig. 5. Neutron production time history for the June 15th flare.

CONCLUSIONS

COMPTEL has proven itself as a very effective tool for the study of solar flare phenomena. The CGRO Target of Opportunity in June of 1991 has already provided some valuable results. Although most of the reported results to-date have concentrated on the June flares, much effort remains in studying the other flare events in the COMPTEL database. Cataloging efforts are already underway, in preparation for a more careful study of all available events.

REFERENCES

1. V. Schönfelder et al., Ap. J. (Suppl.), 86, 629 (1993).
2. C. Winkler et al., Adv. Space Res. 6 113-117 (1986).
3. M. McConnell et al., Adv. Sp. Res., to be published (1992).
4. J. Ryan et al., Adv. Sp. Res., to be published (1992).
5. J. Ryan et al., Compton Centennial Symposium Proc., to be published (1993).
6. G. Rank et al., Compton Centennial Symposium Proc., to be published (1993).
7. J. Ryan et al., these proceedings.
8. G. Kanbach et al., Astr. Ap. Suppl., 97, 349 (1993).
9. V. V. Akimov et al., in Proceedings of the 22nd International Cosmic Ray Conference, (Dublin, Ireland, 1991), 3, 73.
10. M. Varendorff et al., these proceedings.
11. R. Suleiman et al., these proceedings.

THEORETICAL MODELS FOR HIGH-ENERGY SOLAR FLARE EMISSIONS

Reuven Ramaty and Natalie Mandzhavidze[1]
Laboratory for High Energy Astrophysics
NASA/GSFC, Greenbelt, MD 20771

ABSTRACT

We discuss gamma ray production mechanisms and ion transport processes in solar flares. We investigate the implications of the extended GeV gamma ray emission observed from the 11 June and 15 June 1991 flares. We find that this extended emission could be produced by ions trapped in loops, provided there is a suitable combination of size, twist, field convergence and turbulent energy density. We also consider in some detail the possibility of continuous acceleration by Alfvén turbulence. We find that this would require the continuous presence of turbulence with energy density of at least 1 erg cm^{-3}. The strong pitch angle scattering caused by this turbulence leads to anisotropic pion decay emission with much steeper spectrum than observed. We discuss various alternatives, including the possibility of episodal acceleration.

I. INTRODUCTION

High energy solar flare emissions (gamma rays and neutrons) result from the interaction of flare accelerated particles with the ambient solar atmosphere. The photon and neutron production mechanisms are by now quite well understood (e.g. ref. 1). A considerable amount of research has also been carried out on the relevant particle transport processes[2-6]. New interest in these processes has been stimulated by observations of a series of X-class flares in June 1991 with instruments on the COMPTON Gamma Ray Observatory[7-9] (CGRO) and GAMMA-1 (ref. 10). Of special interest are the observations of GeV gamma ray emission that lasted for hours. These observations are raising questions on the nature of the fundamental transport processes (adiabatic motion, pitch angle scattering by plasma turbulence, drifts), as well as on the structure of the magnetic field. In addition, the possibility of particle acceleration to GeV energies over long periods of time has also been brought up. Such acceleration should take place under markedly different physical conditions than the acceleration of the ions responsible for the gamma ray emission observed during the impulsive phase of flares.

In the present paper we first review the photon production processes. We then systematically discuss the transport processes, comparing analytical approximation with Monte Carlo simulations. We investigate the very important role of plasma turbulence in both the transport and the acceleration of particles. Finally, we discuss the 11 June 1991 and 15 June 1991 flares from which extended GeV gamma ray emission was observed. We summarize the arguments in favor and against both long term trapping and continuous acceleration.

---

[1] NAS/NRC Res. Research Assoc.; also at the Inst. of Geophysics, Tbilisi, Georgia

II. PHOTON AND NEUTRON PRODUCTION MECHANISMS

The principal mechanisms that produce high energy photons and neutrons in solar flares are summarized in Table 1. Here we briefly discuss these mechanisms and the resultant emissions.

TABLE 1, High Energy Photon and Neutron Production Mechanisms

Emissions	Processes	Observed Photons or Neutrons	Primary Ion or Electron Energy Range
Continuum	Primary Electron Bremsstrahlung	20 keV - 1 MeV >10 MeV	20keV-1GeV
Nuclear Deexcitation Lines	Accelerated Ion Interactions, e.g. ^{4}He$(\alpha,n)^7$Be* ^{4}He$(\alpha,p)^7$Li* ^{20}Ne$(p,p')^{20}$Ne* ^{12}C$(p,p')^{12}$C* ^{16}O$(p,p')^{12}$O*	Lines at e.g. 0.429 MeV 0.478 MeV 1.634 MeV 4.438 MeV 6.129 MeV	1-100 MeV/nucl
Neutron Capture Line	Neutron Production by Accelerated Ions followed by ^{1}H$(n,\gamma)^2$H	Line at 2.223 MeV	1-100 MeV/nucl
Positron Annihilation Radiation	β^+ Emitter or π^+ Production by Accelerated Ions, e.g. ^{12}C$(p,pn)^{11}$C$\rightarrow^{11}$B+e$^+$+ν p+p$\rightarrow \pi^+..,\pi^+ \rightarrow \mu^+ \rightarrow e^+$ followed by e$^+$+e$^- \rightarrow 2\gamma$ e$^+$+e$^- \rightarrow$ Ps+hν or e$^+$+^{1}H $\rightarrow$ Ps+p Ps $\rightarrow 2\gamma, 3\gamma$	Line at 0.511 MeV Orthopositronium Continuum <511 keV	1-100 MeV/nucl
Pion Decay Radiation	π^0 and π^+ Production by Accelerated Particles, e.g. p+p$\rightarrow \pi^0, \pi^\pm..$ followed by $\pi^0 \rightarrow 2\gamma, \pi^\pm \rightarrow \mu^\pm \rightarrow e^\pm$ e$^+ \rightarrow \gamma$brem, γann. in flight e$^- \rightarrow \gamma$brem	10MeV-3GeV	0.2-5GeV
Neutrons	Accelerated Particle Interactions, e.g. ^{4}He$(p,pn)^3$He p+p$\rightarrow \pi$+n+.. ^{22}Ne$(\alpha,n)^{25}$Mg	Neutrons in Space (10-500 MeV) Neutron Induced Atmospheric Cascades (0.1-10 GeV) Neutron Decay Protons in Space (20-200 MeV)	10MeV-1GeV 0.1-10GeV 20-400MeV

Bremsstrahlung. Interactions of the accelerated electrons with ambient gas in the flare region produce continuum X-ray and gamma ray emission via nonthermal bremsstrahlung. This continuum extends from about 20 keV to over 100 MeV. At the low energy end it merges into the thermal bremsstrahlung produced by hot flare plasma. There is no known high energy cutoff; the highest energy observed[10] bremsstrahlung is around several hundreds of MeV.

Bremsstrahlung production in solar flares in the relevant hard X-ray – gamma ray range was calculated in a thin target model yielding both angle integrated[11] and angle dependent photon spectra[12]. Both of these calculations assumed an ionized ambient gas; for the angle dependent calculation various anisotropic electron distributions were assumed. Bremsstrahlung production by ultrarelativistic electrons in flare loops, assuming that the photons are emitted along the direction of motion of the electrons, was investigated[4,13,14].

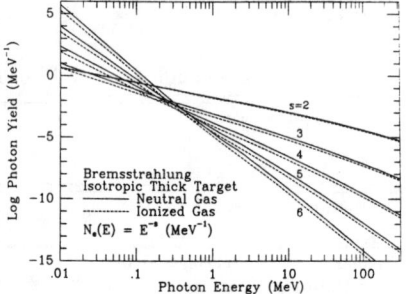

Fig. 1. Bremsstrahlung Production.

In a similar transport calculation[3] the angular distribution of the emitted radiation was also taken into account. A detailed electron transport calculation was carried out[6], however, the bremsstrahlung was only calculated[15] in an approximate manner (electron-electron bremsstrahlung was ignored), and no photon spectrum valid over a broad energy range was given.

In a previous paper[16] we presented the results of accurate calculations of angle integrated thick target bremsstrahlung in the transrelativistic region (0.3 - 1 MeV). Here we extend these calculations over a broad range of photon and electron energies (0.01 - 1000 MeV). We use the non-screened electron-proton[17] and electron-electron[18] cross sections (valid for the entire energy range), and the electron-atom cross section[19] valid in the ultrarelativistic region. Angle integrated thick target bremsstrahlung spectra, for both neutral and ionized ambient gases, are shown in Fig. 1, where the incident electron spectra are assumed to be power laws in kinetic energy. The bremsstrahlung yield in an ionized gas is generally lower than that produced in a neutral gas because of the higher rate of energy loss in the ionized case.

We have used[16] this isotropic bremsstrahlung model to fit the observed 0.3-1 MeV continuum spectra of 10 flares and 6 individual emission episodes during the 6 March 1989 flare. Although the angular distribution of the electrons could be anisotropic, the use of the isotropic model is justified since in this energy range the bremsstrahlung angular pattern is not strongly beamed and Coulomb collisions will nearly isotropize the electrons. We combined the results with data on nuclear line emission, and derived the ratio of the electron flux at 0.5 MeV to the proton flux at 10 MeV. The flux ratio of 0.5 MeV electrons to 10 MeV protons was extensively studied for solar flare particles observed in interplanetary space[20]. For these interplanetary particles, on the average, the 0.5 MeV electron to 10 MeV proton flux ratio is much larger for impulsive flares (in which particles are thought to be accelerated from hot flare plasma near the site of flare energy release) than for gradual flares (in which particles are accelerated from cooler coronal gas). The gamma ray results, pertaining to the particles which interact at the Sun, reveal an even higher electron to proton ratio, regardless of whether the flare is impulsive or gradual. This result suggests that the particles responsible for gamma ray production and the particles observed in interplanetary space from impulsive flares are probably accelerated by the same mechanism. In §III we argue that this mechanism is stochastic acceleration due to gyroresonant interactions with plasma waves.

For many flares, the gamma ray spectrum between about 1 to 8 MeV is dominated by nuclear line emission (see below). Above 10 MeV bremsstrahlung can become important again. There are, however, only two flares (21 June 1980 and 3 June 1982) for which there are published data[21,22] on the continuum below 1 MeV and continuum above 10 MeV extending to around 100 MeV. In Fig. 2 we show the data for these two flares along with calculated isotropic bremsstrahlung spectra fitted to the 0.3 - 1 MeV data. The excellent fit provided by this model to the 21 June 1980 data above 10 MeV may be the consequence of the location of the 21 June 1980 flare close to the limb of the Sun. In the framework of standard loop geometries (§III), the directional bremsstrahlung from limb flares is not too different from the angle integrated emission. On the other hand, the excess between about 20 to 70 MeV predicted by the isotropic model for the 3 June 1982 flare (heliocentric angle 72°) could be evidence for anisotropic emission. However, the discrepancy between the data and the calculated curve in this energy range could also be due to synchrotron losses or a steepening in the spectrum of the radiating electrons. The flattening in the observed spectrum of the 3 June 1982 flare above 70 MeV is most likely due to pion decay emission discussed below.

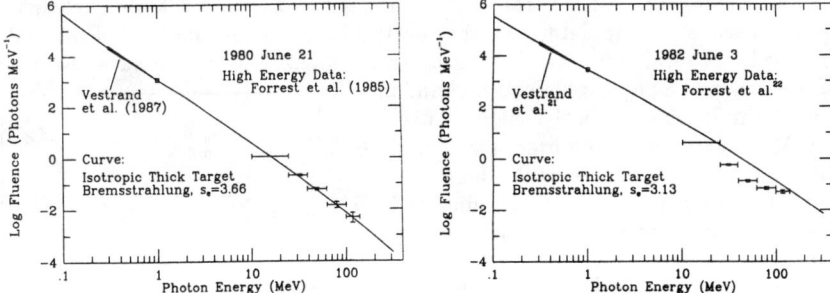

Fig. 2. Isotropic thick target bremsstrahlung fits to the 21 June 1980 and 3 June 1982 continuum.

Line Emission. Nuclear deexcitation lines result from the bombardment of ambient C and heavier nuclei by accelerated protons and α particles, and from the inverse reactions in which ambient hydrogen and helium are bombarded by accelerated carbon and heavier nuclei[23]. Because of their low relative abundances, interactions between accelerated and ambient heavy nuclei are not particularly important. Furthermore, since H and He have no bound excited states, p-p and p-He interactions can also be ignored. However, interactions of α particles with ambient He produce two strong lines, at 478 keV from ^{7}Li and at 429 keV from ^{7}Be. As the shape of the spectral feature resulting from the superposition of these α-α lines is strongly dependent on the angular distribution of the interacting α particles, measurements with good spectral resolution in the energy range 0.4 - 0.5 MeV could turn out to be particularly useful in the study of the anisotropy of the interacting particles. We return to this issue in §III.

The observed gamma ray spectrum of the 27 April 1981 flare[24] has been used to derive abundances of both the ambient gas and the accelerated particles[25]. The derived accelerated particle abundances indicate a very significant enhancement of heavy element abundances, similar to the heavy element enhancement observed in interplanetary particles from impulsive flares[26]. This supports the conclusion mentioned above that the particles responsible for gamma ray production and the particles observed in interplanetary space from impulsive flares have a common origin. The derived ambient gas composition points to enhanced Ne, Mg, Si and Fe abundances relative to C or O. The enhanced Mg, Si and Fe abundances (elements with low first ionization potential, FIP) could be understood in terms of a charge dependent ambient gas transport process from the photosphere to the chromosphere and corona which favors the collisionally ionized, low FIP elements in the photosphere[27]. The enrichment of Ne (a high FIP element) could be due to photoionization by soft X-rays[28]. This interpretation of the Ne enhancement predicts that S should also be enhanced. Both the Ne and S enhancements have been confirmed by observations[29] with the Flat Crystal Spectrometer on SMM. Furthermore, it is possible that the feature at about 2.26 MeV observed[24] from the 27 April 1981 flare contained a significant contribution from the ^{32}S line at 2.230 MeV.

Neutrons. Neutron production in solar flares was studied in detail[5,30,31]. Solar flare neutrons have been observed directly with detectors on spacecraft, and indirectly with detectors on the ground. Solar flare neutrons have also been studied indirectly by observing neutron decay protons in interplanetary space. We have reviewed these observations[32,33]. The recent neutron observations from the June 1991 flares are summarized in Table 3 (§IV). The bulk of the neutrons which move downward to the photosphere are captured on H and ^{3}He in the photosphere. Capture on H produces the 2.223 MeV line. The ratio of the fluence in this line to the 4-7 MeV nuclear deexcitation fluence is used to determine the spectral index of the accelerated ions. We have reviewed this technique recently[16]. Studies of the 2.223 MeV line have also been used to determine the photospheric ^{3}He abundance[34].

Pion Decay Radiation. In the energy range above 10 MeV, along with the bremsstrahlung from primary electrons, there can also be a significant contribution from pion decay radiation. The theory of pion decay emission in solar flares was treated in detail[30,35] and we have reviewed the observations[32,33]. The long duration pion decay emission observed from two flares in June 1991 is discussed in §IV. We show in Fig. 3 the ratio of the angle integrated 4-7 MeV nuclear deexcitation emission[16] to the angle integrated total gamma ray production from π^o decay as a a function of the proton power law spectral index S_p. Both quantities were calculated in an isotropic thick target model. The solid curve (comp 1) refers to a case in which both the ambient medium and the energetic particles have photospheric composition; the dashed curve (comp 2) refers to abundances (both of the gas and the energetic particles) derived from the gamma ray observations of the 27 April 1981 flare discussed above. We point out that, whereas $Q(4-7)$ increases by about a factor of 8 as the composition changes from comp 1 to comp 2, the pion production increases only by about 50%.

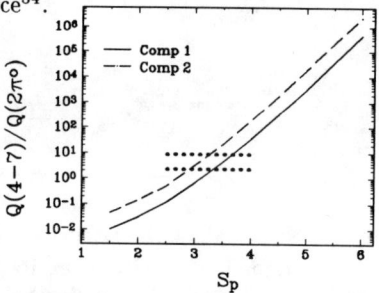

Fig. 3. Nuclear excess to pion production ratio; dots – data, see §IV.

Positrons. Positrons in solar flares result from the decay of radioactive nuclei and charged pions[36]. The contribution from radiaoctive nuclei is closely related to 4-7 MeV nuclear deexcitation emission. The ratio[37] of this positron production to the 4-7 MeV photon production is not strongly dependent on the ion spectrum and the composition. For the comp 1 abundances, it varies from 0.25 to 0.6 when the proton power law spectral index S_p varies from 5 to 3. In addition to the positron production, the 511 keV line flux also depends on the fraction of the positrons which annihilate via positronium[38] and the possible attenuation of the 511 keV line in the solar atmosphere. If the density at the annihilation site is $<10^{15}$ cm^{-3}, about 90% of the positrons will annihilate via positronium, yielding 0.65 line photons per positron. For such annihilation sites we do not expect much attenuation. If the annihilation site is deeper in the atmosphere, the number of line photons per positron increases (but never exceeds 2); however, we also expect more attenuation, especially for flares near the limb.

Data on 511 keV line emission are available for a few flares[25,39,40] (4 and 7 August 1972, 21 June 1980, 1 July 1980, 27 April 1981, 3 June 1982). For the 21 June 1980 flare it was shown[37] that the observed[39] 511 keV line flux is consistent with that expected to accompany the observed 4-7 MeV nuclear deexcitation emission. The bulk of the positrons responsible for the 511 keV line emission in this flare resulted from the decay of radioactive positron emitters[37]. On the other hand, in the 3 June 1982 flare, the 511 keV line emission resulted from positrons from both charged pions and radioactive positron emitters[30]. The 3 June 1982 flare is the only one for which simultaneous pion decay emission and 511 keV line observations were reported.

III. TRANSPORT AND ACCELERATION

There are two strong arguments that suggest that the bulk of the observed gamma ray emission is produced by particles accelerated and trapped in closed magnetic structures, most likely loops. The first argument pertains to relativistic electrons. Gamma ray emission at energies >10 MeV was observed from many disk flares. This gamma ray emission is mostly bremsstrahlung from ultrarelativistic electrons whose radiation pattern is highly collimated along the direction of motion of the electrons. Since it is much more likely that these electrons are accelerated in the corona rather than in the photosphere, in the absence of trapping the electrons would radiate predominantly downwards toward the photosphere because the amount of material above the acceleration is negligible relative to the radiation length of relativistic electrons. In this case radiation would not be observed from disk flares. On the other hand, mirroring in convergent magnetic flux tubes, or pitch angle scattering by plasma turbulence, can reflect the particles and allow them to radiate on their way up in the solar atmosphere.

The other argument follows from the comparison of the number of interacting particles, as derived from the gamma ray observations, with the number of escaping particles from the same flare, obtained from interplanetary observations. This comparison shows that for electrons[41,42] the ratio of escaping to interacting particles (the escape ratio) is less than 1 for all the flares that were studied. For protons, the escape ratio can be both less than or greater than 1; but it is typically less than 1 for impulsive flares[16,43], indicating that at least for these flares the bulk of the protons remain trapped at the Sun. In addition, as we show below, long term trapping of particles in loops provides a natural explanation for the observation of high energy gamma ray emission hours after the impulsive phase of the flare.

a. Transport

The loop model employed in the transport calculations[2,3,5,35] for gamma ray production in solar flares consists of a semicircular coronal segment joined to two radially aligned straight segments extending to the photosphere. In the coronal segment the magnitude of the magnetic field and gas density are constant, while in the subcoronal segments both the field and gas density increase with increasing depth. In the calculations that we discuss below we use the specific model employed in our pion production calculations[35,44]. In this model, the transition between the coronal and subcoronal segments is at 2000 km above the photosphere, and R is the radius of the semicircular coronal segment. The gas density at the photosphere is 3.7×10^{17} cm^{-3}, it decreases exponentially with scale height h_a to the transition, and has a constant value n_c in the coronal segment. The magnetic field decreases linearly in the subcoronal segments, from B_p at the photosphere to B_c at the transition.

<u>Adiabatic Motion</u>. To first order, the motion of the accelerated particles throughout the loop can be described by the conservation of $(1-\mu^2)/B$, where μ is the cosine of the particle's pitch angle. Particles with large pitch angle can mirror many times before they interact or their energy falls below the gamma ray production threshold. On the other hand, the energy of particles with small pitch angles can drop below the threshold before they mirror. The cone containing the velocity vectors of these particles is defined as the loss cone.

The solid curves in Fig. 4 show the amount of matter (expressed in g cm^{-2}) encountered by a particle of initial pitch angle μ_0 traversing the distance from the top of the coronal segment to the mirror point; the dashed curves represent the grammage traversed in the subcoronal segments. In cases 1, 2 and 3 the scale height and coronal density are the same, and B_p/B_c (the mirror ratio) varies.

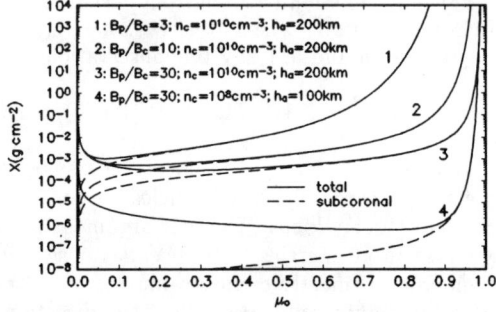

Fig. 4. Grammage from top of the loop to the mirror point.

In case 4 the mirror ratio is high, and the scale height and coronal density are low. The density at the transition is 1.7×10^{13} and 7.6×10^8 cm^{-3} for h_a equal to 200 and 100 km, respectively. As expected, the grammage to the mirror point decreases with increasing mirror ratio (compare cases 1, 2 and 3) and decreasing scale height (compare cases 3 and 4, dashed curves). In fact, in case 4 the density just below the transition is so low that for $\mu_0 < 0.9$ essentially all of the grammage is traversed in the corona. For the other cases, the coronal contribution is important only for large pitch angles.

Using the results of Fig. 4, we can estimate, for various interaction products, the cosine μ_c of the loss-cone half-angle. For 4.438 MeV ^{12}C nuclear deexcitation photons, we assume a typical 40 MeV proton, which will fall below the threshold for line production after traversing ~ 0.7 g cm^{-2}; for pion radiation, we assume a 700 MeV proton, for which the corresponding grammage is ~ 50 g cm^{-2}; and for >10 MeV bremsstrahlung we take a 20 MeV electron, which will lose 10 MeV in ~ 2 g cm^{-2}. Then, in case 2 for example, μ_c is approximately 0.91, 0.95, and 0.92 for 4.438 MeV line production, pion production and >10 MeV bremsstrahlung, respectively. Larger (smaller) values of μ_c will result from larger (smaller) mirror ratios. For pion production, for example, μ_c is approximately 0.81. 0.95 and 0.98 for cases 1, 2 and 3, respectively.

Particles in the loss cone will be removed from the loop on a very short time scale. In Table 2 we show the removal times for nuclear line producing and pion producing protons for three pitch angles outside the loss cones and the loop parameters of case 2. We see that the line producing protons are removed faster than the protons which produce pions. The same is true for relativistic electrons. Thus, if the motion is purely adiabatic, the pion producing protons will execute a very large number of bounces before they interact, and can, in principle, remain trapped in the loop for a long time.

TABLE 2, Time Scales, Adiabatic Motion
($B_p/B_c = 10$; $h_a = 200$ km; $n_c = 10^{10}$ cm^{-3}; $R = 10^9$ cm)

Pitch Angle Cosine	Grammage to Mirror Point (g cm^{-2})	Number of Bounces		Time (s)	
		Nuclear Lines	Pions	Nuclear Lines	Pions
0.1	5.40x10^{-4}	1300	9.3x10^4	4.8x10^3	1.2x10^5
0.5	1.56x10^{-3}	450	3.2x10^4	3.3x10^2	8.3x10^3
0.7	5.16x10^{-3}	135	9.7x10^3	7.2x10^1	1.8x10^3

<u>Pitch Angle Scattering by Plasma Turbulence.</u> The motion of the particles is also influenced by scattering due to plasma turbulence. Plasma turbulence is expected to be present in the ionized coronal segment but not below the transition where the gas is mostly neutral. We consider scattering by isotropic Alfvén turbulence with a Kolmogorov spectrum which extends down to a wave number of 3×10^{-6} cm^{-1}, corresponding to the gyroradius of a 10 GeV proton in a magnetic field of 100 G (e.g. ref. 3). The pitch angle diffusion coefficient is given by $D_{\mu\mu} = \nu |\mu|^{n-1}(1-\mu^2)$, where the scattering rate $\nu \simeq 130 [(\gamma^2-1)^{1/3}/\gamma] W_A$ (s^{-1}), W_A is the total turbulent energy density in erg cm^{-3}, γ is particle Lorentz factor, and $n = 5/3$.

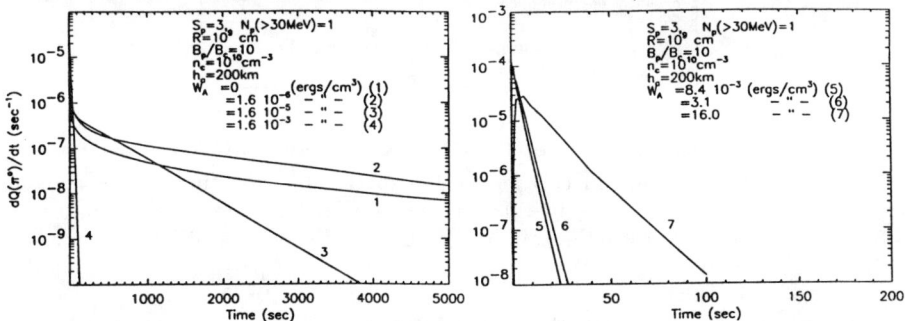

Fig. 5. Neutral pion production rate for various energy densities of plasma turbulence.

We studied the effects of the pitch angle scattering by carrying out Monte Carlo calculations of pion production. In Fig. 5 we show the time dependent rate of π° production in the loop for various values of W_A. Curves 3 - 6 can be approximated by exponentials yielding the characteristic decay times T_d shown by the crosses in Fig. 6. For curves 2 and 7, we estimated the decay time by using the calculations after 1000 s and 10 s, respectively. In the case of no pitch angle scattering (curve 1, $W_A = 0$) the emission decays as a power law. Also shown in Fig. 6 are analytical estimates of the dependence of T_d on W_A for 700 MeV protons (the effective energy for pion production) and the indicated loop parameters (case 2). We distinguish 3 regimes:

(i) Weak scattering. In this regime the transit time of the particles across the loop is much shorter than the pitch angle diffusion time across the loss-cone half-angle, α_c, $\pi R/[v\langle\mu\rangle] \ll \alpha_c^2/\nu$, where $\langle\mu\rangle$ is the average pitch angle cosine in the loss cone, and v is particle velocity. It has been suggested[45,46] that in this weak regime T_d is inversely proportional to ν (or W_A). However, our Monte-Carlo calculations do not agree with the proposed normalization coefficients. We find that in the range $10^{-5} \lesssim W_A \lesssim 10^{-3}$ erg cm^{-3}, $T_d \simeq 0.01/W_A$; at lower values of W_A this approximation breaks down, most likely because particles with relatively small pitch angles are removed from the loop by interactions with matter before they experience significant scattering due to the plasma turbulence.

(ii) Saturated Scattering. The transition to this regime occurs when $\nu \simeq \alpha_c^2 \, v\langle\mu\rangle/(\pi R)$ ($W_A \simeq 0.01$ erg cm^{-3} for the indicated parameters). In this regime the decay time scale is[3] $T_d \simeq \pi R/[v\langle\mu\rangle(1-\mu_c)]$. Using the values of μ_c given above (for case 2), we obtain values of T_d of about 4.2, 2.6, and 1.3 s, for nuclear line, pion and bremsstrahlung production, respectively. It has been suggested[2,3] that saturated pitch angle scattering is responsible for the short decay times of the gamma ray time profiles in impulsive flares (e.g. 21 June 1980).

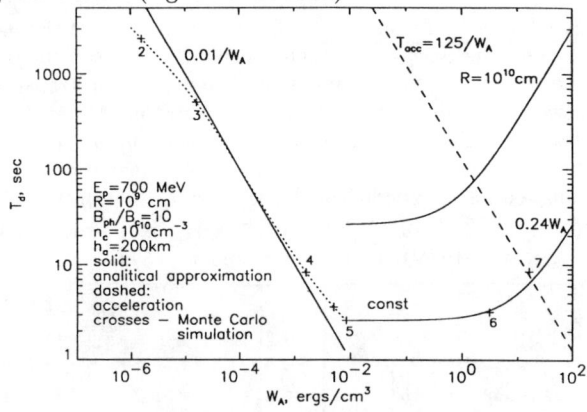

Fig. 6. Trapping and acceleration times of particles in a loop; dotted curve – guide to the eye.

(iii) Strong Scattering. In this regime the spatial diffusion time in the coronal segment is longer than the decay time in the saturated regime. To estimate the diffusive decay time we used the solution of the one dimensional diffusion equation with absorbing boundaries[47]. We found that the number of particles in the loop decays exponentially with characteristic decay time $0.81(\pi R/2)^2/2\kappa_\| = R^2/\kappa_\|$, where $\kappa_\| = (v^2/2\nu)/[(2-n)(4-n)]$ is the spatial diffusion coefficient along the filed lines[48]. Thus, the transition between the saturated and strong scattering regimes occurs when $R^2/\kappa_\| > \pi R/[v\langle\mu\rangle(1-\mu_c)]$. For our parameters ($R = 10^9$ cm), this transition occurs at $W_A \simeq 8.7$ erg cm^{-3}. The solid curves for $W_A > 10^{-2}$ erg cm^{-3} in Fig. 6 represent the analytic decay time[46] (for two values of R),

$$T_d = \pi R/[v\langle\mu\rangle(1-\mu_c)] + R^2/\kappa_\|. \tag{1}$$

This approximation, valid in the saturated and strong regimes, is in good agreement with our Monte-Carlo simulations (see Fig. 6).

Drifts. Particles can also be removed from the loop by drifts. The effects of the drifts have been studied[49] recently employing a magnetic field model that satisfies the force-free equilibrium equation, $\nabla \times \vec{B} = \lambda \vec{B}$, and boundary conditions such that the photospheric magnetic field is concentrated in two spots separated by a distance L_0. The twist exhibited by the resulting loop-like structure is determined by the parameter λ. The particles can drift to the boundaries of the loop as well as into the loss cone. The presence of twist causes some of the particles to drift on closed paths, and these particles can remain trapped in the loop indefinitely[49].

Using results from ref. 50, we plot in Fig. 7 the fraction of the particles that remain trapped in the loop as a function of time for various values of λ, $L_0 = 2 \times 10^9$ cm, and two values of the proton energy E_p. The effects of collisions with the ambient gas and pitch angle scattering due to plasma turbulence have not been taken into account in these calculations. As expected, the low energy protons remain trapped in the loop for very long times, independent of the amount of twist.

In the absence of twist ($\lambda = 0.1$), most of the high energy protons are removed from the loop after about 1 hour. However, for $\lambda = 3.4$ (larger values lead to instabilities) a fraction (6%) of these protons remain trapped indefinitely. On the other hand, because the time scale is proportional[50] to L_0^2, if $L_0 = 10^{10}$ cm, essentially all the high energy protons will remain trapped for at least 8 hours independent of the amount of twist.

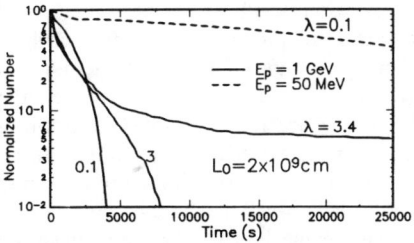

Fig. 7. The effect of twist on particle trapping.

Angular Distributions. An important consequence of particle transport in loops is that the angular distribution of the interacting particles is anisotropic[3-5,13,14]. Evidence for anisotropic gamma ray emission at energies >10 MeV has been provided in solar cycle 21 by the distribution of the observed flare positions on the Sun which showed limb brightening over the distribution expected for isotropic emission[21,51]. This implies that the angular distribution of the interacting electrons is anisotropic. However, during solar cycle 22, >10 MeV emission has been observed from many disk flares, and a statistical analysis[32,33] showed that the hypothesis of isotropic emission cannot be ruled out.

The angular distribution of the interacting ions can be studied using gamma ray line shapes[52,53]. In Fig. 8 (left panel) we show the calculated[52] profiles of the ^{7}Be and ^{7}Li lines (§II), where the arrows indicate the rest energies at 0.429 and 0.478 MeV, respectively. For no pitch angle scattering, the calculated features peak at essentially the rest energies because the angular distribution of the interacting particles peaks tangentially to the photosphere. For saturated scattering, the distribution of interacting particles is downward peaked, redshifting the lines by about 25 keV. In the right panel we show the corresponding count spectrum obtained[52] by folding the calculated spectrum (which also included the 0.511 MeV and other nuclear lines) through the SMM/GRS response. Here, the redshift is much less obvious, but still visible. For limb flares this effect is much less pronounced. Therefore, the observation of the α-α lines from 27 April 1981 flare[52] (heliocentric angle $\sim 91°$) did not allow to distinguish between saturated scattering and no scattering. On the other hand, data on the α-α lines from the 15 November 1991 disc flare (heliocentric angle 18°) obtained with YOHKOH[54] suggest the possibility of downward beaming.

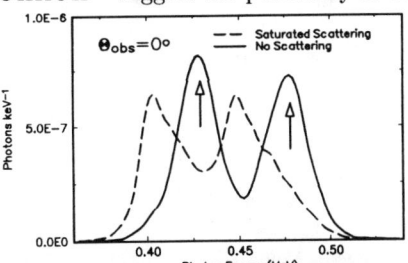 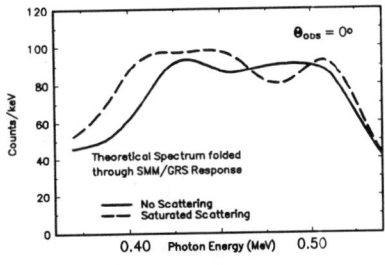

Fig. 8 (from ref. 52). Left panel: calculated spectrum of α-α lines; right panel: detector response to theoretical spectrum; θ_{obs} is heliocentric angle.

b. Acceleration

The particles are most likely accelerated in the corona. This is primarily because pitch angle scattering by plasma turbulence is required for the two most commonly discussed acceleration mechanisms, stochastic acceleration and diffusive shock acceleration[55]. There are arguments for favoring stochastic acceleration over shock acceleration in the impulsive phase of flares. At the site of the impulsive energy release, where the magnetic field is high, the Alfvén speed is expected to exceed the velocity of mass motions. In such an environment magnetoshocks will not develop, thus making shock acceleration an unlikely candidate. Furthermore, it has never been shown how shock acceleration could produce the ^{3}He and heavy element enhancements that are known to characterize both the particles observed in interplanetary space from impulsive flares, as well as the particles which produce the gamma rays. On the other hand, a credible mechanism has now been developed[56] that preferentially accelerates ^{3}He and heavy ions via gyroresonance with shear Alfvén waves. For the same reason, the observed abundance enhancements also favor stochastic acceleration over acceleration in large scale DC electric fields.

Reviews of stochastic acceleration are available in the literature (e.g. refs. 55,57). Here we limit our discussion to acceleration by gyroresonant interactions with Alfvén turbulence. As before, we assume a Kolomogorov spectrum with a cutoff at a wave number corresponding to the gyroradius of a 10 GeV proton. The average rate of energy gain of protons is given by[3] $\langle dE/dt \rangle = 1.10 \times 10^{-6} \, W_A \, V_A^2 \, (\gamma^2 - 1)^{1/3}$, where $\langle dE/dt \rangle$ is in MeV s^{-1}, W_A is in erg cm^{-3}, and V_A is the Alfvén velocity in km s^{-1}. The corresponding mean acceleration time is then given by

$$T_{acc} \simeq \frac{E}{\langle dE/dt \rangle} \simeq 8.5 \times 10^8 \frac{(\gamma - 1)}{(\gamma^2 - 1)^{1/3}} \frac{1}{W_A \, V_A^2} \quad \text{s}. \qquad (2)$$

In particular, for a 700 MeV proton and $V_A = 2000$ km/s, $T_{acc} \simeq 125/W_A$ s. This expression is plotted as a dashed line in Fig. 6.

The ratio T_d/T_{acc} determines the spectrum of the accelerated particles. For proton energies around 700 MeV, $T_d/T_{acc} \simeq 2.5 \, \alpha T$, where αT is the parameter that characterizes the proton spectrum in stochastic acceleration[55]. We found[58] that the observed spectrum of the pion decay emission from the 15 June 1991 flare can be fit with $\alpha T \simeq 0.09$, where we have used relativistically correct proton spectra[59]. This implies that $T_d/T_{acc} \simeq 0.2$. The value of T_d/T_{acc} should not be much smaller, because then the spectrum is too steep to produce pions; on the other hand, the ratio should not be much larger, because then the spectrum will be much harder than the spectra derived from gamma ray observations. $T_d/T_{acc} \simeq 0.2$ requires $W_A \simeq 6$ erg cm^{-3} for $R = 10^9$ cm and $W_A \simeq 0.6$ erg cm^{-3} for $R = 10^{10}$ cm (Fig. 6). For the shorter loop, this energy density implies a short acceleration and decay time. Stochastic acceleration in relatively short loops, therefore, could play a dominant role during the impulsive phase of flares. For the longer loop, the time scales are longer, and therefore such loops are probably not adequate for impulsive phase acceleration. Concerning the extended pion decay emission observed from the 11 June and 15 June 1991 flares that lasted for hours (§IV), even for loop lengths as large as 10^{10} cm, the acceleration and decay times are still quite short (< 100 s). Thus, if we assume continuous stochastic acceleration in loops, the time profile of the emission is determined not by these times, but essentially by the injection of seed particles into the accelerator. It was suggested[60] that pion decay emission from the 3 June 1982 flare with characteristic decay time of 500 s was produced by protons stochastically accelerated during their diffusive transport in a loop. From Fig. 6 we see that, even for a loop length of 10^{10} cm, such diffusive trapping requires $W_A \simeq 16$ erg cm^{-3}, for which T_{acc} is about 10 seconds. As discussed above, this would lead to a very hard proton spectrum, as in fact was found in ref. 60 (S_p of about 1 at relativistic energies and much flatter at lower energies). Such a spectrum is much harder than typical proton spectra derived[16,30,44,58] from gamma ray observations for both impulsive and long duration events, and is in conflict with the neutron observations[61] of the 3 June 1982 flare.

TABLE 3, High Energy Emissions from the June 1991 X-class Flares

Date	Class	Location	Max Soft X-Rays (UT)	Lines	>10 MeV Emission	π Decay Emission	Neutrons
June 4	X12	N30E70	03:39	OSSE[9] 0.51 MeV 2.22 MeV 4.44 MeV 6.13 MeV $\sim$ 7 MeV	OSSE[9]		OSSE[9] Mt.Norikura[63,64] Neut. Mon. Neut. Telsc. Muon Telsc.
June 6	X12	N33E44	01:07	OSSE[9] 2.22 MeV 4.44 MeV 6.13 MeV $\sim$ 7 MeV	OSSE[9]		OSSE[9]
June 9	X10	N34E04	01:43	OSSE[9] 2.22 MeV 4.44 MeV 6.13 MeV $\sim$ 7 MeV COMPTEL[8,65] 2.22 MeV 4-7 MeV	OSSE[9]		COMPTEL[8,65]
June 11	X12	N31W17	02:09	OSSE[9] 2.22 MeV 4.44 MeV 6.13 MeV $\sim$ 7 MeV COMPTEL[8] 2.22 MeV SIGMA[62] $F_{2.2}/F_{4-7}$	OSSE[9] EGRET[7]	EGRET[7]	
June 15	X12	N33W69	08:21	COMPTEL[8] 2.22 MeV 4-7 MeV	GAMMA-1[10]	GAMMA-1[10]	COMPTEL[8]

IV. 11 JUNE AND 15 JUNE 1991 FLARES: TRAPPING VS. CONTINUOUS ACCELERATION

Gamma ray emission was detected from the series of X-class flares that occurred in June 1991 with various instruments on CGRO[7-9], with GAMMA-1 (ref. 10) and with SIGMA/GRANAT[62]. Neutrons were also observed from 4 of these flares with CGRO and from one of them with ground level instruments[63,64] (neutron monitor, neutron telescope and muon telescope on Mt. Norikura). In Table 3 we summarize these observations. We note, however, that for most of these data actual photon fluxes have not yet been presented.

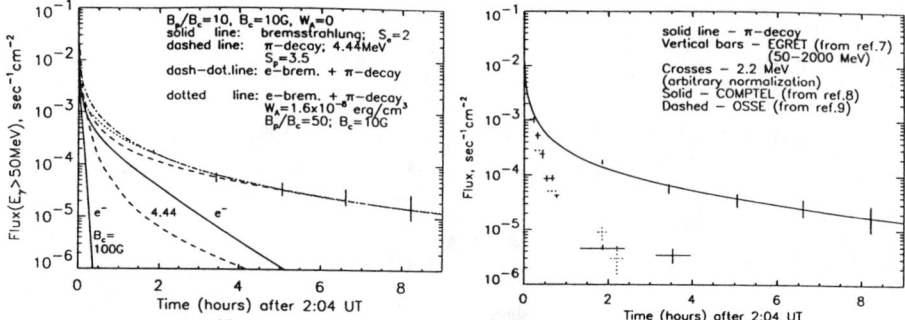

Fig. 9. Calculated[44] and measured gamma ray time profiles from the 11 June 1991 flare.

The remarkable feature of the 11 June and 15 June flares was the unusually long duration of the gamma ray emission. In the case of the 11 June flare 0.05-2 GeV gamma rays were measured with EGRET for 8 hours after the impulsive phase of the flare[7] (Fig. 9). These gamma rays are mostly of pionic origin with some admixture of primary electron bremsstrahlung at relatively low energies (50-70 MeV) during the early period of the observation (Fig. 10). Line emission at 2.22 MeV was also detected for about 2.5 hours and 4 hours following the impulsive phase, with OSSE[9] and COMPTEL[8] (Fig. 9).

Gamma rays of 0.03-3 GeV, also resulting from pion decay, were observed[10] with GAMMA-1 from the 15 June flare during two orbits of the satellite amounting to a total duration of about 2 hours. Between these two orbits 1-10 MeV emission was measured with COMPTEL[8] for about 40 minutes (Fig. 11). The central issue concerning these long lasting emissions is whether they were produced by particles that were continuously accelerated or by particles that remained trapped at the Sun after being accelerated in the impulsive phase of the flares.

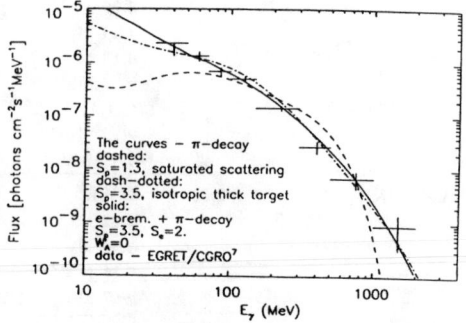

Fig. 10. Measured and calculated spectra of the gamma rays from the 11 June 1991 flare.

11 June 1991. We showed[44] that both the energy spectrum (Fig. 10) and the time profile (Fig. 9) can be fitted with the combination of primary electron bremsstrahlung and pion decay emission produced by particles trapped in coronal magnetic loops. The conditions that allow the long term trapping of the particles are: low level of plasma turbulence and relatively high mirror ratio ($W_A < 2 \times 10^{-8}$ ergs cm^{-3} for B_p/B_c=50) to prevent the fast precipitation of the particles through the loss cones; low coronal magnetic field ($B_c = 10$ G) to prevent synchrotron losses of the high energy electrons; matter density in the coronal part of the loop $n_c < 5 \times 10^{10}$ cm^{-3} to prevent Coulomb and nuclear losses. If these conditions are not satisfied, the emissions would decay too fast and therefore additional acceleration of the particles would be required.

The acceleration could be essentially continuous or episodal. Particles could be accelerated continuously by shocks moving up in the corona or by turbulence in the loops which would have to be constantly regenerated because of the short damping time of the waves. Episodal acceleration would involve several short acceleration periods between which the particles would be essentially trapped. We argued[44] against acceleration by a single shock produced in the impulsive phase because in 8 hours the shock would have moved to about $4R_\odot$ from where most of the particles would not be able to get back to the Sun.

We now examine in more detail the possibility of continuous acceleration by plasma turbulence. As we have seen (§III), the extended time profile of the GeV gamma ray emission requires the presence of a high level of plasma turbulence ($W_A > 1$ erg cm^{-3}) and the continuous supply of seed particles. We calculated the energy spectrum of the pion decay emission produced in the loop in the case of strong pitch angle scattering. As can be seen from Fig. 10, even for a very hard proton spectrum ($S_p = 1.3$), the resulting photon spectrum is not hard enough to fit the data at high energies. This is due to the facts that in the case of strong scattering the emission is mostly produced by particles in the loss cones and the directivity of pion decay emission increases with energy. Therefore, the spectrum in the backward direction (flare heliocentric angle 35°) becomes quite steep at high energies. Thus, we conclude that during most of the EGRET observing period the level of the plasma turbulence in the loop should have been far below the saturation level. This naturally rules out the model of continuous stochastic acceleration of the particles in magnetic loops.

It may still be possible that the particles were accelerated in several episodes separated by time intervals during which the particles are mostly trapped. In this case the level of the plasma turbulence can on average be low enough so that the strong anisotropy mentioned above does not develop. In addition, this model would also allow the weakening of the requirements on the level of plasma turbulence and coronal magnetic field that are necessary if the particles are accelerated only during the impulsive phase. On the other hand, the smooth decay of the gamma ray emission seen in Fig. 9 argues against multiple acceleration episodes, although the absence of variability can be due to the poor time resolution of the gamma ray observations. In any case, if multiple acceleration episodes to GeV energies indeed took place, the fact that no H$_\alpha$ flares occurred during the 8 hours of EGRET observations implies that acceleration to very high energies and production of gamma rays can take place in the absence of optical flares.

Stochastic acceleration, in principle, could occur in other magnetic structures which would not impose a strong anisotropy on the interacting particles. In Fig. 10 we also show the calculated energy spectrum of pion decay emission assuming an isotropic thick target interaction model. As can be seen, the fit is reasonably good, and the fact that the measured spectrum can be explained entirely by pion decay emission without a contribution from electron bremsstrahlung makes this model even more appealing. However, the fact that the spectrum of the accelerated particles remains essentially constant for about 8 hours, while the acceleration efficiency decays by orders of magnitude, is difficult to understand within any model of continuous acceleration. The value of $S_p \simeq 3.5$, derived[62] during the impulsive phase of the flare from the fluence ratio of the 2.22 MeV and the 4.44 MeV lines, is similar to the S_p that provides the fit to the energy spectrum of the pion decay emission measured 2-8 hours later (Fig. 10).

We conclude that although continuous acceleration of the particles cannot be ruled out, the assumption of particle trapping is probably the most natural explanation of the long lasting gamma ray emission from the 11 June flare. The initially suggested[44] trapping model should be further elaborated by taking into account OSSE and COMPTEL data on nuclear line emission after these data become available in terms of actual photon fluxes. The complete analysis should also include data on pion decay emission early in the flare when the EGRET spark chamber was saturated due to high photon fluxes. Such data can become available from OSSE and EGRET/TASC. This will allow us to constrain the parameters of the accelerated particles as well as the parameters of our model. We will also be able to check the specific predictions of particle trapping, in particular the fact that nuclear line emission should decay faster than the pion decay emission (Fig. 9). This follows from the fact that in the case of weak (or no) pitch angle scattering the spectrum of the trapped particles becomes harder with time due to Coulomb losses (§III). Of course this is only true if the effects of drifts (§III) are not significant, because the smaller Coulomb losses at high energies could be compensated by the more rapid escape from the loop due to drifts. Thus, if the comparison reveals that the nuclear lines decay faster than the pion decay emission, this would be a clear indication of particle trapping, while the absence of the effect can not be taken as an argument against trapping. As can be seen from Fig. 9, the currently available data do not allow to make such comparison because there is not sufficient overlap in time between the EGRET and OSSE/COMPTEL data, and the data are not accurate enough. One experimental fact that does provide support to trapping is the finding that, as expected in the trapping model (see Fig. 9), the electron bremsstrahlung decayed faster than the pion decay emission[7].

The other signature of trapping that could in principle allow to distinguish it from continuous acceleration is the fact that due to Coulomb losses the spectrum of the pion decay emission is expected to become harder with time (see Fig. 12b). However, as can be seen, the effect is not very strong and therefore spectral data of much higher accuracy than currently available would be necessary to verify this prediction.

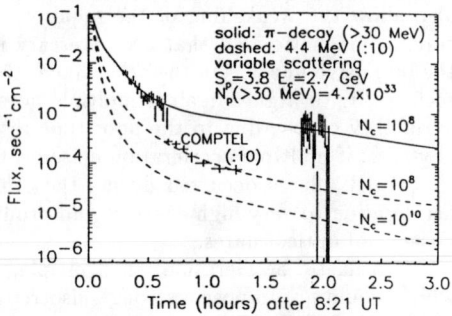

Fig. 11. Measured and calculated gamma ray fluxes for the 15 June 1991 flare[8,58].

15 June 1991. We studied[58] this flare by combining the GAMMA-1 pion decay emission data with the 1-10 MeV COMPTEL data. Since this range is dominated by the prompt nuclear line emission, we assume that these data also represent the time profile of the 4.44 MeV line. We obtained the absolute normalization from the measured 4-7 MeV fluence of (12.1 ± 1.9) ph/cm^2 (M. McConnell, private communication 1993) and the theoretical ratio $F_{4-7}/F_{4.4}=3.7$. The measured time profiles are shown in Fig. 11. No gamma ray observations were made during the impulsive phase of the flare because it occurred during the satellite nights of both GAMMA-1 and CGRO. In Fig. 11 we also show our fits to the GAMMA-1 and COMPTEL data which were obtained assuming that all the particles were injected into the loop instantaneously at the time of maximum of the accompanying soft X-ray emission (08^h21^m UT, according to GOES data). We found that to fit the GAMMA-1 data it is necessary to assume that the precipitation rate decreases with time. This can be caused, for example, by the expansion of the loop (reduction of the loss cones), or by the damping of the turbulence. The time profiles shown in Fig. 11 were calculated assuming that the energy density in the turbulence decays exponentially from an initial level of 1.6×10^{-5} ergs cm^{-3} with a characteristic time constant of 250 s. In order to simultaneously fit the time profiles of nuclear line and pion decay emissions we need a relatively low matter density in the coronal part of the loop (Fig. 11). On the time scales considered here, pion decay emission is practically insensitive to this parameter because the bulk of the pion producing ions interact in the subcoronal part of the loop. On the other hand, because of the much higher importance of the Coulomb losses, a significant fraction of nuclear lines are already produced in the corona.

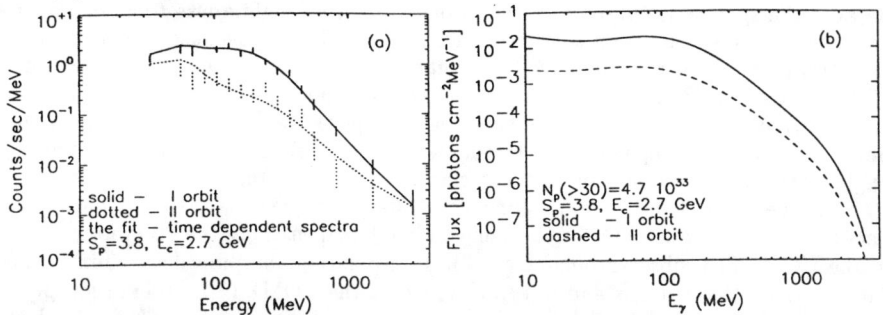

Fig. 12. Measured (GAMMA-1) and calculated energy spectra of pion decay emission from the 15 June 1991 flare (ref. 58). The fits in Fig. 12a were obtained by folding the theoretical spectra shown in Fig. 12b through the GAMMA-1 response and adding the background.

The energy spectrum of the accelerated ions used in Fig. 11 is the one that provides the best fit to the energy spectrum of the high energy gamma rays measured with GAMMA-1 (Fig. 12). We analyzed the spectral data by folding theoretical pion decay emission spectra through the GAMMA-1 response function, adding the background and comparing the result with the data. The theoretical spectra were integrated over the time interval during which the observations were made and the position of the flare was taken into account. Unlike the high energy spectrum from the 11 June flare, which contained a contribution from primary electron bremsstrahlung, the spectrum from 15 June flare was purely pionic and therefore more suitable to study the spectrum of accelerated ions. We tested spectra of pion decay emission resulting from accelerated ions with various spectral shapes, such as power laws, power laws with exponential cutoffs (characteristic energy E_c), Bessel functions and numerical solutions of the Fokker-Planck equation describing stochastic acceleration[59] which, unlike Bessel functions, are valid at all energies. This procedure allowed us to narrow down the range of the possible spectral parameters, namely we can exclude power laws with indexes < 3 and > 4.5, as well as the Bessel function. Among the remaining spectra only the power law with $S_p=3.8$ and $E_c=2.7$ GeV is consistent with GAMMA-1/COMPTEL flux ratio. However, we must note that this spectrum leads to a total time integrated 4.44 MeV flux (540 ph/cm^2) which is larger than any previously reported 4.44 MeV fluence.

The latter problem, as well as the requirement of low coronal density, can probably be overcome if instead of instantaneous injection of particles we assume that the acceleration lasted for several minutes, as in fact was indicated by neutron monitor data[66]. It is also possible that these problems are related to the simplified loop model that we are using, namely that we assume a constant magnetic field in the coronal part of the loop. In a more realistic magnetic field geometry (for example the model based on force-free equilibrium[49], §III), convergence of the magnetic field in the coronal part would allow to keep particles for a longer time in a region with a relatively low matter density, resulting in longer decay times of the nuclear line emission. Convergence of the field lines in the corona would also diminish the precipitation of the particles, and this would alleviate the problem of the very low energy density in the plasma turbulence which we needed to postulate in order keep the particles trapped for a long time for both 11 June and 15 June flares.

Finally, we consider here also the possibility that the particles were continuously accelerated. Unlike the 11 June flare, where we could rule out continuous acceleration because the spectrum obtained in the strong scattering regime was too steep at high energies, here we can obtain a fit with such spectra. This is because the effects of the anisotropy are not very pronounced for a flare near the limb (heliocentric angle of the 15 June flare was 73°). We find that the GAMMA-1 spectral data during the first orbit can be fitted with a spectrum corresponding to a power law proton spectrum with S_p between 3.4 and 3.5 depending on the model (isotropic thick target or saturated pitch angle scattering in a loop). In both cases an exponential cutoff at 2.7 GeV is required.

On the other hand, the 15 June flare offers the possibility of comparing pion decay and nuclear line data for the same time interval, namely the 40 minute interval of the COMPTEL observations (Fig. 11). Interpolating the pion decay flux between the two GAMMA-1 orbits and integrating it over the COMPTEL observation period, we obtain $F_\gamma(\pi) = (2.1 \pm 0.5)$ ph/cm^2. The corresponding ratio $F_{4-7}/F_\gamma(\pi)$ is plotted in Fig. 3, where the theoretical curves are based on isotropic thick target calculations (§II). The possible range of the power law spectral index is 2.9-3.6, depending on the composition. This range is consistent with that obtained from the analysis of the pion decay spectrum. (Note that the spectrum of the pion decay emission practically does not depend on the composition and the ratio of the nuclear line yield to the total energy integrated pion decay emission yield does not depend much on the presence of cutoff in the primary proton spectrum above 1 GeV). We conclude that continuous acceleration cannot be ruled out for this flare.

Observations of line emission at 511 keV could help distinguish between the models. The 4-7 MeV fluence of about 12.1 photons cm^{-2} (Fig. 11) should be accompanied by a 511 keV fluence of about 3 photons cm^{-2} resulting from radioactive nuclei (using a positron to 4-7 MeV photon production ratio of 0.4, and a positronium fraction of 0.9, §II). Even in the case of strong pitch angle scattering, we do not expect much attenuation, because, similar to nuclear deexcitation lines[2], the radioactive nuclei are expected to be produced at relatively high altitudes in the solar atmosphere. In addition, we also expect 511 keV line emission from positrons resulting from π^+ decay. Using a π^+/π° production ratio[37] of 4, we obtain a 511 keV line flux of $4.3f$ photons cm^{-2}, where f takes into account the uncertainty in the positronium fraction and the attenuation. There are no calculation of this parameter for solar flare loop models. We estimate that f will be about 0.7, when there is no scattering. When there is strong scattering, the positrons are produced deep in the atmosphere[35]. Therefore, for a flare near the limb, f should be close to 0. Thus, the predicted 511 keV line fluence is between 3 and 6 photons cm^{-2}, depending on the model. We also expect that the 511 keV line will be accompanied by positronium continuum below 511 keV. The CGRO observing period (0.6-1.3 hours, Fig. 11) is particularly favorable because there should not be much primary electron bremsstrahlung during this late phase of the flare.

V. SUMMARY

We have reviewed the accelerated particle interaction and transport processes relevant to gamma ray production in solar flares. The transport processes that we considered are adiabatic motion, pitch angle scattering by plasma turbulence, spatial diffusion and drifts. These processes affect the gamma ray time profiles, angular distributions and gamma ray line shapes. We also considered stochastic acceleration by the same turbulence that scatters the particles. We point out that, in addition to the energy density in plasma turbulence, the size of the loop in which the particles are accelerated is an important parameter that determines the energy spectrum of the particles, as well as the rise and decay times of the accompanying gamma ray emission. Flares with very impulsive time profiles (both rise and decay times on the order of seconds) require acceleration in small loops ($R \lesssim 10^9$ cm).

We investigated the long term trapping of particles in magnetic loops as a possible explanation of the extended gamma ray emission observed from the 11 June and 15 June 1991 flares. We find that such trapping requires large loops, or smaller but twisted loops, or loops which cover a large area of the solar surface. For all of these structures, the turbulent energy density should be sufficiently low to avoid the fast precipitation of particles through the loss cones. We also investigated the possibility of continuous acceleration of particles over extended time periods. The turbulent energy density required by such acceleration will cause the rapid precipitation of the particles and will lead to a highly anisotropic angular distribution of GeV gamma rays from pion decay. We found that the resultant energy spectrum is inconsistent with observations. However, we cannot rule out the possibility that particles were accelerated during several discrete episodes and remained essentially trapped between them.

The combination of nuclear deexcitation line and pion decay emissions allows us to determine, for the first time, the spectrum of the accelerated protons over a broad energy range (10 MeV-5 GeV). For the two flares mentioned above the spectrum is consistent with a power law of index around 3.5, and for one of them (15 June) there is also a high energy exponential cutoff around 3 GeV. There is as yet no convincing acceleration theory that predicts a single power in kinetic energy over this broad energy range.

We wish to acknowledge Y.-T. Lau and J. A. Miller for useful discussions, V. Akimov and N. Leikov for the collaboration in interpreting the GAMMA-1 data, and J. G. Skibo for help with the bremsstrahlung calculations.

REFERENCES

1. R. Ramaty and R. J. Murphy, Space Science Rev., 45, 213 (1987).
2. X. M. Hua, R. Ramaty and R. E. Lingenfelter, ApJ, 341, 516 (1989).
3. J. A. Miller and R. Ramaty, ApJ, 344, 973 (1989).
4. A. L. MacKinnon and J. C. Brown, Astr. and Ap., 232, 544 (1990).
5. V. G. Gueglenko et al., Solar Phys., 125, 91 (1990).
6. J. M. McTiernan and V. Petrosian, ApJ, 359, 524 (1990).
7. G. Kanbach et al., Astr. and Ap. Suppl., 97, 349 (1993).
8. J. M. Ryan et al., in: Compton Workshop (St. Louis), in press (1993).
9. R. J. Murphy et al., in: Compton Workshop (St. Louis), in press (1993).
10. V. V. Akimov et al., 22nd Internat. Cosmic Ray Conf. Papers, 3, 73 (1991).
11. T. Bai, PhD. Dissertation, Univ. of Md., (1977).
12. C. D. Dermer and R. Ramaty, ApJ, 301, 962 (1986).
13. P. E. Semukhin and G. A. Kovaltsov, 19th Internat. Cosmic Ray Conf. Papers, 4, 106 (1985).
14. V. Petrosian, ApJ, 299, 987 (1985).
15. J. M. McTiernan and V. Petrosian, ApJ, 359, 541 (1990).
16. R. Ramaty, N. Mandzhavidze, B. Kozlovsky, and J. Skibo, Adv. Space Res.(COSPAR), in press (1993).
17. H. W. Koch and J. W. Motz, Rev. Mod. Phys., 31, 920 (1959).
18. E. Haug, Zs. Naturforsch., 30a, 1099 (1975).
19. G. R. Blumenthal and R. J. Gould, Rev. Mod. Phys., 42, 237 (1970).

20. M. B. Kallenrode, E. W. Cliver, and G. Wibberenz, ApJ, 391, 370 (1992).
21. W. T. Vestrand et al., ApJ, 322, 1010 (1987).
22. D. J. Forrest et al., 19th Internat. Cosmic Ray Conf. Papers, 4, 146 (1985).
23. R. Ramaty, B. Kozlovsky, and R. E. Lingenfelter, ApJ Supp., 40, 487 (1979).
24. Murphy, R. J., G. H. Share, J. R. Letaw, and D. J. Forrest, ApJ, 358, 298 (1990).
25. R. J. Murphy, R. Ramaty, B. Kozlovsky, and D. V. Reames, ApJ, 371, 793 (1991).
26. D. V. Reames, ApJ Supp., 73, 235 (1990).
27. J.-P. Meyer, ApJ Supp., 57, 151 (1985).
28. A. Shemi, Mon. Not. Royal Astr. Soc., 251, 221 (1991).
29. J. T. Schmelz, ApJ, 408, 381 (1993).
30. R. J. Murphy, C. D. Dermer, and R. Ramaty, ApJ Supp., 63, 721 (1987).
31. X. M. Hua and R. E. Lingenfelter, Solar Phys., 107, 351 (1987).
32. N. Mandzhavidze and R. Ramaty, Nuclear Physics B, Proc. Suppl., in press (1993).
33. R. Ramaty and N. Mandzhavidze, in: Compton Workshop (St. Louis), in press (1993).
34. X. M. Hua and R. E. Lingenfelter, ApJ, 319, 555 (1987).
35. N. Mandzhavidze and R. Ramaty, ApJ, 389, 739 (1992).
36. B. Kozlovsky, R. E. Lingenfelter, and R. Ramaty, ApJ, 316, 801 (1987).
37. R. J. Murphy and R. Ramaty, Adv. Space Res., 4, No. 7, 127 (1984).
38. C. J. Crannell, G. Joyce, R. Ramaty, and C. Werntz, ApJ, 210, 582 (1976).
39. G. H. Share, E. L. Chupp, D. J. Forrest, and E. Rieger in: Positron Electron Pairs in Astrophysics, eds. M. L. Burns et al. (New York:AIP), p. 15 (1983).
40. R. Ramaty, in: The Physics of the Sun, eds. P.A. Sturrock (Reidel, Dordrecht, 1986) Vol. II, p. 291.
41. B. Klecker et al., 21st Internat. Cosmic Ray Conf. Papers, 5, 80 (1990).
42. E. I. Daibog, Yu. I. Logachev, V. G. Stolpovsky, V. F. Melnikov and T. S. Podstrigach, 21st Internat. Cosmic Ray Conf. Papers, 5, 96 (1990).
43. E. W. Cliver et al., ApJ, 343, 953 (1989).
44. N. Mandzhavidze and R. Ramaty, ApJ, 396, L111 (1992).
45. C. F. Kennel and H. E. Petschek, JGR, 71, 1 (1966).
46. V. G. Gueglenko, L. G. Kocharov, and G. A. Kovaltsov, in: Nuclear Astrophysics, ed. G. E. Kocharov (St. Petersburg), p. 62 (1991).
47. R. Ramaty in: High Energy Particles and Quanta in Astrophysics, eds. F. B. McDonald and C. E. Fichtel (MIT, Cambridge), p. 122 (1974).
48. J. A. Earl, ApJ, 193, 231 (1974).
49. Y.-T. Lau, T. G. Northrop, and J. M. Finn ApJ, in press (1993).
50. Y.-T. Lau and R. Ramaty, this volume.
51. E. Rieger, Solar Phys., 121, 323 (1989).
52. R. J. Murphy, X. M. Hua, B. Kozlovsky, and R. Ramaty, Ap.J, 351, 299 (1990).
53. C. Werntz, F. L. Lang, and Y. E. Kim, ApJ Supp., 73, 349 (1990).
54. M. Yoshimori et al., ApJ Supp., in press (1993).
55. M. A. Forman, R. Ramaty, and E.G. Zweibel, in: The Physics of the Sun, eds. P.A. Sturrock (Reidel, Dordrecht, 1986) Vol. II, p. 249.
56. J. A. Miller and A. Viñas, ApJ, in press (1993).
57. J. A. Miller and R. Ramaty, in: Particle Acceleration in Cosmic Plasmas, eds. G.P. Zank and T.K. Gaiser (AIP, New York, 1992) p. 223.
58. N. Mandzhavidze, R. Ramaty, V. Akimov, and N. Leikov, 23rd Internat. Cosmic Ray Conf. Papers, in press (1993).
59. J. A. Miller, N. Guessoum, and R. Ramaty, ApJ, 361, 701 (1990).
60. J. M. Ryan and M. A. Lee, ApJ, 368, 316 (1991).
61. E. L. Chupp et al., ApJ, 318, 913 (1987).
62. G. Trottet et al., Astr. and Ap. Suppl., 97, 337 (1993).
63. Y. Muraki et al., ApJ, 400, L75 (1992).
64. K. Takahashi et al., 22nd Internat. Cosmic Ray Conf. Papers, 3, 37 (1991).
65. J. M. Ryan et al., in: The Compton Observatory Science Workshop, p. 470 (1992).
66. V. Akimov et al., this volume.

HIGH-ENERGY GAMMA-RAY SIGNATURE OF PROTON ACCELERATION DURING 1991 JUNE 15 SOLAR FLARE.

G.E.Kocharov, E.I.Chuikin, G.A.Kovaltsov and I.G.Usoskin
A.F.Ioffe Physical-Technical Institute, St.Petersburg 194021, Russia

L.G.Kocharov
State Technical University, St.Petersburg 195251, Russia

ABSTRACT

We consider the γ-ray and radio observations for the 06.15.1991 flare. It is shown that these data give the first possibility for precise determination of proton spectrum in 10 MeV – 10 GeV energy range in flare site. The power law γ-ray spectrum up to 1 GeV indicates that the maximum energy in power law primary proton spectrum is $\geqslant$ 10 GeV. The time profiles of cm-radioemission and of γ-ray in 0.8-10 MeV energy band and above 50 MeV coincide. A continuous and simultaneous acceleration of protons and relativistic electrons at the gradual phase of the flare gives a natural explanation of the data.

INTRODUCTION

High energy ($\geqslant$100 MeV) gamma-emission in the solar atmosphere is generated either by neutral pion decay or by ultrarelativistic ($\geqslant$100 MeV) electron bremsstrahlung. The neutral pions are produced in pp- and pα-nuclear reactions by action of accelerated ions with energy $\geqslant$0.2 GeV/n. Thus this radiation let us know information on the characteristics of relativistic particles in the flare site. In this paper we consider observational data on the 1991 June 15 flare [1,2] obtained by the GAMMA-1 telescope of 50-4000 MeV gamma-emission. Here we present the results of our analysis. We also use [3] gamma-observations by GRO/COMPTEL installation in 0.8-30 MeV range and micro-wave radioemission data collected in 'Zimenki' station (Radiophys.Res.Inst.,Nizhny Novgorod, Russia) [4]

GAMMA-EMISSION ON 1991 JUNE 15.

The time profiles of gamma- and microwave-emission during the gradual X12/3B (N33W69) solar flare are shown in Figure 1. It is clear that gamma-observations by both GAMMA-1 (0837-0902 UT) and GRO/COMPTEL (0858-0937 UT) were carried out after the impulsive phase of the flare (about 0815 UT). GAMMA-1 data demonstrate an exponential decay of high energy gamma-emission lasted till about 0850 UT followed by plateau. The exponential decay time was about 9.8 min. Figure 2 shows count rate of GAMMA-1 in two energy ranges. It is seen that there was no significant variations of energy spectrum until 0902 UT when the impact of the South-Atlantic Anomaly became essential. We consider GAMMA-1 count rate after 0902 UT as the upper

© 1994 American Institute of Physics

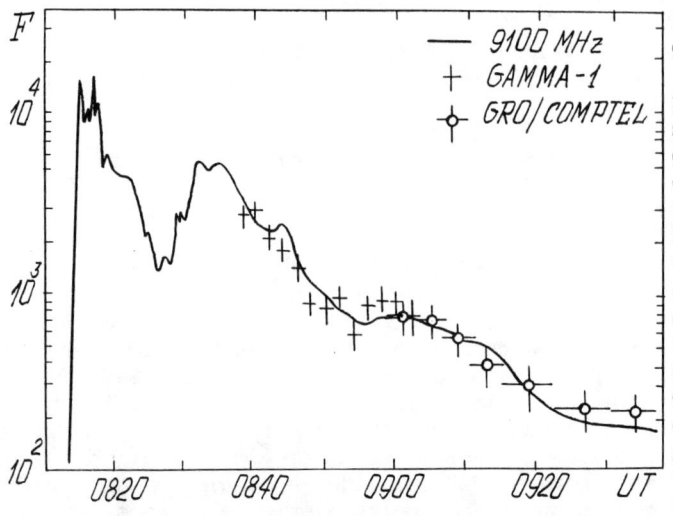

Fig.1. Time profiles of radio (sfu) and γ-emission (arb.units).

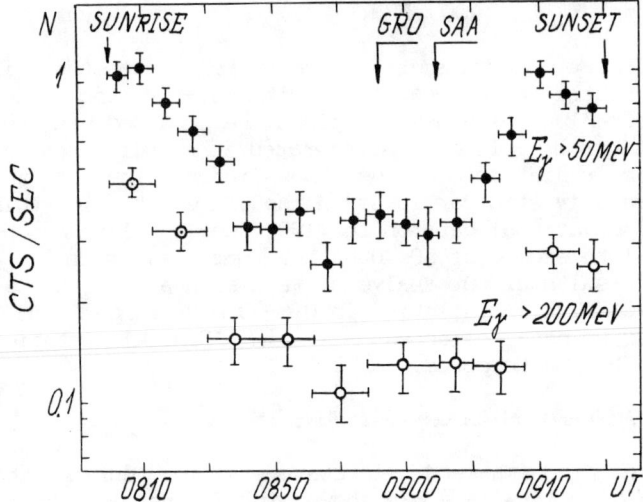

Fig.2. Time history of GAMMA-1 count rate.

limit for solar gamma-emission.
Akimov et.al.[1] proposed that observed high energy gamma-emission was originated from the pion decay. We tested this hypothesis by use of our calculations of nuclear reactions in the solar atmosphere and GAMMA-1 original data on count rate. We used the thick target isotropic model of neutral pion generation in the solar atmosphere. The primary proton energy spectrum above 200 MeV was proposed being double power law up to cutoff energy Em (see Figure 3). Then we simulated the GAMMA-1 count rate by means of known response function. The simulated device response was compared with the observed one by means of the $\chi^2(\nu)$ criterion (the number of energy intervals was $\nu+1$). At first a single power law spectrum (in Fig.3 S1=S2) was analyzed. But it occurred that it is impossible to obtain satisfactory accordance in the whole energy range. However, taking into account high energy range (Eγ >250 MeV) only, we reached a good fit if the primary proton cutoff energy Em is $\geq$10 GeV (see Fig.3). The best fit proton power law spectral index S_2 is equal to 3.5 ($\chi^2(4)=1.8$). The 10% significance level (i.e. $\chi^2(4)<7.8$) corresponds to proton spectral index S2 from 2.9 upto 4.1. The total number of accelerated above 1GeV protons for 0837-0902UT interval was about 10^{28} and roughly independent on S2.

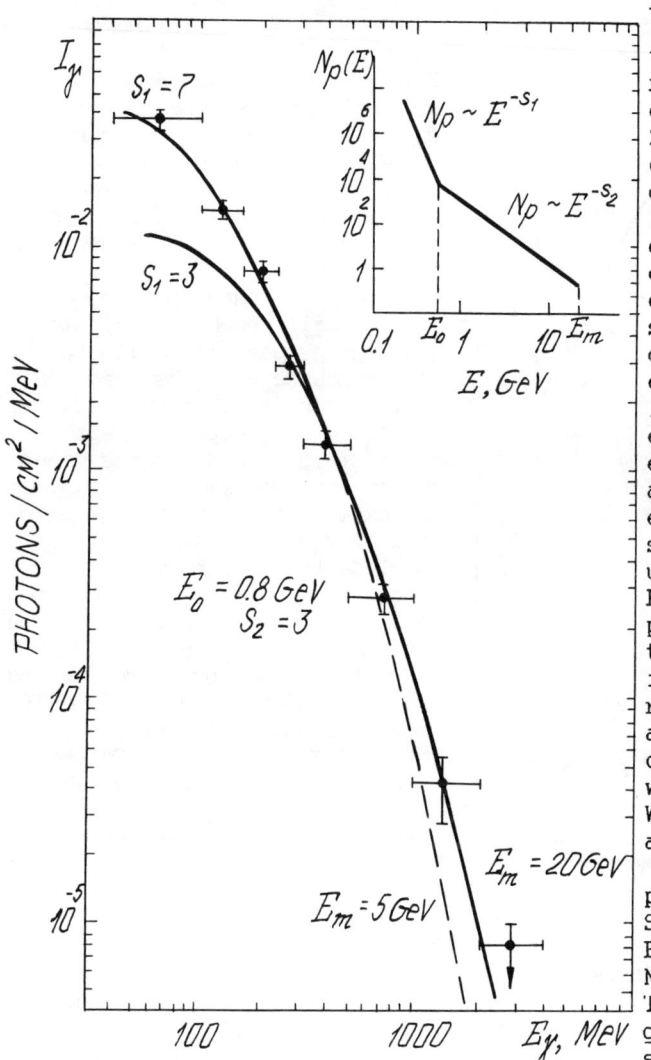

Fig.3. Observed and calculated spectra of γ-emission (0837-0902 UT).

For the single power law proton spectrum the calculated count rate in gamma-ray energy range below 250 MeV is significantly less than the observed one. Usually this fact is explained by additional relativistic electron bremsstrahlung[6-8]. On the other hand the excess of observed lower energy gamma-emission can be explained in the assumption of steeper primary protons spectrum from 0.2 GeV up to certain E_0. Proposing double power law shape of the spectrum ($S_1 > S_2$ in Fig.3) we compared the simulated and observed GAMMA-1 count rates in the whole energy range. We found a good accordance by the proton spectrum parametres: $S_2=3$, $S_1=7$, $E_0=0.8$ GeV, $E_m \geq 10$ GeV and $N_p(>1\text{GeV})=8*10^{27}$. The corresponding gamma-spectrum is shown in Fig.3. In order to visibly demonstrate accordance between calculations and observational data we also plot the experimental points presented in paper[1].

This flare was also observed by the GRO/COMPTEL device which can detect γ-rays in 0.8-30 MeV band[2]. Emission was visible up to 8 MeV only. The neutron capture line 2.2 MeV and other nuclear lines were observed. It means that electron bremsstrahlung is small in comparing with nuclear γ-emission which is produced by 10-100 MeV protons. GRO/COMPTEL observations can be explained under the assumption of the primary proton spectrum being either power law one with spectral

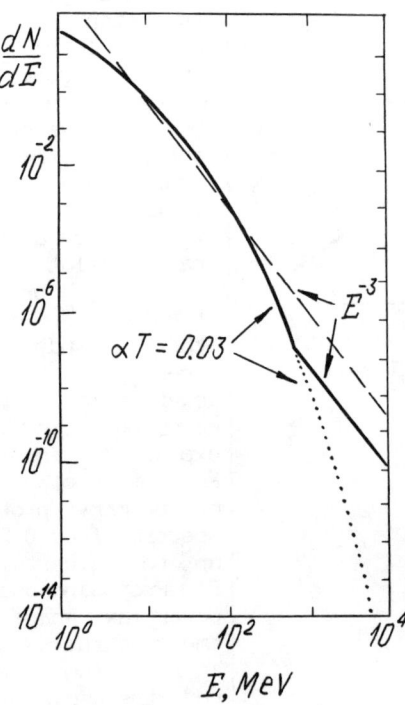

Fig.4. Primary proton spectrum at the Sun.

index ~3_a or Bessel function (αT~0.03)^a. The total number of above 30MeV protons for 0858$_{32}$– 0937 UT interval was about 10^{0}. In the first case one can extrapolate this power law spectrum to above 1 GeV and calculate pion decay gamma-emission. Simulated by this way GAMMA-1 count rate after 0858 UT is much higher than observed one. On the other hand the Bessel function spectrum gives small enough calculated high energy gamma-emission to be in accordance with observed one. It is essential that the Bessel spectrum steepens with energy (e.g.^{o}) and becomes steep enough around 0.8 GeV to explain the excess of 50–250 MeV γ-emission observed by GAMMA-1. Thus we conclude that the primary proton spectrum was a Bessel function upto 0.8 GeV and a power law one with the spectral index S^2~3 from 0.8 upto 10 GeV or more during the whole GAMMA-1 and GRO/COMPTEL observational time (see Fig.4). Both GAMMA-1 and GRO/COMPTEL observations can be explained under this proposition. In this case there is no necessity in additional relativistic electron bremsstrahlung in 50–250 MeV range.

DISCUSSION

According to 'Zimenki' station's observations the flare of 1991 June 15 was characterized by a continuous generation of radioemission which had a complicated multi-impulsive structure. It is seen from Figure 1 that the first pulse of centimeter radio-emission (9100 MHz) at 0815 – 0825 UT corresponds to the impulsive phase of the solar flare. It is followed by two increases corresponding to the extended phase. Pulse increase intervals in the centimeter band correspond to maxima in the decimetre band (950 MHz) during the extended phase of the flare. The decimetre emission may be caused by plasma radioemission and follows continuous multi-impulsive energy release processes^4. The centimetre radioemission can be interpreted as synchrotron emission of electrons with energy $\geqslant$1 MeV accelerated due to that energy release.

There are two possibilities to explain the long duration γ-emission. The first is continuous acceleration of particles during the solar flare. This explanation would be in accordance with the above interpretation of the radio emission. The second one is the only act of acceleration during the impulsive phase of the flare followed by trapping of energetic particles in a magnetic arch. This approach was applied to a number of solar flares (e.g. ^{10}). In Figure 1 we show overlapping time profiles of cm-radioemission and gamma-ray radiation observed by GAMMA-1 and GRO. One can see that the time profiles in the centimeter band and of gamma-ray radiation in nuclear lines and resulting from pion-decay coincide. Energy losses of particles producing these kinds of emission differ significantly one from another and so the observed temporal behaviour of those kinds of emission can not be explained by the primary particle deceleration. However in trapping models the temporal behaviour of the secondary emission is mainly determined by precipitation of particles due to scattering on MHD waves.The exponential decay of emission is typical for trapping models but in general case decay times are different for different kinds of emission because of the dependence of scattering time on particle gyroradius. Thus, in order to explain the observations on the base of the trapping model it is necessary to propose action of some special process which scatters the all kinds of particles with the same efficiency. Under this ad hoc assumption it seems to be possible to propose the only act of acceleration during the impulsive phase of the flare. In this case one has to extrapolate the total number of the primary relativistic protons back to the impulsive phase using an exponential law with decay time of 9.8 min obtained from GAMMA-1 observation. Unfortunately in this case those protons would produce the secondary high energy neutron flux that is in contradiction with Alma-Ata neutron monitor data. That is why we have to turn to the possibility of relativistic proton acceleration after the impulsive phase.

A shock wave which produces type II radioemission is a traditional candidate for the second stage acceleration. The metre II type radioemission$_{11}$ had onset at 0816UT in the impulsive phase and duration about 20 min . Thus one can consider the third possibility proposing relativistic protons being accelerated just before GAMMA-1 observation onset. However in this case all theoretical problems concerning to the trapping arise again. Hence we conclude that the first possibility is the best for the data in hand explanation. It gives a natural explanation of the temporal behaviour of microwave and gamma-ray emission during 0837 - 0937 UT interval, i.e. during about one hour after the impulsive phase. Thus we consider electromagnetic emission observed during the gradual phase of 1991 June 15 to be the result of continuous and simultaneous acceleration of protons and electrons in a wide range of energy. Some stochastic acceleration$_{12}$ mechanism (e.g. acceleration by an ensemble of shock waves exited due to multi-impulsive energy release) seems to be the most suitable for the data in hand explanation. The obtained above two component proton spectrum (Bessel function with a hard power law tail) may be originated from a special structure of accelerated region. Namely more energetic 'kernel' of the acceleration region may be a source of the hard power law tail above 1 GeV and more quite 'halo' may be a source of lower energy protons with Bessel function spectrum.

Recently[13,14] γ-radiation of the 1991 June 11 solar flare were reported. That flare took place at the same active region as the June 15 one. It is seen from June 11 observations that three time intervals and accordingly three time scales exist. The first time scale of about minute characterized the impulsive phase. Then 25 minutes e-folding time was observed during two hours. At last 255 minutes e-folding time of high energy gamma-emission was seen during at least 6 hours. One can propose that three spatial scales correspond to these three time scales, e.g. $\sim 10^9$ cm, $\sim 10^{10}$ cm and $\sim 10^{11}$ cm. For the June 15 flare we have got γ-observations for the second time interval considered above as an interval of extended energy release and acceleration. Note that high energy γ-spectra observed in the second intervals in both flares were similar and so a similar interpretation of these spectra is possible. It was reported[1,2] that high energy gamma-emission on June 15 was seen at the next GAMMA-1 orbit too, i.e two hours later the impulsive phase. It is possible to propose that this emission belong to the third time interval considered in as an interval of trapping of accelerated protons in giant magnetic loop. However in the case of June 15 flare we have no sufficient information to verify this hypothesis.

ACKNOWLEDGEMENTS

We would like to thank Dr.V.F.Melnikov, Prof.T.S.Podstrigach, Dr.E.V.Vashenyuk and Prof.E.V.Kolomiets for placing microvawe and neutron monitor data at our disposal. We are also grateful to Prof. V.Petrosian and Dr.V.F.Melnikov for fruitful discussion.

REFERENCES

1. V.V.Akimov et al., Proc.22nd ICRC,Dublin,3,73 (1991).
2. N.G.Leikov et.al., Astron.Astrophys.Suppl.Ser., 97, 345 (1993).
3. M.McConnell et.al., Report at "Recent Advance in High Energy Astronomy",Toulouse, France (1992).
4. G.A.Kovaltsov, G.E.Kocharov, L.G.Kocharov, V.F.Melnikov, T.S.Podstrigach, I.G.Usoskin and E.I.Chuikin, Izvestiya RAN, Ser.Fiz., in press (1993).
5. V.V.Akimov et.al., Space Sci.Rev., 49, 125 (1988).
6. L.G.Kocharov et.al., Nuclear Astrophysics (Phys.-Tech.Inst., St.Petersburg,1991),p.5.
7. R.Ramaty et.al., Proc.Compton Observatory Science Workshop, NASA CP3137, 480 (1992).
8. N.Z.Mandzhavidze and R.Ramaty,Ap.J.Lett., 396, L111 (1992).
9. J.A.Miller, R.Ramaty and R.J.Murphy, Proc. 20th ICRC, Moscow, 3, 33 (1987)
10. V.G.Gueglenko et.al.,Solar Phys., 125, 91 (1990).
11. Solnechnye dannye, #6 (1991).
12. A.M.Bykov and I.N.Toptyghyn, Izv.AN SSSR,Ser.Fiz., 45,474 (1981).
13. G.Trottet et.al., Astron.Astrophys.Suppl.Ser.,97,337 (1993).
14. G.Kanbach et.al., Astron.Astrophys.Suppl.Ser.,97,340 (1993).

COMPTEL'S SOLAR FLARE CATALOG

R.Suleiman, D.Forrest, M.McConnell, J.Ryan
University of New Hampshire, Space Science Center, Durham, NH
03824

R.Diehl, G.Lichti, G.Rank, V.Schönfelder, A.Strong, M.Varendorff[1]
Max-Plank Institut für Extraterrestrische Physik, D-8046
Garching, FRG

K.Bennett, L.Hanlon, C.Winkler
Space Science Department of ESA/ESTEC, 2200 AG Noordwijk, The
Netherlands

H.Bloeman, W.Hermsen, B.Swanenburg
SRON Leiden, P.B. 9504, NL-2300 RA Leiden, The Netherlands
1 Currently at University of New Hampshire

ABSTRACT

COMPTEL, the imaging gamma-ray telescope, capable of detecting gamma rays in the range of 0.1-30 MeV, is one of four instruments aboard NASA's Compton Gamma-Ray Observatory. The Comptel burst detectors (single Detector Mode) have a field of view of ~2.5π sr. These detectors of COMPTEL permit measurements of energy spectra and time histories of solar flare gamma-ray emission. A search through the Single Detector Mode's data is being conducted. We summarize the preliminary results of this search.

INTRODUCTION

The COMPTEL Solar Flare Catalog was obtained using the data generated in the single detector mode. The BATSE trigger flare list was used as the primary reference source. This catalog includes almost every trigger in the Burst Processor. The particular intensity threshold of solar flares needs no defining. The data from the high range of the Single Detector Mode (June 1991 to February 1992) has been processed and thirty flares have been confirmed. For each processed flare, energy spectrum and several time history plots, were produced for different energy bands. Such plots include (0.6-1) MeV for the 0.847 MeV line from ^{56}Fe, (2-2.4) MeV for the 2.223 MeV line from neutron capture on hydrogen, (3-7) MeV for the 4.44 and 6.13 MeV lines due to ^{12}C and ^{16}O deexcitation, and finally (8-10) MeV.

BURST DETECTOR

The imaging Compton Telescope, COMPTEL, uses 2 out of the 14 NaI cells in the lower (D2) detector (Fig. 1) to accumulate burst spectra upon receiving a trigger signal

52 COMPTEL's Solar Flare Catalog

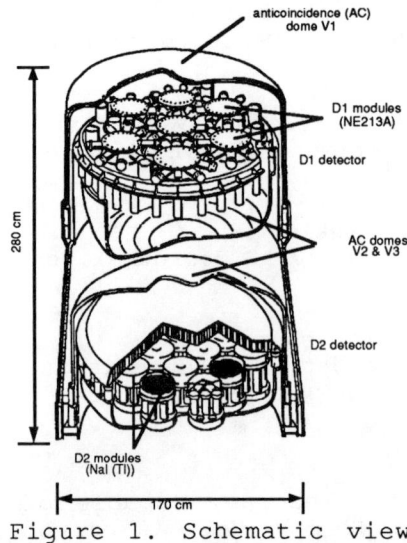

Figure 1. Schematic view of the COMPTON telescope COMPTEL. The burst detectors are shaded in black.

from BATSE. Each NaI(TI) crystal of the Single Detector Mode has a diameter of 28 cm, and a height of 7.5 cm. Each is viewed from below by 7 photomultipliers. The two modules measure energy in two different ranges: low range (0.1 MeV to ~0.6 MeV) and high range (~0.6 MeV to ~10.7 MeV). The threshold (0.1 MeV) in the low range is mostly due to the absorption of gamma-rays in the veto domes and the aluminum sandwich plate sheltering the (D2) assembly (see Fig. 1). The detector modules are supplied with dedicated ADC's and an electronic subsystem (Burst Spectrum Analyzer 'BSA')[1].

DATA and ANALYSIS

An extensive search is being conducted in the data of the BSA. If a trigger is found, then it is cross checked with BATSE and GOES data, to determine if it is a flare. If such a flare shows good statistics then further analysis is performed. The method of finding the background-subtracted flare is to subtract the average of 15 orbits before and after each flare. In the first analysis only the flares within the field of view of COMPTEL telescope mode were processed, but by taking advantage of the wider field of the single detector mode other flares were found. Approximately 30 flares were found and 26 flares have been excluded due to poor statistics.

RESULTS

In figure 2 we display the raw time profiles of 5 solar flares (not dead time corrected).

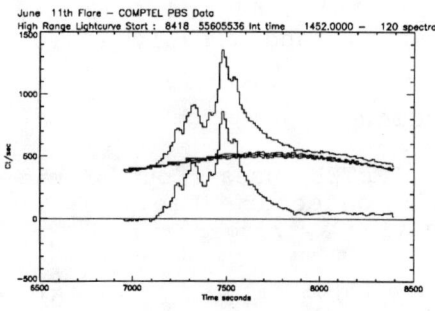
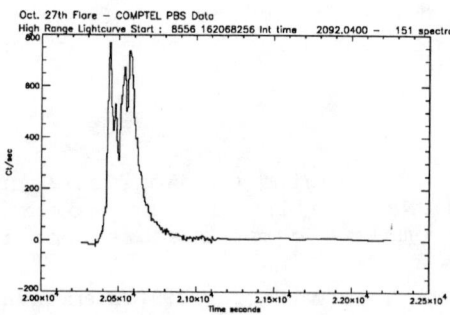

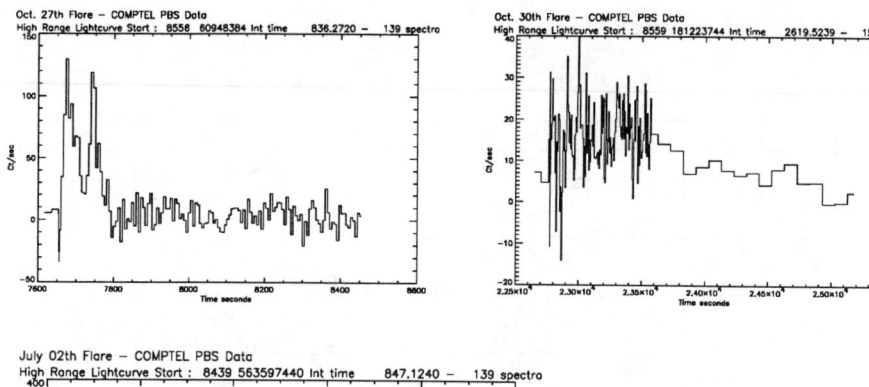
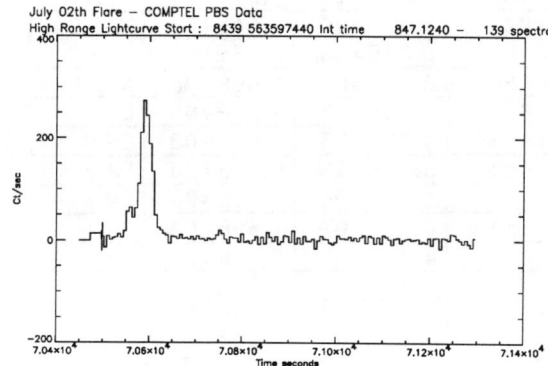

Figure 2. Time profiles of 5 solar flares detected in COMPTEL's burst mode. The first one includes the flare with the background, the middle lines are the background before and after and their average, and the flare minus background.

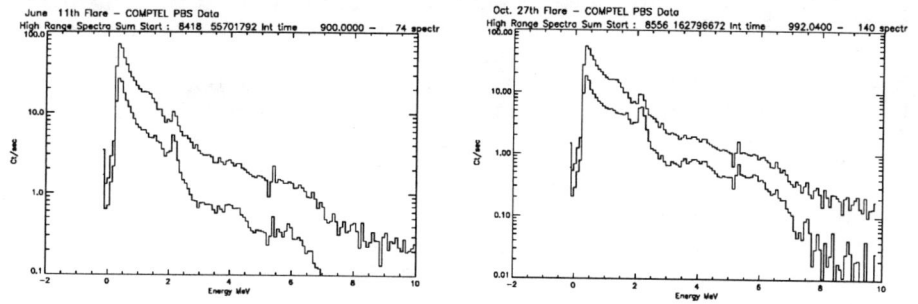

Figure 3. Energy spectra of solar flares. Top plot is the flare with background while the bottom is the flare minus the background.

We show some of the processed solar flares in Table 1. For each flare the onset time and the background subtracted counts in the energy range of 0.6 to ~10 MeV are given.

Table 1. Summary of some of the observed solar flares

Flare Date	Onset time (UT)	Batse Event	Batse's Counts	Comptel's Counts	Class
12/15/91	18:32:20	2102	1.00E+07	8.44E+03	M1.4
02/02/92	09:05:25	2551	1.36E+06	8.60E+03	M2.8
02/04/92	07:53:52	2596	6.65E+04	1.01E+03	C3.7
01/26/92	14:50:25	2511	8.39E+04	5.31E+04	C5.3
07/02/91	19:34:30	684	2.54E+07	1.90E+05	M4.6
07/17/91	06:25:00	786	5.24E+07	5.27E+04	X1.1
06/04/91	03:37:00	387	1.05E+09	1.61E+06	X12.0
06/09/91	01:39:00	419	6.02E+08	7.15E+04	X10.0
06/11/91	01:56:00	485	8.06E+08	3.49E+05	X12.0
06/30/91	02:55:00	665	1.50E+07	3.21E+05	M5.0
10/07/91	10:14:35	1436	4.30E+05	1.65E+03	C9.9
10/14/91	17:33:42	1491	9.50E+06	9.59E+03	M6.6
10/15/91	09:13:20	1498	1.36E+06	1.91E+04	C7.7
10/24/91	02:36:21	1551	6.54E+07	1.59E+04	X2.1
10/27/91	05:38:20	1648	4.29E+08	2.06E+06	X6.1
10/27/91	02:07:30	1643	3.40E+07	1.29E+05	X1.9
10/30/91	06:18:20	1723	7.13E+07	1.76E+05	X2.5
10/30/91	19:11:42	1743	7.30E+06	2.30E+04	M4.3
10/30/91	22:24:10	1749	2.70E+05	1.24E+04	C5.3

SUMMARY

This paper reports on the preliminary results of the construction of the COMPTEL Solar Flare Catalog. The data presented in this paper are from the COMPTEL burst mode. Results from spectral studies and neutron analysis, from COMPTEL's telescope mode, for some of the flares in June and October of 1991, can be found in these proceedings[2,3].

REFERENCES

1. Winkler,C., et. al, Adv. Space Res. 6, 113 (1986)
2. McConnell, et al. , 1992, COMPTEL observations of solar flare gamma-rays, COSPAR E.3-S.5.06.(in press)
3. Rank, G., et al. , 1992, Observations of the 1991 June 11 solar flare with COMPTEL, COMPTON Symposium, Washington University,St. Louis, (October 1992) (AIP conference, in press)

COMPTEL OBSERVATIONS OF GAMMA-RAY FLARES IN OCTOBER 1991

M. Varendorff, D. Forrest, M. McConnell, J. Ryan, R. Suleiman
University of New Hampshire, Institute for the Study of Earth, Oceans and Space,
Durham NH 03824, USA

R. Diehl, G. Lichti, G. Rank, V. Schönfelder
Max-Planck Institut für extraterrestrische Physik, D-8046 Garching, FRG

K. Bennett, L. Hanlon, C. Winkler
Astrophysics Division, Space Science Department of ESA, ESTEC
2200 AG Noordwijk, The Netherlands

B.N. Swanenburg
SRON-Leiden, P.B. 9504, NL-2300 RA Leiden, The Netherlands

ABSTRACT

The COMPTEL experiment on GRO images $0.75 - 30$ MeV celestial gamma-radiation that falls within its 1 steradian field of view. During observation 12 (primary target Cen A) in October 1991 the sun had been in the fov and several solar flares associated with the active region 6891 had been observed. Time profile and energy spectra had been produced, using COMPTEL's primary mode of operation (the telescope mode). Additionally the number of counts received in the D2-single burst detector (the secondary mode of operation) are given. We summarize the preliminary results on all of these flares.

INTRODUCTION

The imaging Compton Telescope, COMPTEL, onboard NASA's Compton Gamma Ray Observatory is a powerful tool for studying both the photon and neutron emission from energetic solar flares[1,2]. Since shortly after its launch in April 1991, COMPTEL has observed a number of solar flares, including the flares on June 9, 11 and 15, 1991[3,4].

In COMPTEL's imaging telescope or 'double scatter' mode ($0.75 - 30$ MeV), it measures both positions and spectra of cosmic γ-ray sources that fall within its 1 steradian field of view. A pulse shape discrimination and time of flight measurement allows the separation between γ-rays and neutrons. In addition, in burst or 'single detector' mode COMPTEL accumulates independent $0.1-1.1$ MeV and $1-10$ MeV spectra in two of its lower NaI detectors.

In COMPTEL's 'double scatter' mode, a photon which Compton-scatters in one of the seven upper D1 detectors, is then detected in one of the lower fourteen high-Z D2 detectors. In the simplest case of a single Compton scatter in D1 and complete absorption in D2, the possible γ-ray source positions lie on a circle of radius $\bar{\varphi}$ around the direction of the scattered photon, with

$$\cos\bar{\varphi} = 1 - \frac{1}{\epsilon_2} + \frac{1}{\epsilon_1 + \epsilon_2}, \quad (1)$$

where ϵ_1 and ϵ_2 are the energy deposits measured in the upper (D1) and lower (D2) detectors, respectively (in units of the electron rest-mass). In the case of solar flares, the position of the source is known and the location information is used for event selection to suppress the background. This also allows us to compile a spectrum, where events from the sun with only partly absorption in the lower D2-detector are suppressed, leading to a nearly diagonal response.

A descritption of the 'single mode' together with a catalog of spectra obtained from solar flares can be found in these proceedings[5].

DATA and ANALYSIS

In October 1991 the sun was in a very active phase and produced several X-class flares. Comptel registered 9 solar flares in its 'single mode' in October 1991. During observation period 12 from October 17 to 31, 1991, the sun was in the field of view of the telescope. The instrument was pointed towards the Cen-A region on the sky. BATSE registered several hundred solar flares during this time. For the stronger ones we produced time profiles of the COMPTEL data. We used the flare onset time and duration measured by BATSE, to select an appropriate time window. Only telescope events which satisfy the optimum event selection criteria, and which were consistent with the solar position, were used.

If the time profile showed evidence for the presence of a flare signal, an image was produced, to verify, that the signal is coming from the direction of the sun.

RESULTS

In Figure 1 we display the raw time profiles (not yet deadtime corrected) of 5 solar flares.

Two of the flares, 24.10.91 2:36:21 and 27.10.91 5:36:23, were so intense, that the instrument showed severe deadtime effects up to 90%. For the flare on 24.10.91 16:53:22s the instrument was shut off shortly after the onset of the flare because of a SAA passage. The flare on 30.10.91 6:16:18 occured just ~200 seconds after the SAA pass of the satellite. In figure 2 we show the energy loss spectra of two solar flares. For information on the background in the energy histograms, the measured countrate 15 orbits earlier and 15 orbits later is given. These orbits can be used as a good first estimate of the background count rate. For the flare of 27.10.91 5:39:35s the real background will be lower, because of the mentioned deadtime effect.

Table 1. Summary of the observed solar flares

Flare (date)	onset time (UT)	class	duration (s)	telescope src/bkg
24.10.	2:37:10	X2.1	110	181/ 5
24.10.	16:54:15	M3.2	25	74/ 35
27.10.	2: 7:30	X1.9	140	610/ 65
27.10.	5:39:35	X6.1	375	614/ 50
30.10.	6:16:18	X6.1	1650	1357/ 600

We summarize our results in Table 1. For each flare the onset time and duration of the impulsive phase in the energy range from 0.72 to 30 MeV is given. Furthermore the total number of counts and the estimated number of background counts in the 'telescope mode' during the impulsive phase of the flares are given. A

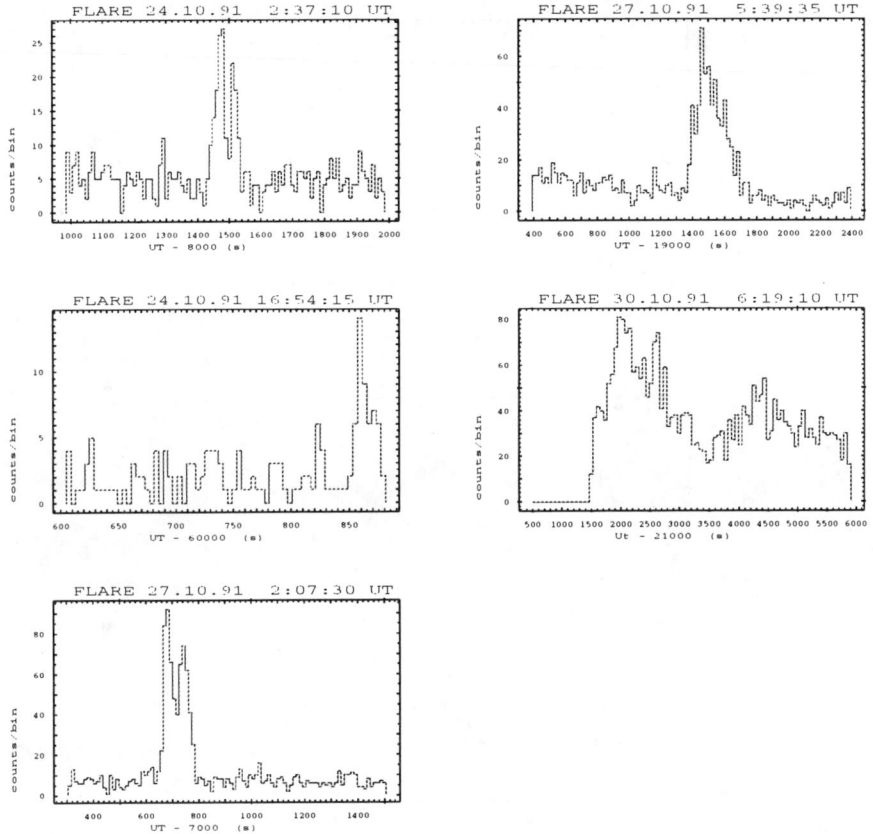

Figure 1. Time profiles of solar flares detected in COMPTEL telescope mode.

more detailed study of the background is still in progress. This should allow us to better define the extent of the γ-ray emission.

SUMMARY

This paper reports on the preliminary results of the analysis of 10 solar flares during an observation in Ocotber 1991. For the time profiles and spectra only data from COMPTEL's telescope mode were used. The total number of counts from solar flares using the single detector mode are given as well. Results from spectral studies and analysis from neutrons from solar flares in June 1991 and a catalog of preliminary spectra and countrates from the single detector mode data can be found in these proceedings.

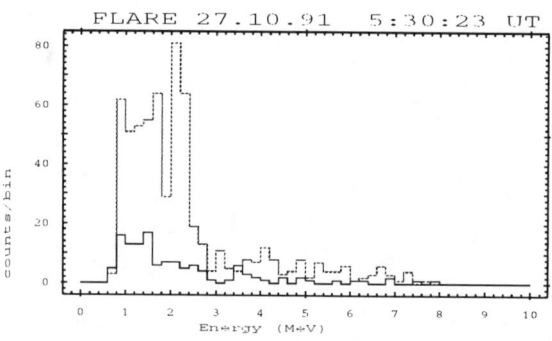
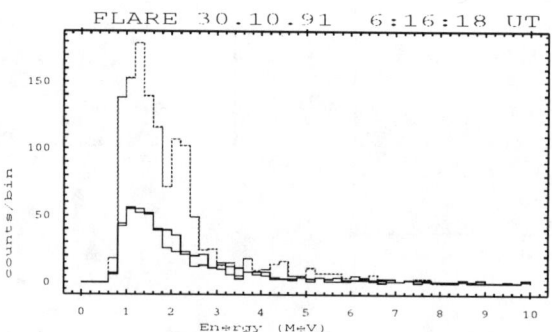

Figure 2. Energy loss spectra of solar flares with background spectrum overlayed.

REFERENCES

1. Schönfelder, V. et al. , Ap. J. Suppl. (in press, 1993).
2. Ryan, J. et al. , Data Analysis in Astronomy IV, ed. V. diGesu et al. , Plenum Press, New York 261 (1992).
3. McConnell, M. et al. , COSPAR E.3-S.5.06 (1992).
4. Rank, G., et al. , COMPTON Symposium, Washington University, St. Louis, (October 1992).
5. Suleiman, R., et al. , Conference on High Energy Emission from Solar Flares, Waterville Valley, N.H., USA, (March 1993, in press).

A Search for Low-Energy Protons in a Solar Flare from October 1992: Preliminary Results

T. Metcalf, D. Mickey, R. Canfield, and J.-P. Wülser
Institute for Astronomy, University of Hawaii

ABSTRACT

We give preliminary results from the first use of the University of Hawaii's new Imaging Vector Magnetograph (IVM) to search for linear polarization in the H-alpha spectral line during solar flares. Such polarization has previously been interpreted as impact polarization from 100 keV protons impacting the chromosphere[1,2]. The new data set has several advantages over previous data. First, the field of view is substantially larger than that used by Metcalf et al.[2], and, second, the temporal resolution (16 s) is a factor of two better than that previously obtained. We show a preliminary comparison between the flare Hα polarization and hard X-rays observed with the Compton Observatory.

INTRODUCTION

Hénoux et al.[1] recently obtained observations of linear polarization in the Hα line during three solar flares, which they interpreted as impact polarization from a hecta-keV proton beam colliding with chromospheric hydrogen. They discussed several characteristics of impact polarization which allow its identification. First, for a vertical proton beam, the polarization should be directed towards the center of the solar disk and, second, the polarization should be strongest at the limb and weakest at disk center (for a truly vertical beam, the polarization would vanish at disk center). Hénoux et al. have identified the observed polarization as impact polarization based on its direction, which pointed along the line from the observation point to disk center in all three flares (to within ±20 degrees).

Metcalf et al.[2] also observed linear polarization in the Hα line using the Stokes Polarimeter at Mees Solar Observatory[3] and presented observations of a flare on June 14, 1991. The Stokes Polarimeter has significantly better sensitivity to the polarization than Hénoux's instrument, but did not have as large a field-of-view. The June 14, 1991 flare was co-observed with the Compton Observatory and the dataset was sufficient to demonstrate the non-thermal nature of the Hα polarization. Hénoux et al.[1] did not have hard x-ray observations.

The data presented here use the University of Hawaii's new Imaging Vector Magnetograph (IVM) at Mees Solar Observatory. This instrument combines the best characteristics of the instruments used before by Hènoux et al.[1] and Metcalf et al.[2]. The IVM has about the same sensitivity to the polarization signal as the Stokes Polarimeter, since it was designed as a vector magnetograph, but it uses a CCD camera for imaging rather than the raster scan used by the Polarimeter.

Hence the IVM has both a good temporal cadence (16 s) and a large field-of-view (4 arc minutes).

Our goal in this study is to determine whether the Hα polarization observed during flares is due to the impact of 100 keV protons on the chromosphere. For this preliminary study, we discuss the direction of the linear polarization in the flare of October 23, 1992 (0058 UT). In a future study, we plan to examine the dependence of the polarization on the heliocentric angle and the temporal correlation between the linear polarization in Hα and Compton Observatory observations of the high energy proton population. These may be sufficient to prove or disprove the hypothesis that the linear polarization is impact polarization from a proton beam.

DATA

The Imaging Vector Magnetograph uses a 27 cm reflecting telescope. Wavelength selection is accomplished with a pre-filter and a servo-controlled Fabry-Perot etalon with a spectral range 5000-7000 Å, millisecond settling time, and 70 mÅ bandpass. Polarimetric analysis is done with liquid crystal retarders. The 512 x 512 Tektronix CCD detectors allow us to obtain 1.2" instrumental resolution (0.6" pixels) and a 4.6' by 4.6' field of view. For each polarization image, the IVM uses four CCD images to compute the linear polarization. In the current setup, the IVM observes the Hα line in two wavelengths: line center and 1Å to the blue side of the line.

Figure 1 shows the hard x-rays observed with the BATSE[4] instrument during the flare of October 23, 1992 and Figure 2 shows the flare at Hα line center at the time of the hard X-ray peak (00:58:54 UT). The underlying image is the Hα line center intensity and shows the enhanced Hα emission during the flare. The short line segments show the magnitude and the direction of the linear polarization. Most of the strong polarization is clearly directed towards the center of the solar disk, as predicted by the impact polarization hypothesis. However, there is a distinct component which is directed perpendicular to the disk center direction.

This perpendicular component can also be understood in the context of impact polarization since the presence of higher energy protons and/or horizontally moving photoelectrons would give this signature. These populations would give a smaller polarization signal than 100 keV protons, and this may explain why the perpendicular component is primarily observed at the edges of those Hα kernels with strong polarization directed towards disk center. The high energy protons and/or photoelectrons may be present in the kernels as well, but their effect is swamped by the 100 keV protons. This remains speculation at this stage in the analysis, however.

Figure 3 shows a polar histogram of the number of IVM pixels showing linear polarization in each direction at the time of the hard x-ray peak (00:58:51 UT). There are clearly two components to the the polarization, one directed

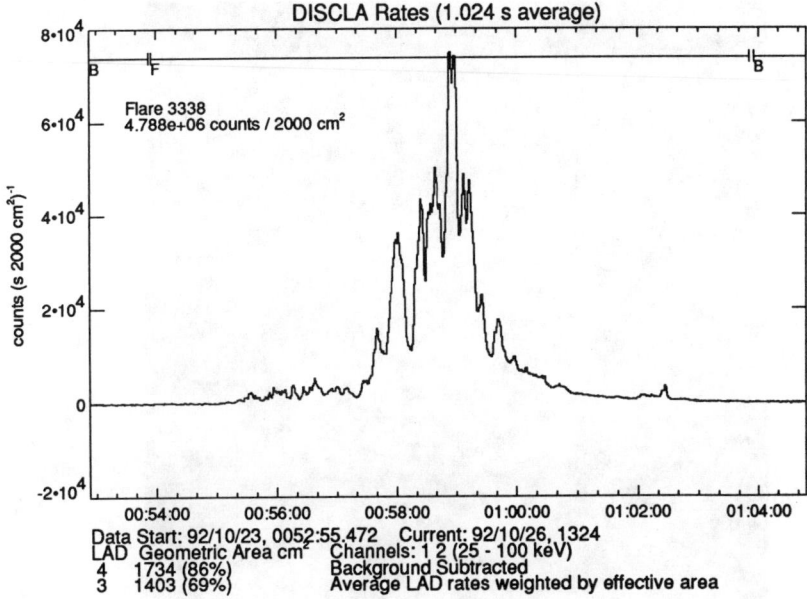

Figure 1: The BATSE hard x-ray light curve (25-100 keV) for the flare of October 23, 1992.

towards the center of the disk and one directed perpendicular to this. This is consistent with impact polarization from a nearly vertical beam of particles. The 20 degree offset from the disk center direction could be an indication of the tilt of the magnetic field line in the chromosphere along which the beam is moving.

Figure 4 shows the polarization at 01:02:06 UT, after the hard x-ray burst. At this time, the polarization has spread out quite suddenly. The extended polarization was not observed in the previous IVM polarization image at 01:01:51 UT. Could this be an injection of protons into a magnetic field line not previously involved in the flare? At no time was there bright unpolarized Hα emission from this site.

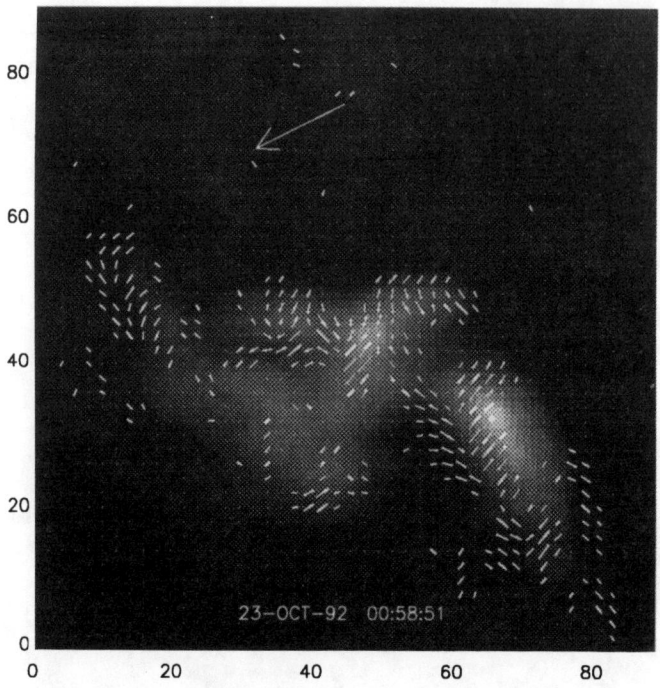

Figure 2: The underlying image is the flare seen at line center in Hα at the time of the peak in the hard x-rays (00:58:51 UT). The short line segments show the magnitude and the direction of the linear polarization (binned by 2's), and the arrow shows the direction to disk center. The line segments are 2.4 arcseconds apart.

CONCLUSIONS

Our preliminary conclusions are:
- The new Imaging Vector Magnetograph at the University of Hawaii is ideal for observations of linear polarization in the Hα line. It has significantly better temporal and spatial resolution than achieved previously.
- Significant linear polarization was observed in the October 23, 1992 flare.
- The linear polarization was preferentially directed 20 degrees from the direction to disk center.
- The Hα polarization peaked at the time of the hard x-ray burst maximum, though the polarization did not decay as rapidly as the hard x-rays.
- The component of the observed polarization directed towards disk center is consistent with the predictions of impact polarization by approximately 100 keV protons.

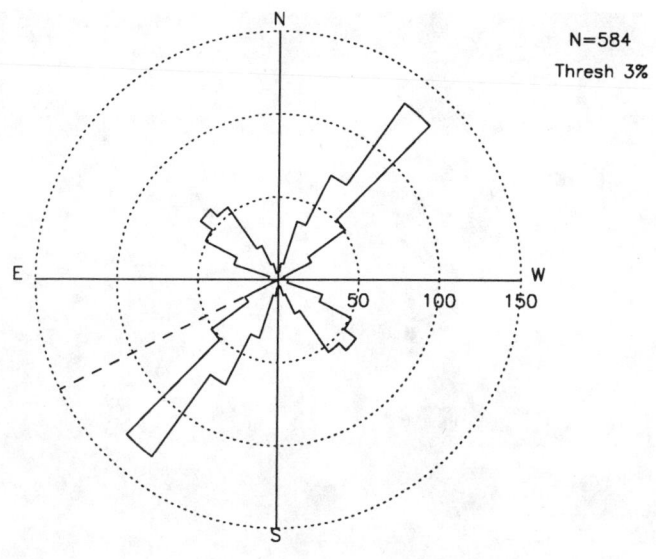

Figure 3: A polar histogram of the number of IVM pixels with linear polarization in each direction at the time of the hard x-ray peak (00:58:51 UT). Only pixels with linear polarization above 3% are plotted. The radial dashed line shows the direction of disk center.

REFERENCES

1. Hénoux, J. C., Chambe, G., Smith, D., Tamres, D., Feautrier, N., Rovira, M., and Shal-Bréchot, S., Ap. J. Supp., **73**, 303 (1990).
2. Metcalf, T. R., Wülser, J.-P., Canfield, R. C., and Hudson, H. S., in "The Compton Observatory Science Workshop", 536 (1992).
3. Mickey, D. L., *Solar Phys.*, **97**, 223 (1985).
4. Fishman *et al.*, in *Proceedings of the Gamma Ray Observatory Science Workshop*, ed. W. N. Johnson (1989).

64 A Search for Low-Energy Protons in a Solar Flare

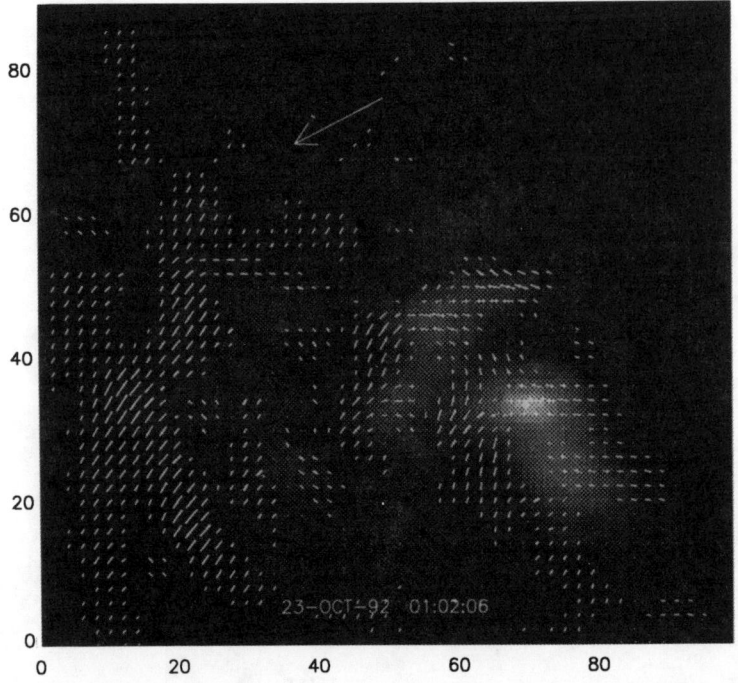

Figure 4: An Hα flare image at 01:02:06 UT with the linear polarization vectors superimposed. The arrow shows the direction of disk center.

A CORRELATION BETWEEN 4-8 MEV GAMMA-RAY-LINE FLUENCE AND >50 KEV X-RAY FLUENCE IN LARGE SOLAR FLARES

E.W. Cliver
Space Physics Division (GPSG), Phillips Laboratory, Hanscom AFB, MA 01731-3010 USA

N.B. Crosby
DASOP, Observatoire de Paris, Section d'Astrophysique, 92195, Meudon, France

B.R. Dennis
Laboratory for Astronomy and Solar Physics, Code 682.2, NASA/Goddard Space Flight Center, Greenbelt, MD 20771 USA

ABSTRACT

For large flares observed by the Solar Maximum Mission (SMM) satellite from 1980-1982, we find a reasonably good correlation between 4-8 MeV gamma-ray-line (GRL) fluences and >50 keV hard X-ray fluences. We find no compelling evidence for a distinct population of large hard X-ray flares that lack commensurate GRL emission. Our results are consistent with the acceleration of the bulk of the ~100 keV electrons and ~10 MeV protons (i.e., the populations of these species that interact in the solar atmosphere to produce hard X-ray and GRL emissions) by a common process in large flares of both long and short durations.

INTRODUCTION

Forrest[1] found that the >300 keV electron bremsstrahlung continuum and 4-8 MeV GRL emissions of solar flares were correlated down to the GRL detection threshold of the Gamma Ray Spectrometer[2] (GRS) on SMM. Thus Forrest[1] and Chupp[3] argued that all flares with >300 keV continuum could be GRL flares, given a sensitive enough detector. Bai[4] criticized this inference because, in his picture, protons and the bulk of relativistic electrons are both accelerated by a second-step process that operates only in GRL flares. From Bai's viewpoint, the correlation found by Forrest[1] was not surprising but was instead an expected result of second-step acceleration. Bai[4] argued that there exists a population of large hard (e.g., >50 keV) X-ray flares with steep spectra ("non-GRL flares" that result from a primary or "first-step" acceleration mechanism) in which a second-step process does not operate to accelerate particles to high energies. Such events should weaken any correlation in a plot of >50 keV emission vs. 4-8 MeV line emission for a sample of large flares.

To test Bai's[4] contention, we compared flare 4-8 MeV GRL fluences with >50 keV fluences to see if the correlation reported by Forrest[1] could be extended to lower X-ray energies. To conduct this test, we made use of data reduction/analysis programs recently completed by the Hard X-ray Burst Spectrometer (HXRBS) team that enables one to readily determine X-ray fluences. The ratio of the flare bremsstrahlung continuum emission produced by accelerated electrons to the GRL emission produced by protons provides a measure of the electron to proton (e:p) ratio of interacting particles. Cane, McGuire, and von Rosenvinge[5] were the first to show that the e:p ratios of solar energetic particle (SEP) events observed in space following flares are ordered by the flare duration, with impulsive flares having higher e:p ratios. Thus we also examined the effect of flare duration on the e:p ratio of the particles that interact at the sun to produce X-ray and GRL emission to see if a similar relationship held.

The analysis is described in Section 2; results are discussed in Section 3.

ANALYSIS

Fluence Calculation. The HXRBS detector[6] consists of a CsI(Na) scintillation spectrometer with a large anticoincidence shield. The HXRBS Event Catalog[7] contains 7045 events for the three years 1980-1982. As the first step in our procedure for obtaining fluences, we required that an event have a detectable flux in Channel 3 as reported in the HXRBS Catalog. From February 1980 - December 1982, the low energy cut-off for this channel increased from 49 keV to 63 keV. Of the events with a signal in Channel 3, we selected those having durations >200 s and/or peak count rates, integrated over all channels, of >100 c s^{-1}. Non-solar events and events flagged as having "noisy data" were not considered. Each selected HXRBS event was broken down into discrete time intervals by the automated procedure described in Crosby, Aschwanden, and Dennis[8]. Then the integral count rate above pre-event background for each interval was deconvolved to approximate the incident photon flux, which was assumed to have a power-law spectrum of the form $E^{-\gamma}$, using conversion factors generated by modeling the detector response to an incident flux with such a spectrum. A least-squares spectral fit for each interval was performed using an automated procedure, and the fit parameters were stored in a "summary file". There were 2878 such events during 1980-1982. Only flare intervals with power-law slopes greater than 1.1 or less than 7.0 were used in the fluence calculations. For $\gamma < 1.1$, the integral of the X-ray spectrum diverges, while values of $\gamma > 7.0$ may reflect a thermal spectrum and are, in any case, unreliable because of the relatively poor energy resolution of the CsI(Na) detector. To ensure that we considered only detected >50 keV emission and were not merely integrating background noise from erroneous spectral fits both early and late in flares when the counting rate is low, we followed the procedure of Crosby, Aschwanden, and Dennis[8] and only considered those intervals for which the calculated value of the thick-target energy in >50 keV electrons exceeded the value of this parameter averaged over all intervals that met the above criteria. The >50 keV fluence value for a given event was then obtained by summing the contributions from all valid intervals.

As a check on the accuracy of the HXRBS fluences obtained by the above method, we compared our >50 keV fluences with preliminary 40-140 keV fluences measured by the GRS on SMM (Vestrand, private communication, 1992) for a sample of large flares observed from 1980-1982. The result of the HXRBS-GRS comparison is shown in Figure 1. The plot contains nearly all HXRBS events with >50 keV fluences ≥5000 photons cm^{-2} for which the peak of the burst was observed, and a decreasing fraction of such well-observed events for smaller fluences.

The circled data points in Figure 1 indicate events affected by pulse pile-up[9]. The presence of pulse pile-up in an event is revealed by a comparison of the outputs from the two X-ray detectors on GRS. One of these detectors has an additional iron filter to block lower energy photons and, therefore, is less susceptible to pulse pile-up distortion of counting rates. Any difference in the output of the two detectors for a common energy range can be attributed to a greater degree of pulse pile-up in the detector without this filter. In terms of their level of "shielding", the two GRS X-ray detectors bracket the HXRBS X-ray detector, one being more heavily shielded and one less so. Thus an indication of pile-up in the GRS X-ray detectors indicates that the output from the HXRBS detector may also be affected, especially because the HXRBS detector is larger than the GRS detectors. The fact that the circled data points in Figure 1 generally lie above the least-squares fit line is consistent with the relative susceptibilities of the HXRBS and GRS detectors to pulse pile-up. The cause of the discrepancy between GRS and HXRBS fluences for the two data points flagged with question marks remains to be determined.

As shown in Figure 1, there is good agreement between the two fluence measures, especially when the circled data points are ignored. The dashed line in Figure 1 is the least squares fit to the "good" data points (not circled or flagged with a "?"); it can be used to correct HXRBS fluences for pulse pile-up affected events for which the GRS 40-140 keV fluence

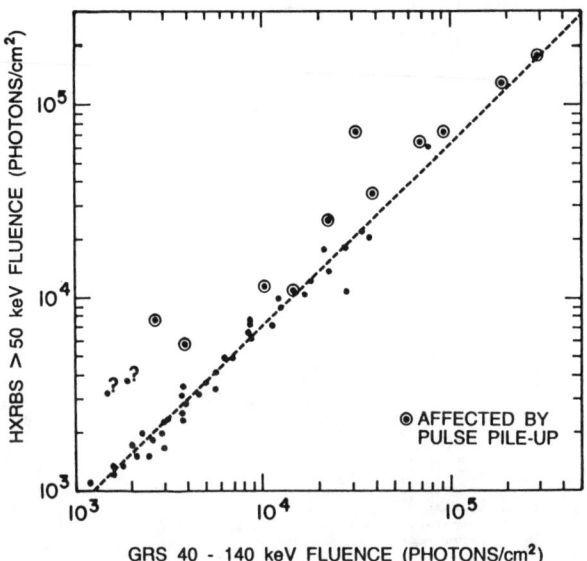

Fig. 1. Plot of HXRBS >50 keV fluence vs. GRS 40-140 keV fluence for large flares.

(from the more heavily shielded detector) has been determined. The assumption underlying any such correction is that the GRS X-ray detector with the additional filter is not affected by pulse pile-up. This assumption may not be valid for intense events, particularly those with a soft spectrum, and for such events the corrected >50 keV fluence will only be an upper limit. We note that the ratio of the GRS fluences to the SMM fluences from the dashed-line fit is ~1.6, corresponding to a power-law energy spectrum with an exponent of ~ -3.2 (cf., ref. 10). Such a flat spectrum results, at least in part, from the tendency of big flares to have harder spectra[11].

4-8 MeV Fluence vs. >50 keV Fluence. Figure 2 is a plot of 4-8 MeV fluence vs. >50 keV fluence (corrected for pulse pile-up as necessary and when possible) for all HXRBS summary file events occurring from 1980-1982 with >50 keV fluences ≥500 photons cm^{-2}. The 4-8 MeV fluences were taken from Cliver et al.[12] The "x" data points in this figure are for non-GRL flares; the values of the ordinates for these points correspond to a 4-8 MeV fluence upper limit of ~0.5 photon cm^{-2} (plotted between 0.35 - 0.9 photon cm^{-2} because of space limitations), the nominal detection threshold of the GRS for GRL emission. The data points with horizontal lines drawn through them indicate that a correction for pulse pile-up (using Figure 1) has been applied to the >50 keV fluence. A relationship similar to that depicted in Figure 2 has been found to exist between the GRS 40-140 keV fluence and the 2.2 MeV neutron capture line fluence[13]. Figure 3 shows an updated version (from Vestrand[14]) of the correlation obtained by Forrest[1] between the >300 keV fluence and the 4-8 MeV fluence for all flares with >300 keV emission observed by GRS during 1980-1985. The plots in Figures 2 and 3 are similar in appearance. In both cases the scatter increases for lower energies, although to a greater degree in Figure 2. There is no "population" of large >50 keV fluence events that lack detectable GRL emission in Figure 2. There are two outliers, labelled with their dates, that fall below the general trend of the data. One of these was an event on 1982 July 12 event that lacked detectable GRL emission; the other outlier occurred on 1981 October 7. For both of the outlying events pulse pile-up effects were severe, particularly so for 1982 July 12. For that

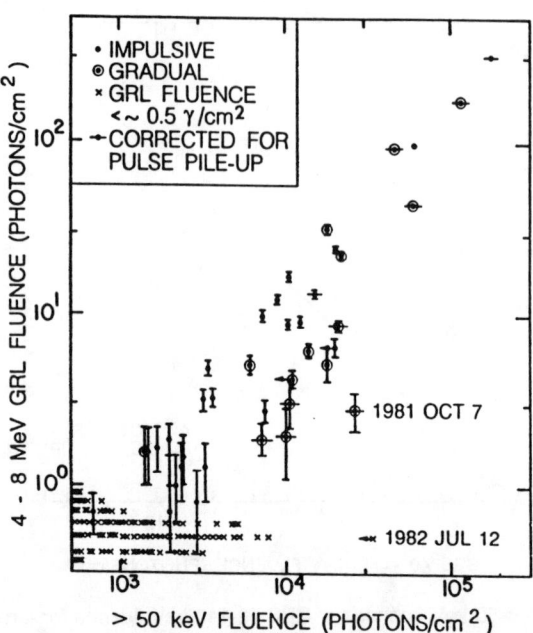

Fig. 2. Plot of GRS 4-8 MeV GRL fluence vs. HXRBS >50 keV fluence, 1980-1982.

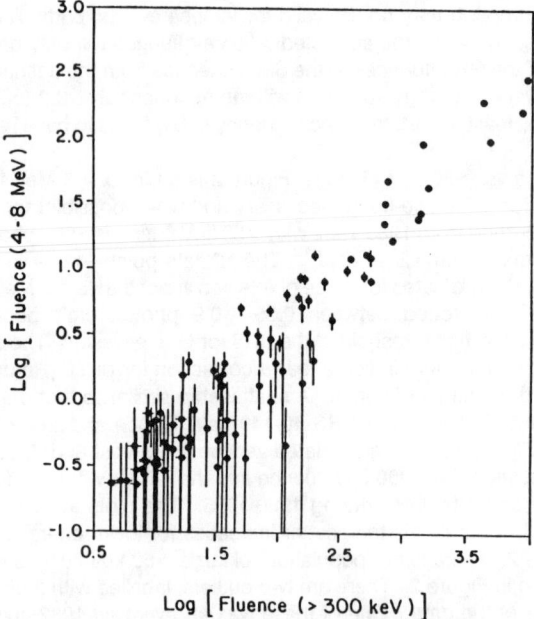

Fig. 3. Plot of 4-8 MeV excess vs. >300 keV electron bremsstrahlung continuum fluence for flares observed by the GRS, 1980-1985 (from Vestrand[14]).

event, the 40-140 keV profile from GRS was so distorted that any fluence value obtained would have been meaningless (Vestrand, private communication, 1993). A leftward pointing arrow on the data point for the July 12 event indicates that the plotted value is an upper limit. Such arrows are also used for two other piled-up events for which the 40-140 keV fluence was not determined. As noted in Section 2.1, however, even the data points corrected for pulse pile-up represent, in a sense, upper limits because the 40-140 keV fluence obtained from the more heavily shielded X-ray detector on GRS that is used to obtain a corrected >50 keV fluence via Figure 1 may also be distorted by pulse pile-up. The circled data points in Figure 2 represent long duration flares following the classification scheme of Cliver et al.[12] (cf., refs. 4, 5). As can be seen in the figure, there is a tendency for these events to have higher e:p (i.e., >50 keV bremsstrahlung fluence : 4-8 MeV GRL fluence) ratios than do the impulsive flares; their data points tend to lie to the right of the trend line, thereby increasing the scatter. (The large (>3000 photons cm^{-2}) non-GRL flares, indicated by "x" data points, are also characteristically gradual events.) The difference in e:p ratios between the gradual and impulsive flares is not great, about a factor of two in the medians, and may be due to the relative sensitivities of the HXRBS and GRS detectors late in long duration events when GRL fluxes fall below the detection threshold. For comparison, Kallenrode, Cliver, and Wibberenz[15] found a difference of a factor of $\lesssim 10$ between the average e:p ratios of SEPs from gradual and impulsive flares. For SEP events, however, the difference is in the opposite direction with higher e:p ratios observed in SEP events associated with impulsive flares. The small, possibly instrumental, difference that we find between e:p ratios of interacting particles from gradual and impulsive flares is consistent, to first order, with the recent result of Ramaty et al.[16] who showed that the ratio of the numbers of interacting 0.5 MeV electrons to 10 MeV protons is independent of flare duration.

There is evidence for a class of impulsive γ-ray flares, called electron-dominated events[17], in which line emission is missing or muted. While such events might be representatives of the population of large "first-step" non-GRL flares argued for by Bai (1986) in the two-step scenario, an identification of the two groups is problematical because the bremsstrahlung continuum in electron-dominated events extends beyond 10 MeV (up to 60 MeV in certain cases) and the spectra exhibit a tendency to flatten with increasing energy. There were eight such flares in the total sample, 1980-1989, of GRS flares that were intense enough to be spectrally analyzed[17]. Three of these flares were observed during the 1980-1982 period we considered (Rieger, private communication, 1990): 1980 June 4, 1980 June 29, and 1982 June 15. Each of these flares had >50 keV fluence in the range from 2-4 x 10^3 photons cm^{-2}; thus, their data points lie in the lower left-hand side of Figure 2 where the scatter is greatest. Because of the subtraction technique used to determine the nuclear excess[14], the line emission in these events is over-estimated[18]. However, even if we reduce the GRL emission observed in these events from the deduced values of ~ 2-5 photons cm^{-2} to the GRS instrumental background of ~0.5 photons cm^{-2}, the altered data points remain within the scatter and our basic result - the apparent correlation of >50 keV and 4-8 MeV fluences for large flares observed from 1980-1982 - is not changed.

DISCUSSION

The correlation in Figure 2 that we find between >50 keV fluences and 4-8 MeV GRL fluences suggests that the bulk of ~100 keV electrons and ~10 MeV ions (needed to produce >50 keV continuum and 4-8 MeV GRLs, respectively) are accelerated in a common acceleration process in large flares (cf., refs. 19, 20). This would be the simplest explanation for the correlation. We note, in particular, the absence of a well-defined population of flares with large >50 keV fluences but without detectable GRLs. In the picture of Bai[4], such events would be those in which the second-step process was not operating. The two high fluence events that are deficient in GRL emission (1981 October 7 and 1982 July 12) have characteristics (delay of high energy X-rays, at least for 1981 October 7, and type II association) that Bai and Dennis[21] and Bai[4] reported for "normal" GRL flares. The anomalous position of the 1982 Jul 12 event is presumed to result primarily from pulse pile-up in the HXRBS detector. In addition, both the 1982 Jul 12 and 1981 Oct 7 events were gradual flares and there is a tendency in Figure 2, that may be an instrumental effect, for gradual events to have larger e:p ratios than impulsive events. We note that gradual events, in general, contribute much of the scatter in the correlation plot in Figure 2. Again, while these events appear to be, as a group, slightly deficient in GRL emission, they exhibit spectral delays and are highly associated with type II emission. Thus there is little evidence that they constitute a separate "class" of event. The common acceleration process we propose for ~100 keV electrons and ~10 MeV ions in large flares could still be a second-step process, following an initial injection as envisioned by Bai[4] (cf., refs. 22, 23), but such a second-step process must dominate electron acceleration down to energies $\lesssim$100 keV rather than the >200 keV level suggested by Bai[4].

Acknowledgement. We thank Tom Vestrand for providing preliminary GRS X-ray fluences for comparison with the HXRBS fluences.

REFERENCES

1. D.J. Forrest, in Positron-Electron Pairs in Astrophysics, ed. M.L. Burns, A.K. Harding, and R. Ramaty (American Institute of Physics, New York, 1983), p. 3.
2. D.J. Forrest, et al., Solar Phys., 65, 15 (1980).
3. E.L. Chupp, Ann. Rev. Astron. Astrophys., 22, 359 (1984).
4. T. Bai, ApJ, 308, 912 (1986).
5. H.V. Cane, R.E. McGuire, and T.T. von Rosenvinge, ApJ, 301, 448 (1986).
6. L.E. Orwig, K.J. Frost, and B.R. Dennis, Solar Phys., 65, 25 (1980).
7. B.R. Dennis, et al., The Complete HXRBS Event Listing, 1980-1989, NASA Technical Memorandum 4332 (1991).
8. N.B. Crosby, M.J. Aschwanden, and B.R. Dennis, Solar Phys., 143, 275 (1993).
9. S.R. Kane and H.S. Hudson, Solar Phys., 14, 414 (1970).
10. W.T. Vestrand, D.J. Forrest, E.L. Chupp, E. Rieger, and G.H. Share, ApJ, 322, 1010 (1987).
11. B.R. Dennis, Solar Phys., 100, 465 (1985).
12. E.W. Cliver, D.J. Forrest, H.V. Cane, D.V. Reames, T.T. von Rosenvinge, R.E. McGuire, S.R. Kane, and R.J. MacDowall, ApJ, 343, 953 (1989).
13. W.T. Vestrand, Phil. Trans. Royal. Soc. London, Series A, 336, 349 (1991).
14. W.T. Vestrand, Solar Phys., 118, 95 (1988).
15. M.-B. Kallenrode, E.W. Cliver, and G. Wibberenz, ApJ, 391, 370 (1992).
16. R. Ramaty, N. Mandzhavidze, B. Kozlovsky, and J. Skibo, Adv. Space Res. (in press, 1993).
17. E. Rieger and H. Marschhauser, in Max91/SMM Solar Flares: Max91 Workshop No. 3, eds., R.M. Winglee and A.L. Kiplinger (Boulder, CO, 1991), p. 68.
18. F.-W. Bech, J. Steinacker, and R. Schlickeiser, Solar Phys., 129, 195 (1990).
19. D.J. Forrest and E.L. Chupp, Nature, 305, 291 (1983).
20. S.R. Kane, E.L. Chupp, D.J. Forrest, G.H. Share, and E. Rieger, ApJ (Letters), 300, L95 (1986).
21. T. Bai and B.R. Dennis, ApJ, 292, 699 (1985).
22. T. Bai and P.A. Sturrock, Ann. Rev. Astron. Astrophys., 27, 421 (1989).
23. N. Mandzhavidze and R. Ramaty, Nuclear Phys. B, Proc. Suppl. (in press, 1993).

TRAPPING OF PROTONS IN TWISTED MAGNETIC LOOPS

Yun-Tung Lau and Reuven Ramaty
Laboratory for High Energy Astrophysics
NASA/Goddard Space Flight Center
Greenbelt, MD 20771

ABSTRACT

We study the evolution of an ensemble of protons in a loop-like magnetic structure derived from force-free equilibrium. We take into account the effects of the twist of the magnetic field. Protons whose velocity vectors are initially in the loss cone are removed on very short time scales. Subsequently, protons are removed by drifting either into the loss cone or to the assumed boundary of the structure. We find that in the absence of collisions with gas and pitch angle scattering due to plasma turbulence, for sufficiently large loops, protons can remain trapped for long periods of time independent of the amount of twist. For smaller structures, the long term trapping implied by high energy gamma ray observations requires a significant amount of twist.

INTRODUCTION

The observation[1,2] of high energy gamma rays from pion decay from two flares in June 1991 has raised questions regarding the lifetime against drifts of GeV protons in solar flare magnetic loops. In a previous paper[3] it was shown that in a magnetic structure derived from force-free equilibrium, the twist of the magnetic field lines can significantly reduce the particle loss due to drifts. Here we use the same model to study the time evolution of a distribution of initially isotropic protons in such a loop-like structure. Our calculations ignore collisions with gas and pitch angle scattering by plasma turbulence in the loop.

ANALYSIS

The plasma pressure in the corona is negligible compared to the magnetic field pressure. Thus we may use magnetic field models that satisfy the force-free equilibrium equation, $\nabla \times \mathbf{B} = \lambda \mathbf{B}$. We solve this equation in a cube extending to ± 1 in each direction from the origin. The normal component B_n vanishes on the boundaries, except for the $z = -1$ face (the photosphere), where $B_n = B_z$ has two concentrated flux regions of opposite sign. These regions are separated by a distance equal to 1 (the normalized loop size L_0), with B_z decreasing like a Gaussian ($\sigma = 0.1$) from the center of each region. We also normalize the magnetic field to B_0, the maximal photospheric magnetic field in the concentrated flux regions. The parameter λ determines the amount of twist in the field lines.

For a typical solution with the above boundary conditions, the field lines form a loop, going from one flux region to the other. For $\lambda \sim 3$, the twist angle of the field lines between the two ends of the loop is about 2π. The mirror ratio B_0/B_c (B_c being the coronal field at the top of the loop) is between 16 and 30. More details about this model were given previously.[3]

Because the gyroradius of the protons is small compared to a typical 10^8 cm loop thickness, the first adiabatic invariant (the magnetic moment of the gyrating particle) is well conserved. We may then use guiding center theory[4] to describe the particle motion. The dimensionless guiding center velocity is,

$$d\mathbf{R}/dt = V_{\parallel}\hat{\mathbf{b}} + \mathbf{V}_D. \qquad (1)$$

Here $\mathbf{R}$ is the position of the guiding center and the unit vector $\hat{\mathbf{b}} = \mathbf{B}/B$. The dimensionless parallel velocity is, $V_{\parallel} = \pm(1 - B/B_m)^{1/2}$, where $B_m = B/\sin^2\theta$ is the magnetic field at the mirror point, and θ is the pitch angle. Note that B_m is a constant along the trajectory of the particle. The other term $\mathbf{V}_D$ in Eq. (1) contains the drifts due to the gradient and the curvature of the magnetic field. Its expression has also been given before.[3]

The guiding center equation has been integrated[3] for the above equilibrium model. It was found that when the twist parameter $\lambda > 3.28$ (or when the twist angle is $\gtrsim 2\pi$), there are particles trapped indefinitely in the twisted loop.

When the bounce motion due to mirroring of the particles is much faster than the drift motion, the second adiabatic invariant, $J = p \oint (1 - B/B_m)^{1/2} ds$, is also conserved.[5] Here $p = \gamma m_0 v$ is the particle momentum. The integral is to be taken along a field line from one mirror point (where the integrand vanishes) to the other, and back again. For a given B_m, J = constant defines a surface in space. The particles drift along these surfaces of constant J.

Because of the fast bounce motion, we may average Eq. (1) over a bounce period to obtain the bounce averaged equation,[5]

$$d\mathbf{r}_{\perp}/dt = (c/eBT) \nabla J \times \hat{\mathbf{y}}, \qquad (2)$$

where the bounce period $T = (1/v) \oint (1 - B/B_m)^{-1/2} ds$, and the position vector $\mathbf{r}_{\perp}$ is on a cross-sectional plane, which is now taken as the x-z plane in our model; $\hat{\mathbf{y}}$, essentially parallel to $\hat{\mathbf{b}}$, is the unit vector normal to this plane. For a given B_m (i.e., for a particle of given initial pitch angle and initial position in the x-z plane), the vector $\mathbf{r}_{\perp}$ traces out curves of constant J in the x-z plane.

Upon normalizing Eq. (2) to dimensionless form, we have,

$$d\hat{\mathbf{r}}_{\perp}/d\tau = (1/\hat{T}) \nabla \hat{J} \times \hat{\mathbf{y}}, \qquad (3)$$

where the normalized quantities are $\hat{\mathbf{r}}_{\perp} = \mathbf{r}_{\perp}/L_0$, $\tau = t/T_0$, $\hat{J} = J/pL_0$, and $\hat{T} = Tv/L_0$, and $T_0 = (L_0/v)^2 \Omega$, with $\Omega = eB_c/\gamma m_0 c$. Note that the particle energy and the loop length appear in T_0 only.

We assume an ensemble of particles with random initial pitch angles θ on the x-z plane. We further assume that the initial positions of these particles are uniformly distributed on the upper half of this plane. For each particle we evaluate B at its initial position and the conserved quantity B_m. Next we trace the field line from that position by solving numerically $d\boldsymbol{\rho}/ds = \mathbf{B}/B$, where $\boldsymbol{\rho}$ is the position vector and s the field line length. If B_m exceeds the maximal B on that field line (which occurs at the footpoint of the field line), then this particle is in the loss cone and is removed. If B_m is less than this maximal B, then we use Eq. (3) to trace out the curve of constant J on the x-z plane. At sufficiently close time intervals, we perform the same test and remove the particle if it has drifted to a field line such that it is now in the loss cone. We also remove the particle when the curve of constant J intersects the boundaries of the box. When the twist is large enough, there are particles for which the curves of constant J are closed. These particles are trapped indefinitely in the loop.

We put initially N_0 (=400) particles on the upper half of the x-z plane. Then we count N, the number of particles remaining in the box as a function of time. In Fig. 1 we show N/N_0 as a function of t/T_0 for three values of λ. In the case of small twist ($\lambda = 0.1$) N/N_0 practically vanishes after $t = 3.5 T_0$. In contrast, for $\lambda = 3.4$ (larger values lead to instabilities), a fraction (6%) of the protons remain trapped indefinitely.

For $B_c = 40$ G and $L_0 = 2 \times 10^9$ cm, $t = 3.5T_0$ corresponds to 69 min for 1 GeV protons. Thus, in the absence of twist most of these protons will be removed in about an hour. However, when the loop is sufficiently twisted, a small fraction of the protons remain trapped. On the other hand, for a larger loop, for example $L_0 = 10^{10}$ cm, $T_0 = 8.2$ hours, and thus 1 GeV protons will remain trapped for long times independent of the amount of twist. For lower energy protons, e.g. 50 MeV, the same B_c and $L_0 = 2 \times 10^9$ cm, $T_0 = 4.89$ hours. These protons, therefore, will not be affected by the drifts for long times, independent of the amount of twist.

We wish to acknowledge N. Mandzhavidze and T. G. Northrop for useful comments.

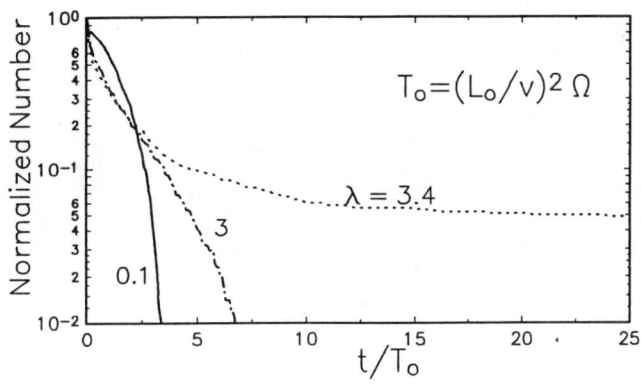

Figure 1: The time evolution of the normalized particle number N/N_0.

REFERENCES

[1] V. V. Akimov, et al., 22nd Internat. Cosmic Ray Conf. **3**, 73 (1991).
[2] G. Kanbach, et al., Astr. and Ap. Suppl., 97, 349 (1993).
[3] Y.-T. Lau, T. Northrop, and J. M. Finn, Ap. J., in press (1993).
[4] T. G. Northrop, 1963, *The adiabatic motion of charged particles*, (New York, Interscience Publishers).
[5] T. G. Northrop, and E. Teller, Phys. Rev. **117**, 215 (1960).

LONG DURATION EVENTS AND EXTENDED ACCELERATION

SOLAR FLARE NEUTRONS AND GAMMA RAYS

Richard E. Lingenfelter
University of California, San Diego, La Jolla, CA 92093-0111

ABSTRACT

We briefly review the rich variety of gamma-ray line and neutron emission that has been observed from solar flares over the past two decades. We then discuss what we are learning from these observations about the nature, interactions and confinement of the flare-accelerated ions that cause the emission, about the structure and properties of the flare region, and about the elemental and isotopic abundances in the photosphere and chromosphere.

1. INTRODUCTION

Observations of neutron and gamma-ray-line emission from solar flares are providing important new information on particle acceleration in solar flares and on the flare process itself. These gamma-rays and neutrons are produced directly by nuclear interactions of the flare-accelerated protons and heavier ions with ambient gas in the solar atmosphere. Thus they yield the most direct information available on the total number, energy spectrum, time dependence, and angular distribution of the accelerated ions in the flare region, and they can provide unique new information on the structure of the flare region and the elemental and isotopic (e.g. ^{3}He/H) abundance ratios in the solar photosphere and chromosphere.

The possibility of detecting neutrons and gamma rays from solar flares was first suggested by Biermann, Haxel and Schluter[1] in 1951, and Morrison[2] in 1958, respectively. It was subsequently shown, in detailed calculations of the expected fluxes by Lingenfelter and Ramaty[3] in 1967 that the principal gamma-ray lines should be those at 2.223 MeV from neutron capture on ^{1}H, at 0.511 MeV from positron annihilation, and at 4.438 and 6.129 MeV from deexcitation of nuclear levels in ^{12}C and ^{16}O, plus broad emission around 70 MeV from pion decay. They also showed that measurements of the 2.223 and 4.438 or 6.129 MeV line fluences from flares could give the first measure of both the spectrum and total number of accelerated particles in flares, since the ratios of these line fluences depend strongly on the particle energy spectrum. They further showed that ratios of the flux in deexcitation lines from various nuclei could provide independent measures of the elemental abundance in the flare region. At the same time, calculations of the expected neutron production also showed[3-5] that measurement of the time dependence of the solar flare neutron flux alone could give a second, independent measure of the particle spectrum and number.

Gamma ray line emission was first observed by Chupp et al.[6] with a detector on OSO-7 from the solar flare of 4 August 1972 at the predicted line energies of 0.51, 2.2, 4.4 and 6.1 MeV. These and other weaker lines have since been observed from dozens of flares by detectors on HEAO-1[7], HEAO-3[8], HINOTORI[9-11], SMM[12-14] and most recently on GRO[15-17]. Neutrons were first observed from the flare of 21

June 1980 and from several subsequent flares with the SMM detector[12,18] and their energy spectra have recently measured from a number of flares with instruments on GRO[15-17]. Relativistic neutrons have also been observed[19,20] from several flares with ground based cosmic-ray neutron monitors. In addition, protons from the decay of lower energy, flare neutrons were detected[21,22] in interplanetary space by ISEE-3. The general characteristics of many of these gamma-ray flares are listed in two recent catalogs[13,23].

These observations have stimulated extensive theoretical calculations of gamma ray line[24-30] and neutron[26,31-35] production by flare particle interactions, and comparisons of the flare observations with these calculations are providing much important new information. The wealth of observables, including time and angular dependent fluxes of at least a dozen gamma-ray lines, plus bremsstrahlung, Compton scattering, pion-decay and positronium gamma-ray continua, together with time, energy and angular dependent fluxes of escaping neutrons, neutron-decay protons and positrons, have all allowed the construction of very complex models of solar flare particle interaction geometries with an equally large number of variables, including elemental and isotopic abundances of both ambient and accelerated ions, angular and time dependent accelerated particle spectra, flare region region density and magnetic field convergence geometries, pitch-angle scattering mean free paths and energy densities of magnetohydrodynamic-turbulence.

2. ACCELERATED PARTICLES

In general, comparisons of these calculations with observations have shown[25] that the nuclear interactions are caused primarily by accelerated particles that remain trapped in the magnetic fields of the flare region and interact as they slow down in the solar atmosphere, rather than by accelerated particles that eventually escape into interplanetary space. For if the escaping particles were responsible for the observed gamma ray line emission, they would have been greatly enriched in spallation products, such as D, T, Li, Be and B, which were not observed[36,37] in the flare particles in interplanetary space. These comparisons have also shown, however, that the longer duration (>100 s) emission of higher energy gamma rays (>10 MeV) and neutrons (>300 MeV) is produced either by particles that remain trapped[38] on coronal loops or by particles that are further accelerated[26] by shocks as they escape into interplanetary space with only negligible interaction in the solar atmosphere.

In particular, these comparisons have provided the first direct determination of the numbers and energy spectra of accelerated particles in flares. The number and energy spectrum of the accelerated particles are generally determined[26,30,31] from measurements of the integrated flux, or fluence, in the 2.223 MeV line and the 4–7 MeV band. The bulk of the emission in the 4–7 MeV band is from the two strongest nuclear deexcitation lines in ^{12}C and ^{16}O, as has been shown[39,40]. Measurements of the fluence, rather than the flux, are needed because the time histories of the two lines differ significantly. The time history of the nuclear deexcitation line emission in the 4–7 MeV band follows almost instantaneously the time dependence of the accelerated particle interaction rate while that of the 2.223 MeV line is delayed by the neutron thermalization and capture times.

Determinations of the accelerated particle numbers and energy spectra from these fluence ratios, however, depend strongly on the location of the flare on the solar disk[31,32,41–43], since the 2.223 MeV line emission is attenuated by Compton scattering, especially for flares occuring near the limb of the Sun. The 2.223 MeV line is much more strongly attenuated than the other major lines because it is produced predominantly by the capture of those neutrons that penetrate deep into the photosphere ($n > 10^{16}$ H cm^{-3}) where their capture lifetime becomes less than that of decay. The broad band 4–7 MeV emission, on the other hand, is produced[27] much higher in the solar atmosphere from deexcitation of nuclear levels excited by the less penetrating charged particles.

Such limb darkening of the 2.223 MeV line, has been calculated[31] in detail, considering accelerated ion angular distributions, spanning the range from downward beaming, expected from some acceleration models[44] to magnetic mirroring[45], and accelerated ion energy spectra that are either Bessel functions in momentum per nucleon, expected[46] from stochastic Fermi acceleration, or power laws in energy per nucleon, expected[47] from shock acceleration in the non-relativistic limit. As we discuss below, the angular distribution of the interacting ions have recently been determined directly in loop models[27–30,33].

The limb darkening of the 2.223 MeV line emission, resulting from the attenuation of the line, produced by neutron capture deep in the solar atmosphere, is most evident when compared with the excess 4–7 MeV emission from nuclear deexcitation, produced by the incident ions much higher in the atmosphere where the attenuation is negligible. The calculated[31] angular dependence of this line emission from the solar atmosphere, relative to that from nuclear deexcitation lines in the 4–7 MeV band, is shown in Fig. 1, together with the measured ratios of the 2.223 MeV line to excess 4–7 MeV fluences from 15 flares between 4 August 1972 and 9 July 1982. As can be seen, the measured ratios show a strong limb darkening of as much as a factor of 30 for flares ranging from the disk to the limb of the Sun.

We also see from Fig. 1, that the accelerated ion spectra can all be characterized by a relatively narrow range of spectra, either Bessel functions with $0.015 < \alpha T < 0.04$, or power laws with $3 < s < 4$. The range of spectral indices, however, still depends somewhat on the assumed abundances, particularly that of helium, and Ramaty et al.[30] find ranges of $0.01 < \alpha T < 0.03$, or $3.5 < s < 4.5$, for the same flares, assuming an enhanced helium abundance. The range of αT is very similar to that of the Bessel function spectra with $0.014 < \alpha T < 0.036$, that were found[48] to fit the measured spectra of protons in the interplanetary medium. The power law indices, on the other hand, are significantly steeper than the mean of 2.4 ± 0.7 fitted[48] to the same sampe of protons. As we discuss below, studies[35] of the observed neutron spectra should enable us to clearly determine whether Bessel function or power law spectra best describe the interacting ions.

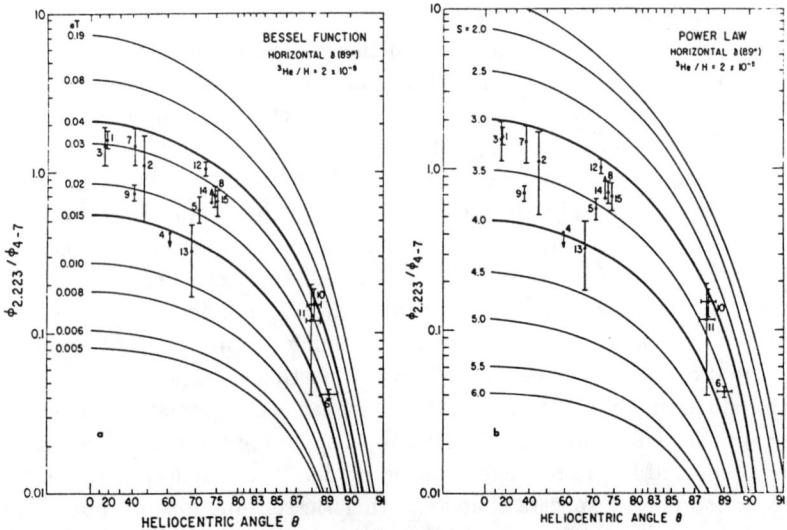

Fig. 1. Calculated[31] angular dependence of 2.223 MeV neutron capture line fluence from the solar atmosphere relative to the nuclear deexcitation line fluence in the 4–7 MeV band, together with the measured ratios of the 2.223 MeV line to excess 4–7 MeV fluences from 15 flares between 4 August 1972 and 9 July 1982.

Very similar spectra were also found for both the accelerated protons trapped at the Sun and those in the interplanetary medium for two of the three flares for which direct comparisons can be made, 7 and 21 June 1980. In the 3 June 1982 flare, on the other hand, the spectrum of the protons in the interplanetary medium was best fit[49,50] by a power law rather than a Bessel function, but that power law was much flatter, $s = 2.1\pm0.3$, than that of 3.3 ± 0.2 required[30,31] by the gamma ray fluence ratios.

These spectral differences are correlated with the two different classes of flares, suggested by a variety of studies[51,52]. These studies indicate that gamma-ray and proton flares can be grouped into two classes: impulsive flares characterized by hard (>30 KeV) X-ray emission lasting only about 10^2 sec and long duration flares with hard X-ray emission lasting for 10^3 sec or longer. In terms of these classes, the observations suggest that the accelerated particles in impulsive flares have a Bessel function like spectrum which could result from stochastic acceleration[46] and that the escape of such particles into interplanetary space is usually energy independent, whereas, in the long duration flares further acceleration by shocks occurs during escape of the particles from the flare.

The number of flare-accelerated protons of energy greater than 30 MeV that escaped into the interplanetary medium also differs for these two classes. As can be seen in Fig. 2, the fraction of accelerated protons that escaped from the Sun,

$F_{esc} = N_{ipm}/(N_{ipm}+N_{sun})$ varies[31] from 0.95 for the 4 August 1972 flare to <0.003 for the 1 Jul 1980 flare.

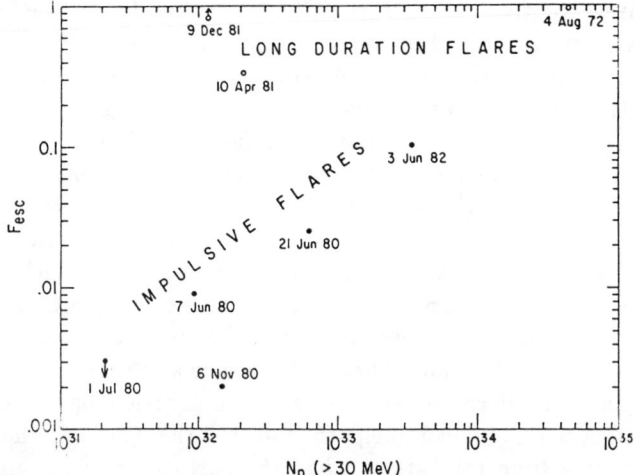

Fig. 2. Estimated[31] fractions of flare-accelerated protons of energy greater than 30 MeV that escaped into the interplanetary medium compared to the total number of such protons that were accelerated in flares, showing the difference between impulsive and long duration flares.

For the impulsive flares (7 and 21 June, 1 July, 6 November 1980 and perhaps 3 June 1982), the number of accelerated protons escaping into the interplanetary medium is substantially less than that remaining trapped at the Sun producing gamma rays. For the long duration flares (4 August 1972 and 10 April 1981), on the other hand, the number of escaping protons is comparable to, or greater than, that remaining at the sun. This can also be seen from observations[53] of the long duration flare of 9 December 1981 which had no detectable gamma-ray line emission. Interplanetary proton measurements suggest that about 9.6×10^{31} protons (> 30 MeV) with a power-law index $s = 3.5$ escaped from this flare. For such a proton spectrum, the upper limit on the 2.2 MeV gamma ray fluence of < 3.7 photons cm^{-2} from this flare at W16N12 implies that $< 1.9 \times 10^{31}$ protons (> 30 MeV) remained trapped at the sun, giving an escape fraction of > 0.83.

For impulsive flares, there may be a strong correlation between the escape fraction and the total number of accelerated protons. If this correlation is supported by further observations, it will place another important constraint on models of the acceleration and escape of particles in solar flares. For long duration flares, on the other hand, the escape fraction is apprently quite independent of the total number of accelerated protons. Recent calculations show[30] that these general relationships are also essentially independent of the assumed spectral form and abundance, although estimated escape fractions can increase by about a factor of 2 with enhanced helium abundance.

The observed time dependence of nuclear deexcitation line fluxes also enables us to study the magnetic mirroring and pitch-angle scattering of the accelerated ions in the flare region. To make such studies, Monte-Carlo and other numerical simulations have recently been developed[27-30,33,34] that follow the propagation of individual accelerated ions throughout a solar flare magnetic loop, including the effects of mirroring of the ions in the convergent flux tubes and their MHD pitch-angle scattering by MHD turbulence. These simulations calculate the depth distribution of nuclear deexcitation line production, the time dependence of the line production, and the escape probability of the line photons from the solar atmosphere. Comparisons of such simulations with SMM measurements[54,55] of the time dependence of the deexcitation line fluxes show[27] that the inclusion of mirroring and pitch-angle scattering can lead to time profiles which are in good agreement with observations. This suggests that the decaying portions of the time profiles of gamma-ray emission observed from solar flares could be governed by transport and not by acceleration. These calculations also show that the gamma-ray emission time scales decrease with decreasing magnetic loop length and with decreasing MHD scattering mean free path, and that there are minimum emission time scales resulting from saturation, when the particles are scattered into the loss cone as fast as they are removed by interactions, so that the time cannot be further decreased by reducing the mean free path.

In particular, for impulsive flares, such as the 21 June 1980 flare, the observed[54] time profiles place[27] an upper limit of about 3×10^9 cm on the length of the loop, but for more extended flares, such as the 3 June 1982 flare[55], the loop length could be as large as 5×10^{10} cm. The decaying portions of the two pulses of the 21 June 1980 flare can both be fit[27] by a mean free path close to that for saturation, so that the time profiles depend only weakly on the energy density in MHD turbulence. These simulations also suggest that a magnetic field that increases below the transition region as the pressure to the 1/5 power provides a good fit to the data. This corresponds to a factor of ~ 20 variation in the magnetic field between the transition region and the photosphere, for the assumed atmospheric model. The bulk of the nuclear deexcitation line emission in such a model is thus produced in the lower chromosphere.

Because the energetic (>30MeV) neutrons produced by flare accelerated ions are not produced isotropically, but tend to be produced predominantly in the direction of motion of the incident ions, measurements of the neutron flux from solar flares can also provide unique information on the angular and energy distributions of the accelerated ions in flares. To understand these measurements, Monte Carlo simulations[32-35] have been made of the angular and energy distribution of neutrons escaping from the solar atmosphere, expected from thick-target interactions of flare accelerated ions with a range of incident angular distributions and energy spectra. These simulations also provide time-dependent fluxes of those neutrons that survive to a distance of 1 AU without decaying, in order to directly compare with the time-dependent observations.

Recent calculations[35] of the time dependences of the neutron flux at a distance of 1 AU from the Sun show: 1) how the different accelerated ion spectral forms (Bessel function and power law) can be distinguished and how the accelerated ion spectrum can thus be determined from the neutron flux measurements, independent of the degree of pitch-angle scattering experienced by the ions, and 2) how the amount of pitch-angle scattering can also be determined from the same neutron flux measurements, independent of the accelerated ion spectra.

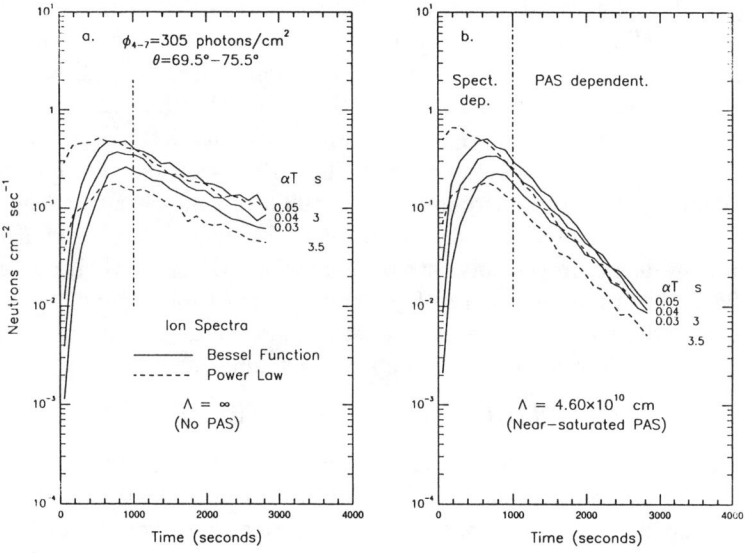

Fig. 3. Calculated[35] time dependences of the neutron flux at a distance of 1 AU from the Sun resulting from interactions of accelerated ions with either Bessel function (αT) or power law (s) spectra in the bracketing cases: 3a) without pitch-angle scattering (pas) of the ions, and 3b) with nearly saturated pitch-angle scattering. These fluxes are from an assumed flare at a heliocentric angle of $\sim 72°$ and they are normalized to a constant 4–7 MeV nuclear deexcitation line fluence.

As can be seen in Fig. 3, in the first 1000 seconds after the impulsive phase of a flare, the shape of the time dependence of the neutron flux at a distance of 1 AU depends very strongly on the form of the ion spectrum and only very weakly on whether or not there is any significant pitch-angle scattering of the ions. Independent of the degree of ion pitch-angle scattering, ions with power-law spectra produce much harder neutron spectra than ions with more exponential-like Bessel function spectra, for the same nuclear deexcitation gamma-ray line fluence. Therefore, we see that the neutron flux at 1 AU from ions with a power-law spectrum (Fig. 3 dashed lines), rises very quickly in the first 100 seconds to within a factor of 2 of the peak flux, while the neutron flux from ions with a Bessel function spectrum (solid lines), rises much more slowly to the peak flux. Thus, from measurements of the neutron flux time dependence in the first 1000 seconds, we should be able to distinguish between power-law and Bessel-function spectral forms and clearly determine the ion spectrum, independent of the degree of ion pitch angle scattering.

Moreover, as we also see in Fig. 3, after the first 1000 seconds of a flare, the time dependence of the neutron flux at 1 AU no longer depends on either the form or the index of the ion spectrum and instead depends very strongly on the degree pitch-angle scattering experienced by the ions. Quite independent of the ion spectrum, we see (Fig. 3a) that without ion pitch-angle scattering, all of the neutron fluxes after the first 1000 seconds fall off with essentially the same relatively slow exponential ($\sim$ 1200 sec) time decay, determined by the extended production of neutrons by unscattered ions mirroring and slowly losing their energy in the upper atmosphere of the Sun. On the other, with nearly saturated pitch-angle scattering, we see (Fig. 3b) that irrespective of the ion spectrum, all of the neutron fluxes after the first 1000 seconds fall off with a very similar but much more rapid exponential ($\sim$ 600 sec) time decay, as the ions are more quickly flushed out of magnetic loop by pitch-angle scattering which causes them to mirror deeper in the solar atmosphere where they lose their energy more rapidly. Thus, from measurements of the neutron flux time dependence after the first 1000 seconds, we should be able to determine the degree of ion pitch angle scattering, quite independent of the the ion spectrum.

3. ELEMENTAL AND ISOTOPIC ABUNDANCES

Elemental abundances of both the accelerated ions and the ambient gas in the flare region can be determined from the analysis of the nuclear deexcitation lines observed in the flare spectra. Monte Carlo simulations have been made[56–58] of the expected fluxes in various nuclear lines, incorporating detailed calculations[59,60] of the excitation cross sections and kinematics and assuming a range of accelerated ion spectra. SMM measurements[61] of the spectrum of the 27 April 1981 flare have provided the most sensitive measure of these abundances, so far.

These analyses have shown[59,60] that the best-fit abundances of the ambient gas where the accelerated ions interacted in this flare are significantly enriched in C, N, O, Ne, Mg, Si, and Fe, compared to H and He in either photospheric or coronal abundances[62]. Moreover, the abundances of the accelerated ions are even more enriched in these elements compared to these same abundances and they appear to have compositions similar to those of the ^{3}He-rich flares.

The measured[61] spectrum of the 27 April 1981 flare, compared to that calculated for the best-fit model abundance determined in the most recent analysis[58] of this flare is shown in Fig. 4. Also shown in Fig. 4b is a Monte Carlo simulation of the spectrum that could be measured with high-resolution Ge spectrometer, showing how many more lines could be seen and how much better the lines could be resolved.

The gamma ray line observations from solar flares also offer[41–43,63] the most direct means of determining isotopic abundance of ^{3}He in the solar photosphere. Flare produced neutrons diffusing into the solar photosphere are radiatively captured on hydrogen to form deuterium and the time dependence of the 2.223 MeV gamma rays resulting from this capture provides a sensitive measure of the ^{3}He abundance, if it exceeds 10^{-5} that of H. For with such abundances the (n,p) reaction on ^{3}He, which has a cross section 1.6×10^4 times that of H(n,γ), can compete effectively for the capture of neutrons. The ratio of capture cross sections is independent of energy since both are proportional to $1/v$.

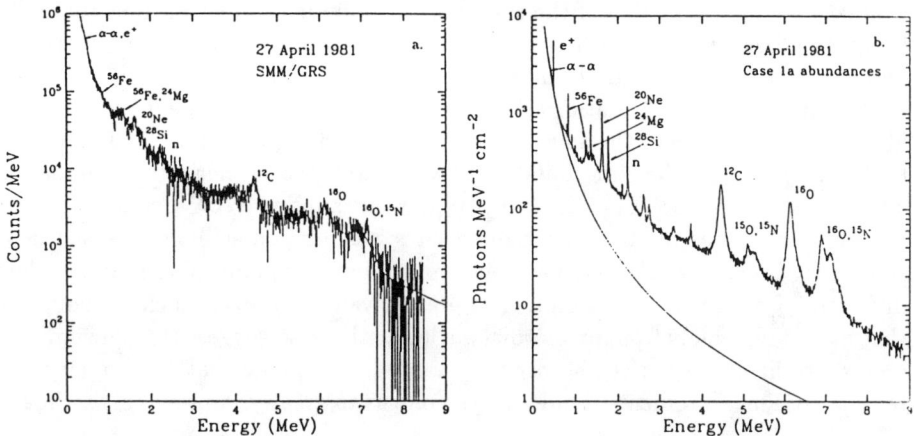

Fig. 4. Gamma-ray spectrum of the 27 April 1981 flare, measured[61] by SMM, compared to that calculated for the best fit model abundance. Also shown 4b) is a Monte Carlo simulation of the much better resolved spectrum that could be measured with high-resolution Ge spectrometer.

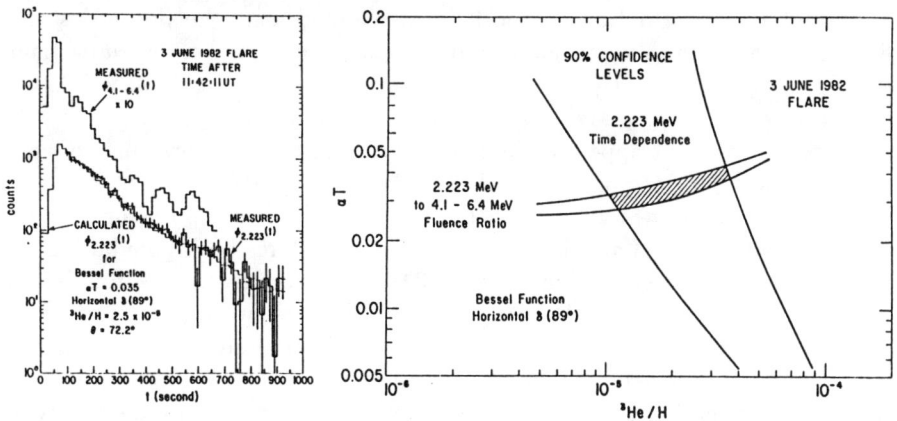

Fig. 5. Measured[64] time dependences of the delayed 2.223 MeV neutron capture line, compared to that calculated[63] for that flare, assuming a photospheric ^{3}He/H of 2.5×10^{-5}. The accelerated ion spectrum is assumed to be a Bessel function with an αT of 0.035, which interacts to produce neutrons with a production time dependence coincident with that measured[19] for the prompt nuclear deexcitation line contribution in the 4.1–6.4 MeV band. Also shown in 5b) is the best-fit determination of the photospheric ^{3}He/H abundance and the accelerated ion spectral index αT.

SMM measurements of the time-dependent neutron capture line emission from the flare of 3 June 1982, from which 2.2 MeV line emission was observed[64] for nearly 1000 s, have provided the most sensitive measure of the solar photospheric ^{3}He/H ratio, so far. The expected variation of this time dependence from a flare at the observed heliocentric angle of 72° was determined by Monte Carlo simulations[63], as a function of the assumed ^{3}He abundance, for neutrons produced by flare-accelerated ions with a range of incident Bessel function energy spectra and a mirroring angular distribution.

Comparing these calculations with the measured time dependence of the 2.223 MeV line emission from the solar flare of 3 June 1982, gives[63] a best-fit ^{3}He/H of $(2.3\pm1.2)\times10^{-5}$ at the 90% confidence level for the photospheric ^{3}He abundance. The previous estimate[65] of $(3.4\pm1.7)\times10^5$, for the outer convective zone of the Sun from measurements of ^{3}He/^{4}He in meteorites was high enough to suggest that a significant amount of ^{3}He may have been mixed into the photosphere by turbulent diffusion[66] from the solar interior. The present value, however, is close enough to that expected[67] solely from primordial nucleosynthesis to suggest that such mixing seems less likely. But the photospheric value is still somewhat preliminary and recent GRO measurements should now enable us to greatly improve this estimate.

4. SUMMARY

Comparisons of solar flare gamma ray and neutron observations with the results of Monte Carlo simulations of accelerated ion interaction models have yielded important new information on the flare accelerated ions, the properties of the flare region and the solar elemental and isotopic (^{3}He) abundances. In particular, studies of the angular dependence of attenuation of the 2.223 MeV line emission relative to the 4–7 MeV band from nuclear deexcitation, suggest[30,31] that the accelerated ions trapped at the Sun have a relatively narrow range of energy spectra. Moreover, comparison of the spectra and total numbers of accelerated protons trapped at the Sun with those of flare-accelerated protons observed[48] in the interplanetary medium show that for impulsive flares both populations seem have very similar spectra, suggesting that they are accelerated medium is essentially energy independent. For long duration flares, however, the protons may undergo further acceleration by shocks as they escape, altering their spectrum to a power law. Such comparisons also suggest that for long duration flares, the fraction of accelerated protons that escape into the interplanetary medium is close to unity, while for impulsive flares, this fraction can be much less than unity and seems to be roughly correlated with the total number of protons accelerated. Recent calculations[35] also show that measurements of the time dependent flux of neutrons that escape from the Sun should enable us to clearly distinguish between the different spectral forms (e.g. Bessel function and power law) for the accelerated ions interacting in the flare.

Analyses of the nuclear deexcitation lines observed in the flare spectra show[59,60] that the abundances of both the ambient gas and accelerated ions interacting in the flare are significantly enriched in C, N, O, Ne, Mg, Si, and Fe, compared to H and He in either photospheric or coronal abundances[62]. Studies of the time dependence of the 2.223 MeV capture gamma-ray line emission, suggest[63] that the ^{3}He/H ratio in the solar photosphere to be $(2.3 \pm 1.2) \times 10^{-5}$. This ratio is less than that previously estimated[65] and it suggests that there is no significant mixing into the photosphere of ^{3}He made in the solar interior.

Acknowledgements. We thank NASA for support under grant NAG5-1597.

REFERENCES

1. L. Biermann, O. Haxel, and A. Schluter, Z. Naturforsch., **6A**, 47 (1951).
2. P. Morrison, Il Nuovo Cim., **7**, 858 (1958).
3. R.E. Lingenfelter, and R. Ramaty, High Energy Nuclear Reactions in Astrophysics, ed. B.S.P. Shen (New York: W.A. Benjamin Inc., 1967), p.99.
4. R.E. Lingenfelter, et al., J. Geophys. Res., **70**, 4077 (1965).
5. R.E. Lingenfelter, et al., J. Geophys. Res., **70**, 4087 (1965).
6. E.L. Chupp, et al. Nature, **241**, 333 (1973).
7. H.S. Hudson, et al. Astrophys. J., **236**, L91 (1980).
8. T.A. Prince, et al. Astrophys. J., **255**, L81 (1982).
9. M. Yoshimori, et al. Solar Phys., **86**, 375 (1983)
10. M. Yoshimori, et al. J. Phys. Soc. Japan, **54**, 487 (1985).
11. M. Yoshimori, et al. J. Phys. Soc. Japan, **54**, 1205 (1985).
12. E.L. Chupp, Ann. Rev. Astron. Astrophys., **22**, 359 (1984).
13. E.W. Cliver, et al. Astrophys. J., **343**, 953 (1989).
14. E.L. Chupp, Astrophys. J. Supp., **73**, 213 (1990).
15. J. Ryan, et al. Compton Symposium, (New York: AIP, 1993), in press.
16. G. Share and R. Murphy, Compton Symposium, (New York: AIP, 1993), in press.
17. G. Kanbach, et al., Compton Symposium, (New York: AIP, 1993), in press.
18. P.P. Dunphy, and E.L. Chupp, 22nd ICRC Papers, **3**, 65 (1991).
19. E.L. Chupp, et al. ApJ, 318, 913 (1987).
20. K.R. Pyle and J.A. Simpson, 22nd ICRC Papers, **3**, 53 (1991).
21. P. Evenson, P. Meyer, and K.R. Pyle, Astrophys. J. **274**, 875 (1983).
22. P. Evenson, et al., Astrophys. J. Supp., **73**, 273 (1990).
23. G.E. Kocharov, Yu. Kartavyh, and G.A. Kovaltsov, Ioffe Inst. Preprint 1461 (1990).
24. R. Ramaty, K. Kozlovsky, and R.E. Lingenfelter, Space Sci. rev., **18**, 341 (1975).
25. R. Ramaty, R.E. Lingenfelter, and K. Kozlovsky, in Gamma Ray Transients and Related Astrophysical Phenomena, (New York: AIP, 1982), p. 211.
26. R.J. Murphy, C.D. Dermer and R. Ramaty, Astrophys. J. Supp, **63**, 721 (1987).
27. X.-M. Hua, R. Ramaty, and R.E. Lingenfelter, Astrophys. J., **341**, 516 (1989).
28. J.M. Ryan and M.A. Lee, Astrophys. J., **368**, 316 (1991).
29. N. Mandzhavidze, and R. Ramaty, Astrophys. J., **389**, 739 (1992).
30. R. Ramaty, et al., Adv. Space Res., (1993) in press.
31. X.-M. Hua, and R.E. Lingenfelter, Solar Phys., **107**, 351 (1987).

32. X.-M. Hua, and R.E. Lingenfelter, Astrophys. J., **323**, 779 (1987).
33. V.G. Guglenko, et al. Solar Phys, **125**, 91 (1990).
34. V.G. Guglenko, et al. Astrophys. J. Supp., **73**, 209 (1990).
35. R.E. Lingenfelter, X.-M. Hua, B. Kozlovsky, and R. Ramaty, Compton Symposium, ed. N. Gehrels (New York: AIP, 1993), in press.
36. R.E. McGuire, et al., 15th ICRC Papers, **5**, 54 (1977).
37. R.E. McGuire, et al., 16th ICRC Papers, **5**, 61 (1979).
38. N. Mandzhavidze, and R. Ramaty, Astrophys. J., **396**, L111 (1992).
39. I.A.Ibragimov and G.E. Kocharov, Sov. Astron. Lett., **3**, 221 (1977).
40. R. Ramaty, B. Kozlovsky, and A.N. Suri, Astrophys. J., **214**, 617 (1977).
41. H.T. Wang, and R. Ramaty, Solar Physics, **36**, 129 (1974).
42. Kanbach, G., et al. 14th ICRC Papers, **5**, 1644 (1975).
43. Kanbach, G., et al. 17th ICRC Papers, **10**, 9 (1981).
44. S. Colgate, Astrophys. J. **221**, 1068 (1978).
45. E.G. Zweibel and D. Haber, Astrophys. J., **264**, 648 (1983).
46. R. Ramaty, in Particle Acceleration Mechanisms in Astrophysics, ed. J. Arons, C. Max and C. McKee (New York: AIP, 1979), p. 135.
47. M.A. Forman, R. Ramaty, and E.G. Zweibel, in The Physics of the Sun, ed. P.A. Sturrock (Dordrecht: Reidel, 1986), v2, p. 249.
48. R.E. McGuire, et al., 17th ICRC Papers, **3**, 65 (1981).
49. F.B. McDonald, and M.A.I. van Hollebeke, Astrophys. J., **290**, L67 (1985).
50. M.A.I. van Hollebeke, et al., Astrophys. J., Supp., **73**, 285 (1990).
51. H.V. Cane, et al., 19th ICRC Papers, **4**, 88 (1985).
52. T. Bai, Astrophys. J., **308**, 912 (1986).
53. E.W. Cliver, et al. 18th ICRC Papers, **10**, 334 (1983).
54. D. Forrest, in Positron and Electron Pairs in Astrophysics, ed. M.L. Burns et al. (New York: AIP, 1983). p. 3.
55. E.L. Chupp, et al. Astrophys. J., **318**, 913 (1987).
56. R.J. Murphy, et al. 19th ICRC Papers, **4**, 249 (1985).
57. R.J. Murphy, et al. 19th ICRC Papers, **4**, 253 (1985).
58. R.J. Murphy, et al. Astrophys. J., **371**, 793 (1991).
59. R. Ramaty, B. Kozlovsky and R.E. Lingenfelter, Astrophys. J., Supp., **40**, 487 (1979).
60. R.J. Murphy, Ph.D. Dissertation, Univ. of Maryland, 206 pp. (1985).
61. R.J. Murphy, et al. Astrophys. J., **358**, 290 (1990).
62. E. Anders and N. Grevesse, Geochim. Cosmochim. Acta, **53**, 197 (1989).
63. X.-M. Hua, and R.E. Lingenfelter, Astrophys. J., **319**, 555 (1987).
64. T. Prince, et al. 18th ICRC Papers, **4**, 79 (1983).
65. J. Geiss, Space Sci. Rev., **33**, 201 (1982).
66. E. Schatzman, and M. Maeder, Astron. Astrophys., **96**, 1 (1981).
67. J. Yang, et al. Astrophys. J., **281**, 493 (1982).

NEUTRON AND GAMMA-RAY MEASUREMENTS OF THE SOLAR FLARE OF 1991 JUNE 9

J. Ryan, D. Forrest, J. Lockwood, M. Loomis, M. McConnell, D. Morris,
W. Webber
Space Science Center, Morse Hall, University of New Hampshire,
Durham, NH 03824

K. Bennett, L. Hanlon, C. Winkler
Astrophysics Division, European Space Research and Technology Center, Noordwijk,
The Netherlands

H. Debrunner
Physikalisches Institut, University of Bern, 3012 Bern, Switzerland

G. Rank, V. Schönfelder
Max-Planck Institut für Extraterrestrische Physik, 8046 Garching, Germany

B.N. Swanenburg,
SRON-Leiden, 2300 RA Leiden, The Netherlands

ABSTRACT

The COMPTEL Imaging Compton Telescope on-board the Compton Gamma Ray Observatory measured significant neutron and γ-ray fluxes from the solar flare of 9 June 1991. The γ-ray flux had an integrated intensity (> 1 MeV) of ~30 cm^{-2}, extending in time from 0136 UT to 0143 UT, while the time of energetic neutron emission extended approximately 10 minutes longer, indicating either extended proton acceleration to high energies or trapping and precipitation of energetic protons. The production of neutrons without accompanying γ-rays in the proper proportion indicates a significant hardening of the precipitating proton spectrum through either the trapping or extended acceleration process.

INTRODUCTION

Active Region 6659 produced powerful flares throughout its transit across the disk from 2 June 1991 to 16 June 1991. On 9 June the region produced an X10/3B class flare that was observed by all instruments on the Compton Gamma Ray Observatory. At this time the active region was located at N34E04 in heliographic coordinates. The flare was impulsive in nature, with a duration on the order of 8 minutes and exhibiting two intense spikes of γ-ray emission > 600 keV. The Sun was positioned approximately 15° off the telescope axis, well within the instrument field-of-view (FOV).COMPTEL measures γ-rays by two means. These are the "telescope" and "burst" modes. The two modes of operation are described elsewhere[1]. The "telescope" mode relies on the Compton scattering process, whereby the γ-ray Compton scatters in a D1 detector (liquid scintillator) and then scatters again in a separate D2 (NaI) detector. In measuring the solar γ-ray flux, the scattering angle, deduced from the Compton formula and the direction of the scattered γ-ray must be consistent with a solar origin for subsequent analysis. The "burst" mode utilizes two of the D2 detectors as omnidirectional spectrometers. Neutrons are only measured in the "telescope" mode, but here the Compton scatter is replaced by an elastic neutron-proton scatter in the organic liquid scintillator in the D1 detector. The energy estimate of the recoil neutron is performed by measuring the time-of-flight of the neutron from D1 to D2. The solar γ-ray and neutron measurement capabilities of the instrument are described in greater detail by Ryan et al.[2].

OBSERVATIONS

The raw omnidirectional count rates of individual detector types are shown in Figure 1. This X10/3B flare was impulsive in nature with a duration on the order of 8 minutes and exhibiting two intense spikes of γ-ray emission. The GOES X-ray flare started at 0134 UT peaking at 0143 UT. The γ-ray onset occurred at 0136 UT and the impulsive phase lasted until ~ 0142 UT. Satellite sunrise precedes the flare only by a few minutes, thereby allowing a long observation of any extended flare emissions, such as neutrons. The count rates in the D1 and D2 detecting systems with thresholds of ~ 60 and ~ 350 keV, respectively, show how the flare behaved in hard X-rays and γ-rays. The slow rise and fall of the D2 counts rate after the impulsive phase arises from excursions in geomagnetic latitude and is not related to the flare. Also shown is the count rate in the Anticoincidence detection system (threshold ~100 keV). These detectors reject charged cosmic ray particles but are also sensitive to large thermal X-ray fluxes. At the X10 level, the X-ray flux falling upon the large Anticoincidence detectors results in pulse pile-up and large dead time effects for the whole instrument. The Anticoincidence rate qualitatively traces the magnitude of the dead time effect, which reaches a value of approximately 75% during the impulsive phase. The dead time effect improves to the ~ 50% level by 0155 UT.

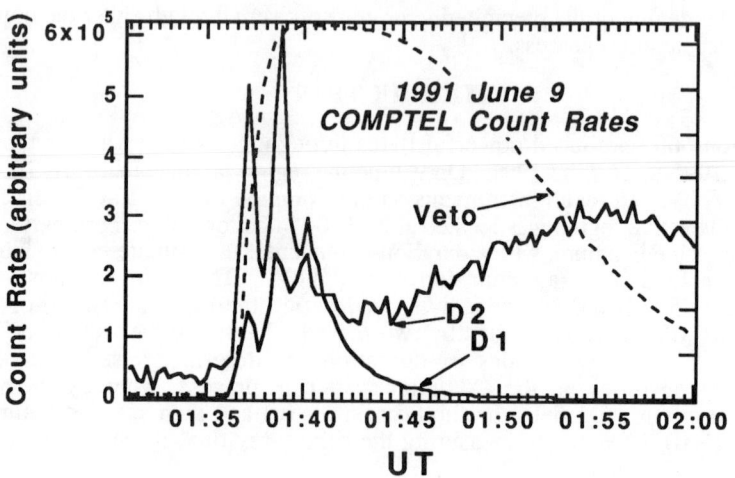

FIG. 1: Raw COMPTEL housekeeping data.

During the impulsive phase the γ-ray emission spectrum is obtained by selecting the telescope events consistent with the Sun's location, as shown in Figure 2. From the γ-ray events in the telescope mode, although not shown here, an image of the sun can be constructed[3]. The strong line from deuterium formation is seen at 2.2 MeV. Other lines are also present with less statistical significance. No background spectrum has been subtracted, but few photons are expected, since the background is suppressed by the large signal-to-noise ratio large.

Solar neutron events from 0155 UT to 0222 UT were selected from the data and reprocessed. This time selection avoids the troublesome period around the impulsive phase with large dead times and other instrumental effects. The time 0155 UT corresponds to a 50 MeV neutron produced at the flare start 0136 UT or a 60 MeV neutron from the flare maximum at 0139 UT. Some selection bias, therefore, exists for neutrons detected shortly after 0155 UT (real time). The selected data were subjected to similar geometric constraints as were the γ-rays, i.e., the origination direction of the neutron event must be consistent with the solar direction (± 10°). The scatter angle (ϕ) of the individual neutron events was restricted to 20° to improve the signal-to-noise ratio. The energy of each neutron is used to compute its production time at the Sun. Shown in Figure 3 in addition to the flux > 600 keV is the live time corrected and background subtracted intensity-time profile of the neutrons as produced at the Sun but plotted at the production time plus the light travel time over 1 AU (507 s). Therefore, photons and neutrons produced concurrently are plotted at the same time value even though the neutrons arrive later.

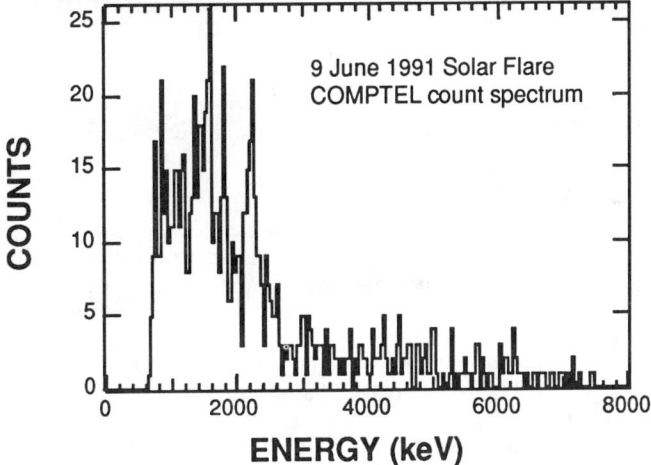

FIG. 2: *The raw telescope mode count spectrum from the 9 June 1991 solar flare.*

The background was estimated from the measured neutron flux ~ 24 hours later when the same orbital-geophysical conditions were reproduced. The on-board instrument software, however, was configured differently from the flare observation, being less restrictive in accepting neutron events. This resulted in a uniform 36%

increase in the neutron background count rate. The background rate was scaled downward by this factor in order to provide a representative background for the flare orbit on 9 June. The background corrections were successfully tested on similar background orbits to ensure that a null result is obtained. Identical data cuts then were made on both the flare data and the prescaled background data. Live time fractions of 60% to 100% were applied to the data as a function of real time from 0155 - 0220 UT.

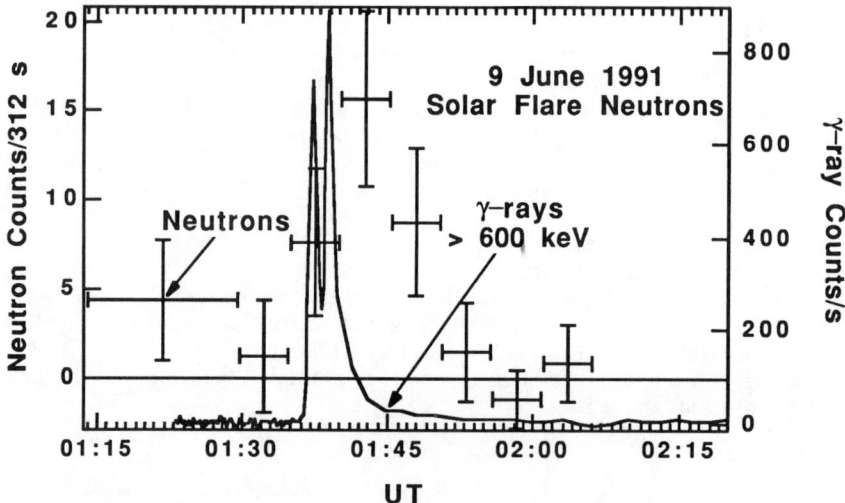

FIG. 3: Neutron and γ-ray emission-time profiles for the 9 June 1991 solar flare, plotted at the time corresponding to a photon arrival time.

DISCUSSION

The most important feature to note in Figure 3 is the time coincidence of the onsets of neutron and the γ-ray production. The neutron intensity-time profile is expected to be smoother and somewhat broader than that of γ-rays because of the neutron energy resolution which maps into an error (FWHM) in production time at the Sun (~ 1-3 minutes, dependent on energy). Although the γ-rays and neutrons are observed to start simultaneously, there is evidence that the neutron emission persists for ~ 10 minutes after the > 600 keV γ-ray flux has subsided.

This extended emission is evidence for a significant evolution (hardening) of the progenitor proton spectrum, arising from either additional acceleration or differential trapping and precipitation of protons. As seen in Figure 2 a large fraction of the γ-rays > 1 MeV are of nuclear origin. With no evolution of the proton spectrum, we would expect the > 600 keV γ-ray flux to follow the neutron production profile. However, a hardening of the spectrum would enhance the neutron emissivity with respect to that of the nuclear γ-rays. It should be noted, though, that the primary electron bremsstrahlung component has not yet been separated from the nuclear component, so that part of the decay of the flux > 600 keV could result from the decay of the pure electron component of the γ-ray flux.

The trapping scenario is consistent with the > 50 MeV γ-ray flux detected by EGRET after 0145 UT[4]. The net neutron count rate from 0136 UT to 0150 UT is positive at the 4.2 σ significance level. The absolute neutron flux is uncertain at this time due to the difficulty in computing the effective area of the instrument under the conditions of the flare and the data restriction used in the analysis. Work continues on this effort. The uncertainty of the instrument response manifests in the relative neutron flux among the three 5 minute intervals of neutron production. The assignment of a production time to an individual neutron is not affected by this uncertainty, only the intensity of the neutron emission. Thus, the intensity-time profile in Figure 3 may change shape when neutron fluxes are computed, but the duration of the event will remain unchanged.

In all other respects the 9 June 1991 was purely impulsive in nature, but the prolonged energetic particle activity as manifest in neutrons and γ-rays > 50 MeV establishes this flare as one of a class of *long duration* events, similar to that of 3 June 1982. Two models for describing this phenomenon have been put forth[5,6]. Ryan and Lee[5] propose that a steadily hardening proton spectrum is produced by proton trapping in a *turbulent* coronal magnetic loop with second order stochastic acceleration, while Mandzhavidze and Ramaty[6] posit that the protons responsible for extended high energy emission such as this arise in the impulsive phase with differential trapping and precipitation in a *relatively quiet* coronal loop. In either scenario, it is the diffusion of protons out of the loop which precipitate and produce the high energy emission. The data of the 9 June 1991 flare do not distinguish between the two models. With the great sensitivity of the instruments on the Compton Gamma Ray Observatory it is likely that other flares exhibiting this behavior will be observed or discovered in the data. By studying these flares we hope to gain a greater understanding of the processes of proton acceleration and transport in solar flares.

REFERENCES

1. Schönfelder *et al.*, Astrophys. J. (Suppl.), **86**, 657.
2. Ryan, J.M. et al., in *Data Analysis in Astronomy IV*, Ed. V. Di Gesù et al.,New York, Plenum Press (1992).
3. McConnell, M. et al., Adv. in Sp. Sci., to be published.
4. Kanbach, G., priv. communication, (1992).
5. Ryan, J.M., Lee, M.A., Astrophys.J., **368**, 316 (1991).
6. Mandzhavidze, N., Ramaty, R., Astrophys. J. (Letters), **396**, L111 (1992)

EGRET OBSERVATIONS OF EXTENDED HIGH-ENERGY EMISSIONS FROM THE NUCLEAR LINE FLARES OF JUNE 1991

E.J. Schneid
Grumman Corporate Research Center, Bethpage, NY 11714

K.T.S. Brazier, G. Kanbach, C. von Montigny, H.A. Mayer-Hasselwander
Max-Plank Institut fur Extraterrestrische Physik, 8046,
Garching bei Munich, Germany

D.L. Bertsch, C.E. Fichtel, R.C. Hartman, S.D. Hunter, D.J. Thompson
NASA/Goddard Space Flight Center

B. L. Dingus, P. Sreekumar
Universities Space Research Association, NASA/GSFC, Code 662,
Greenbelt, MD 20771

Y.C. Lin, P.F. Michelson, P.L. Nolan
Stanford University, Stanford, CA 94305

D.A. Kniffen
Hampden-Sydney College, P.O. Box 862, Hampden Sydney, VA 23943

J.R. Mattox
COMPTON Science Support Center, Computer Sciences Corporation,
Greenbelt, MD 20771

ABSTRACT

EGRET onboard the Compton Gamma Ray Observatory observed four energetic X-type solar flares during June 1991. Two of the these flares were in the EGRET spark chamber field of view and were observed to have high energy gamma ray emission lasting hours after the impulsive phase of the flare. Measurements of all four flares were obtained by the EGRET large NaI spectrometer.

INTRODUCTION

Shortly after the launch of the Compton Gamma Ray Observatory (CGRO) on April of 1991, a period of intense solar activity occurred during the May, June, July time period. Two X-type flares occurred in the first week of June prompting the action of making the Sun a target of opportunity for CGRO and therefore all CGRO instruments were pointed at the Sun when three additional X-type flares occurred during the second week. EGRET was able to observe four of these flares but was turned off during the fifth flare. All of these flares whose nuclear excitation lines in their spectra and had extended emission of high energy radiation. During the June 11, 1991 flare, gamma-rays, with energies up to several GeV were observed in the EGRET spark chamber up to 8 hours after the impulsive phase of the flare. The information for these flares are listed in Table 1. For EGRET, an azimuth angle of zero corresponds to viewing along the CGRO long axis.

Table I
Energetic X-Type Flares of June 1991

Date 1991	Time (UT) Maximum	Active Region	Type	H-α Location	Angles to CGRO Axis Zenith	Azimuth
June 4	0339	6659	X12	N30E70	105°	0°
June 6	0107	6659	X12	N33E44	106°	2°
June 9	0143	6659	X10	N34E04	15°	12°
June 11	0209	6659	X12	N31W17	14°	6°
June 15	0821	6659	X12	N33W69	12°	9°

The EGRET instrument consists of a spark chamber with interleaved tantalum plates to convert gamma rays to electron position pairs, and to image the trajectories of the pair. A NaI scintillation spectrometer measures the energy of pair as it emerges form the bottom of the spark chamber telescope. A large anticoincidence shield to reject charged particle events surrounds the spark chamber but does not extend to cover the NaI spectrometer. The NaI spectrometer (76x76x20 cm) has a special burst/flare mode for recording gamma ray bursts and solar flares. In this mode a pulse height spectrum is accumulated for all NaI events from 1 MeV to approximately 200 MeV. Spectra are routinely accumulated every 32.75 seconds but when activated by a BATSE burst trigger, EGRET accumulates 4 sequential spectra with integration times variable from 1 to 16 seconds. Details of the instrument and calibration can be found in Hughes et al.[1], Kanbach et al.[2], and Thompson et al[3].

RESULTS

The NaI spectrometer measured gamma-ray emission lasting thousands of seconds after the impulsive phase while the spark chamber was able to measure gamma-ray emission lasting hours after the peak of the flare. Figure 1 shows the higher energy (20 - 200 MeV) signals observed in the NaI spectrometer of the four flares seen by EGRET. The two flares on June 4 and 6 were very energetic and were observed to have substantial signal in the NaI even though the radiation was attenuated by the bulk of the CGRO satellite. The flare related gamma-rays were still very strong when earth occultation occurred and still could be seen after occultation.

The June 9 and June 11 flares were in the field of view of the spark chamber enabling measurement of gamma-rays from 50 to greater than 1 GeV to be obtained. For the June 11 flare, gamma ray emission were observed up to 8 hours after the impulsive phase (Kanbach et al.[4]). Figure 2 shows the June 11 time profile observed by the spark chamber and NaI for two energy region 50-150 MeV and >150 MeV. The NaI data covers only the period lasting a few thousand seconds after the impulsive phases. The NaI data appears to rise and then decay. This rise may be due to other contributions to the NaI signal-such as solar neutrons arriving at the spacecraft. Work is in progress to try a resolve this issue.

The dashed line is a prediction by (Mandzhavidze and Ramaty[5]) to fit the spark chamber data and assumes that high energy protons are trapped in coronal loops. The spark chamber spectra could be well fit by a bremsstrahlung component and neutral pi meson decay component consistent with the theory of trapped high energy protons

(Mandzhavidze and Ramaty[5]). Pitch angles, Bremsstrahlung contributions, and other parameters could be adjusted to get a better fit but would not account for a peaking in the NaI data.

NaI spectra for the extended period show clear evidence of nuclear excitation lines caused by energetic protons. Figure 3, 4, 5, 6 show the pulse height spectra for the June 4, 6, 9 and 11 flares respectively. Individual lines can not be resolved but enhancements at 2.0 - 2.4 MeV (the neutron capture region) and at 4.0 - 7.0 MeV (proton excitation of carbon, oxygen, etc. region) can be clearly observed. Each figure contain three spectra for the time intervals 100-300, 300-700, 700-1000 seconds after the first peak of the flare. Even for the June 4 and June 6 flares where the gamma radiation was attenuated and modified by passage through CGRO, the line features can be still observed. The gamma ray line features for all flares decay with increasing time; whereas, the high energy photons (>20 MeV) show enhancements in flux during the extended emission period (see also figure 1). This could indicate an increase of high energy gamma-ray production at the Sun or possibly the arrival of energetic neutrons at the CGRO location. Detailed analyses of these spectra will be discussed by Bertsch et al.[6].

CONCLUSIONS

The EGRET instrument with its high energy capabilities in the spark chamber and NaI spectrometer is providing new information on the high energy particles associated with energetic solar flares. These June flares contribute strongly to data base required to interpret the energetics and mechanisms for the extended emission.

ACKNOWLEDGMENT

The EGRET team gratefully acknowledges support form the following: Berndeministerum fur Forschung und Technologie, grant 50 QV 9095 (MPE); NASA grant NAS65-1742 (HCS); NASA grant NAG 5-1605 (SU); and NASA contract NAS 5-31210 (GAC).

REFERENCES

1. Hughes, E.B., et al., IEEE Trans. Nucl. Sci., NS-27, p 364 (1988)
2. Kanbach, G., et al., Proc of the Gamma Ray Observatory Workshop, p 2-1 (1989)
3. Thompson, D.J., et al., ApJ Suppl. Ser., 89, 629 (1993)
4. Kanbach, G., et al., Astron. Astrophys. Suppl. Ser. 97, 349 (1993)
5. Mandzhavidze, N., and Ramaty, R., ApJ 396, L111, (1992)
6. Bertsch, D.L., et al., to be published (1993)

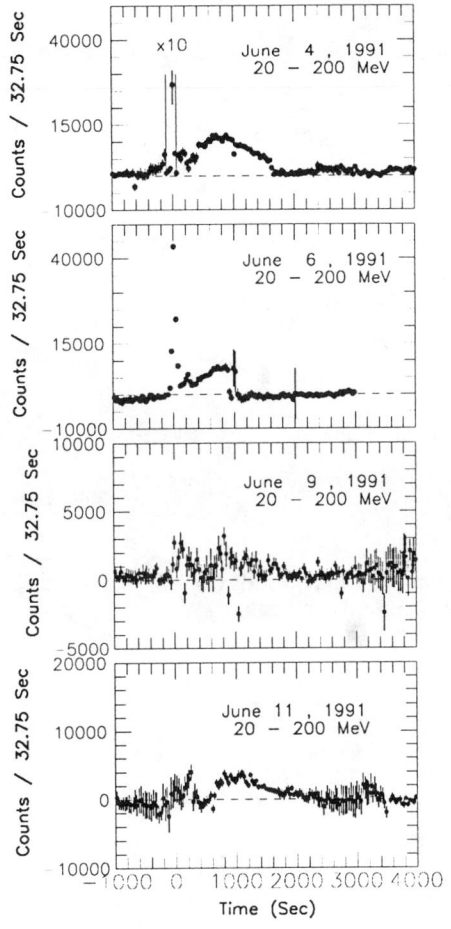

Fig 1 Time profiles of high energy flare related signals observed in the NaI as a function of time (seconds) since the first peak of the flare. These signals may result from a combination of gamma rays, neutrons, and charged particles. The June 4 impulsive peak has been reduced by a factor of 10 for plotting purposes.

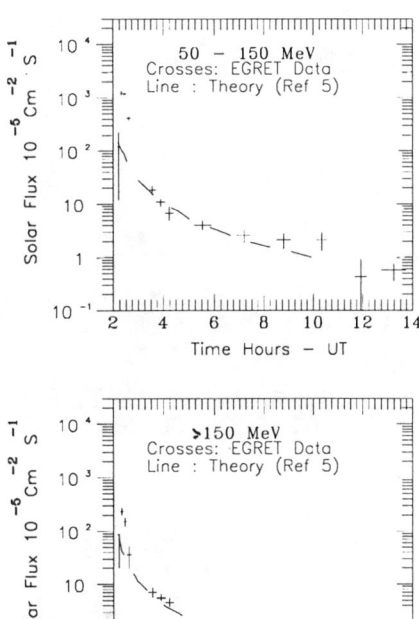

Fig 2 The time profiles of the (50-100 and >150 MeV) extended high energy emissions for the June 11 flare. Data for times earlier than 0300 UT were obtained from the NaI detector and later than 0300 UT were obtained from the spark chamber. The dashed curves are predictions for protons trapped in a coronal loop.

98 EGRET Observations of Extended High-Energy Emissions

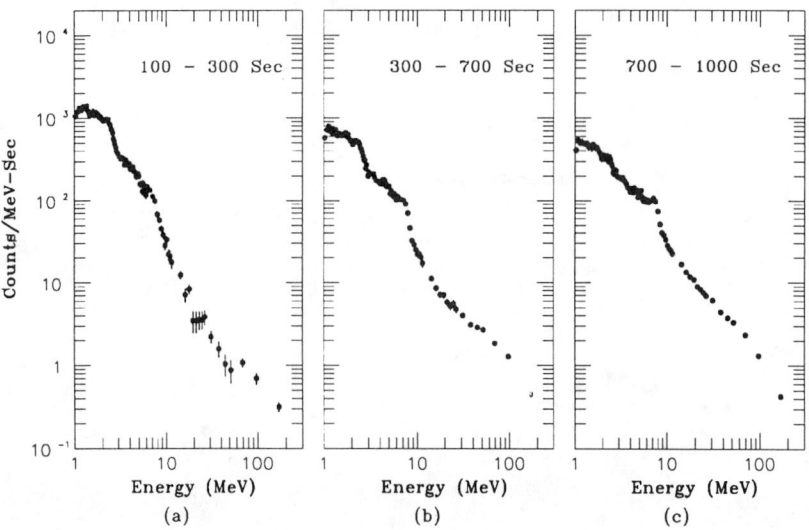

Fig 3   The June 4, 1991 NaI flare related spectra for the periods (a) 100-300, (b) 300-700, (c) 700-1000 seconds after the first impulsive peak.

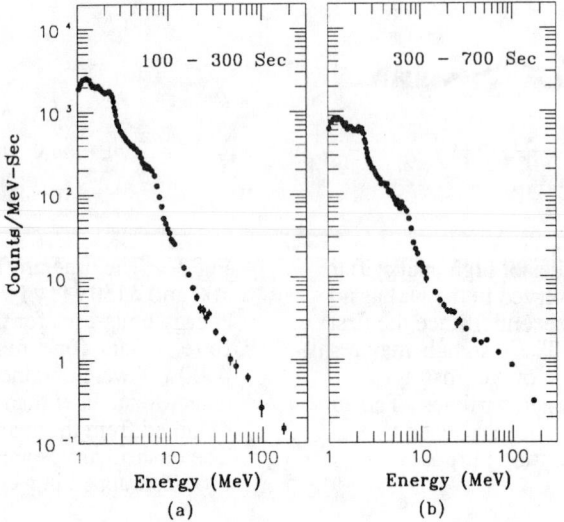

Fig 4   The June 6, 1991 flare related spectra for the periods (a) 100-300, (b) 300-700 seconds after the first impulsive peak.

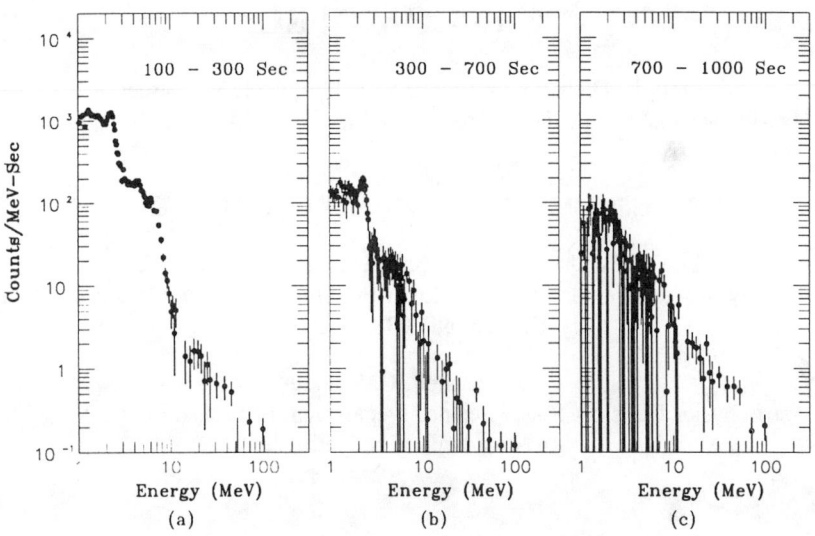

Fig 5 The June 9, 1991 flare related spectra for the periods (a) 100-300, (b) 300-700, (c) 1000 seconds after the first impulsive peak.

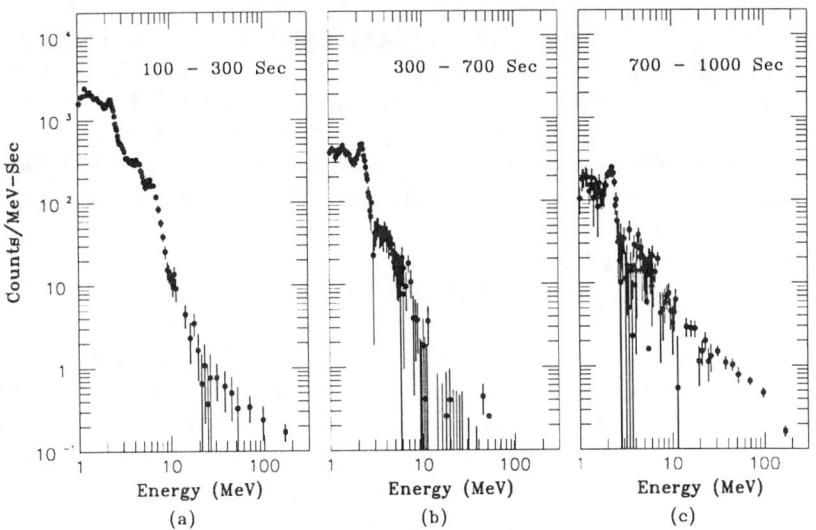

Fig 6 The June 11, 1991 flare related spectra for the periods (a) 100-300, (b) 300-700, (c) 700-1000 seconds after the first impulsive peak.

Observations of the 1991 June 11 solar flare with COMPTEL

G. Rank, R. Diehl, G. G. Lichti, V. Schönfelder, M. Varendorff[*]
Max-Planck Institut für Extraterrestrische Physik, D-8046 Garching, Germany

B. N. Swanenburg, R. van Dijk[**]
SRON-Leiden, P.B. 9504, NL-2300 RA Leiden, The Netherlands

D. Forrest, J. Macri, M. McConnell, M. Loomis, J. Ryan
University of New Hampshire, Institute for the Study of Earth, Oceans and Space,
Durham NH 03824, U.S.A.

K. Bennett, C. Winkler
Astrophysics Division, Space Science Department of ESA/ESTEC, NL-2200
AG Noordwijk, The Netherlands

ABSTRACT

The COMPTEL instrument onboard the Compton Gamma-Ray Observatory (CGRO) is sensitive to γ-rays in the energy range from 0.75 to 30 MeV and to neutrons in the energy range from 10 to 100 MeV.

During the period of unexpectedly high solar activity in June 1991, several flares from active region 6659 were observed by COMPTEL. For the flare on June 11, we have analyzed the COMPTEL telescope data, finding strong 2.223 MeV line emission, that declines with a time constant of 11.8 minutes during the satellite orbit in which the flare occurs. It remains visible for at least 4 hours. We obtained preliminary values for the 2.2 MeV and 4-7 MeV fluences. Neutrons with energies above 20 MeV have been detected and their arrival time at the Earth is consistent with the γ-ray emission during the impulsive phase.

INTRODUCTION

COMPTEL is a useful instrument for studying solar flares. It is able to measure γ-rays in the energy range from 0.75 MeV to 30 MeV and to derive the direction of the incoming radiation. When operating in its full-telescope mode, COMPTEL uses a double scattering of a γ-quantum to get spectral and directional information. In this case, Compton scattering of the γ-ray is the fundamental process.

---

[*]Present adress: University of New Hampshire, Institute for the Study of Earth, Oceans and Space, Durham NH 03824, U.S.A.
[**]Also: Astronomical Institute "Anton Pannekoek", University of Amsterdam, The Netherlands

In addition, COMPTEL has the unique capability of detecting individual neutrons in the energy range from approximately 10 MeV to 100 MeV. The measurement is similar to that of γ-rays, but for neutrons hard-sphere scattering is the basic mechanism. The COMPTEL data undergo a pre-selection onboard the satellite, where a possible neutron signal is separated, mainly by a discrimination in pulse shape, and telemetered in a special neutron event stream (for details see Ryan et al., 1992; McConnell et al., these proceedings).

To avoid background of charged particles, the whole instrument is shielded by anti-coincidence plastic scintillators. To suppress additional γ-ray and neutron background originating in the Earth's atmosphere and the instrument itself, only events being compatible with the direction of the Sun are accepted. This uses the imaging capability of COMPTEL, as it has been demonstrated earlier for the June 11 solar flare (Rank et al., 1992).

The remaining background shows orbital variations due to changes in rigidity and spacecraft orientation relative to the Earth. However, 15 and 16 orbits before and after the flare, the orbital parameters are reproduced quite precisely. Therefore, these data can be used to generate a model background.

The interpretation of the data is very difficult due to several effects which are caused by the enormous X- and γ-ray flux during the flare: There are (1) severe deadtime effects, (2) the occurence of multi-hit processes in the telescope and (3) an overflow of the limited event buffers and the telemetry capability.

OBSERVATIONAL RESULTS

On June 11, 1991, the active solar region 6659 has produced a huge flare, being accompanied by a prominence. Optically it was classified as a 3B event and in soft X-rays as a X-12 GOES event. X-ray emission started at 1:56 UT, reached its maximum at 2:09 UT and faded away at about 2:20 UT. Orbital sunrise of CGRO had taken place at 1:48 UT, a few minutes before the flare onset. Hence, the temporal evolution of the flare could be studied for a whole orbital period.

<u>Analysis of COMPTEL γ-ray measurements:</u>

A time history of the June 11 solar flare can be obtained from COMPTEL telescope data by correcting the measured count rate for background and live-time effects. The γ-ray emission starts at about 1:58 UT, shows a first peak at 2:00 UT and reaches its maximum at 2:05 UT. This indicates a faster rise to the maximum than in soft X-rays. After the maximum, the γ-ray emission in the 2.0-2.5 MeV range shows exponential decline with a time constant of 11.8 minutes during the flare orbit. In the next two orbits a line signal is still remaining. This indicates the production of the 2.223 MeV line and therefore the presence of neutrons for at least 4 hours after the impulsive phase (McConnell et al., 1992; Rank et al., 1992).

To obtain preliminary values for fluences, we set tight restrictions on the incoming direction of the γ-rays in order to get a photo-peak response for the detectors. In this case, the effective areas from the calibration can be used. They have to be corrected for data restrictions (some restrictions are different from the calibration standards) and for the incomplete module combination (two modules of the lower detector were turned off). The remaining effective areas are in the order of 6 cm^2, slightly depending on energy.

The fluences were calculated for the whole flare orbit, beginning at the onset of the flare (2:00 UT) and ending at the orbital sunset (3:00 UT). For the 4-7 MeV range we obtain a fluence of $\phi_{4-7} = 138\pm8$ γ cm^{-2}. For the 2.223 MeV line the fluence in the range from 2.1 to 2.4 MeV is given by $\phi_{2.2} = 114\pm18$ γ cm^{-2}. This leads to a fluence ratio of $\phi_{2.2} / \phi_{4-7} = 0.83\pm0.14$.

Assuming the same acceleration conditions as they have been studied by Hua and Lingenfelter (1987), a preliminary estimate of the spectral hardness of the accelerated proton spectrum can be made. In the case of a power law spectrum, as it would appear for non-relativistic shock acceleration, we get a spectral index of s = 3.6. For second order stochastic acceleration we would expect a Bessel function with $\alpha T = 0.02$.

Analysis of COMPTEL neutron measurements:

The analysis of the neutron data is performed in the same way as it has been done for the γ-rays. However, some effects complicate the analysis:

During the most intense part of the flare, pulse pile-up effects of the γ- and X-ray flux produce a strong signal which is not related to incoming neutrons. These events are arriving at the Earth when the electromagnetic signal is expected. But calculating their emission time at the Sun, they are emitted long before the flare onset. Data restrictions on the energy thresholds of the detectors, combined with tighter windows on the pulse shape discrimination and the time of flight from the upper to the lower detector remove the bulk of the artifical counts during the impulsive phase. However, a few counts still remain. These are scaled up by huge live time factors producing single spikes before the onset of the flare. As they are not related to neutrons they were excluded in the figures.

A phase of extremely low count rate follows this artificial signal. It is caused by the dead-time effects due to the huge X- and γ-ray flux. The dead-time situation is improving when the neutrons arrive. Nevertheless, a live-time correction has been applied to the neutrons in the same manner as was done for the γ-rays.

In figures 1 to 4 the time histories of the neutrons are displayed. The count rates in the four energy ranges 10-20 MeV, 20-40 MeV, 40-60 MeV and 60-100 MeV are calculated back to their emission time at the Sun. A phase of negative values appears in each picture. It is caused by a suppression of counts beneath the background level during the impulsive phase when the dead time is great. The start and end time of each plot is defined by the satellite orbit.

In the lowest energy range almost no signal can be found in the flare orbit. Neutrons with these energies are travelling so slow that they arrive at the end of the satellite orbit when the neutron background from the Earth's atmosphere makes a detection difficult (most recent analysis performed after the Workshop has shown that there is an excess during the next two satellite orbits. This would be in very good agreement with the extended signal of the 2.223 MeV line).

For the higher energy ranges, especially for the 40-60 MeV range, a significant signal can be found. The events are measured at times when neutrons from the impulsive phase are expected to reach the Earth.

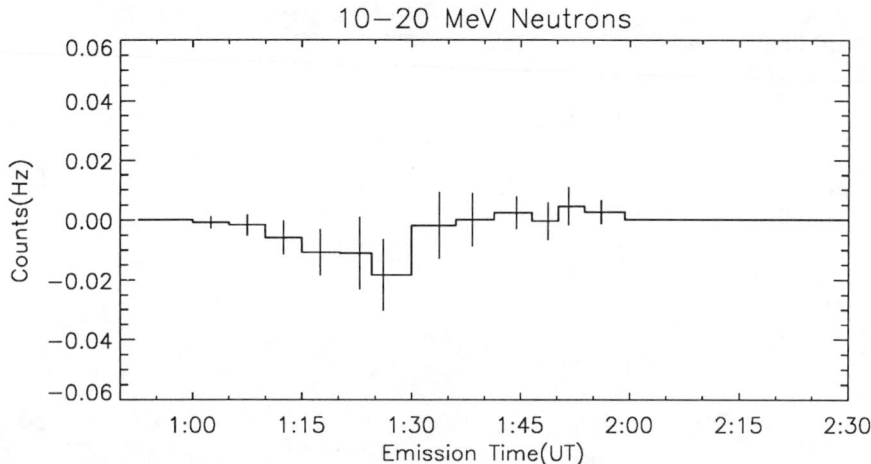

Fig. 1: Time history of the count rate of neutrons in the energy range from 10 MeV to 20 MeV for their calculated emission time at the Sun. The γ-ray event at the sun peaks at about 1:55 UT.

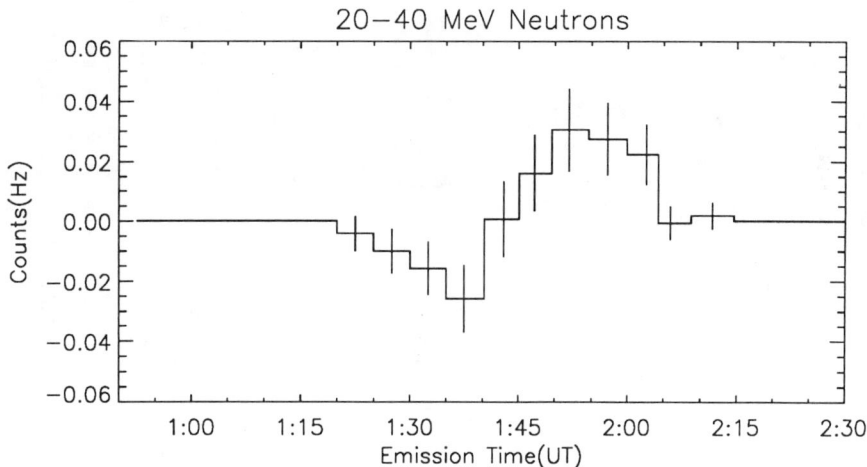

Fig. 2: Time history for neutrons in the energy range from 20 MeV to 40 MeV (see also Fig. 1).

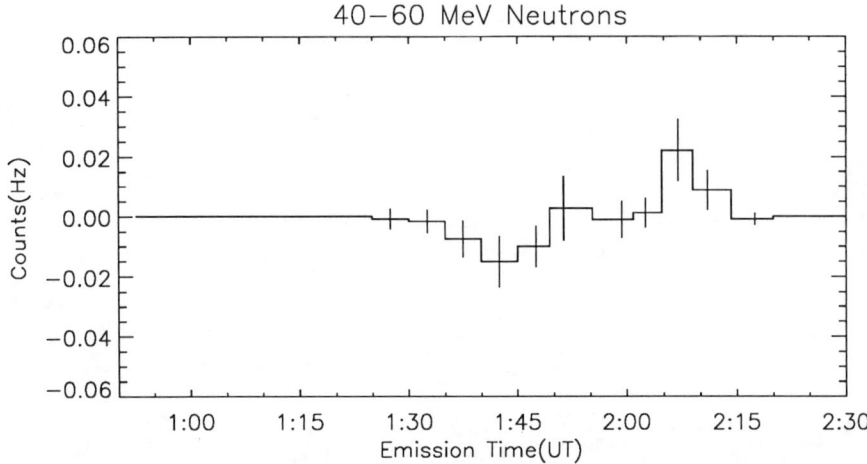

Fig. 3: Time history for neutrons in the energy range from 40 MeV to 60 MeV (see also Fig. 1).

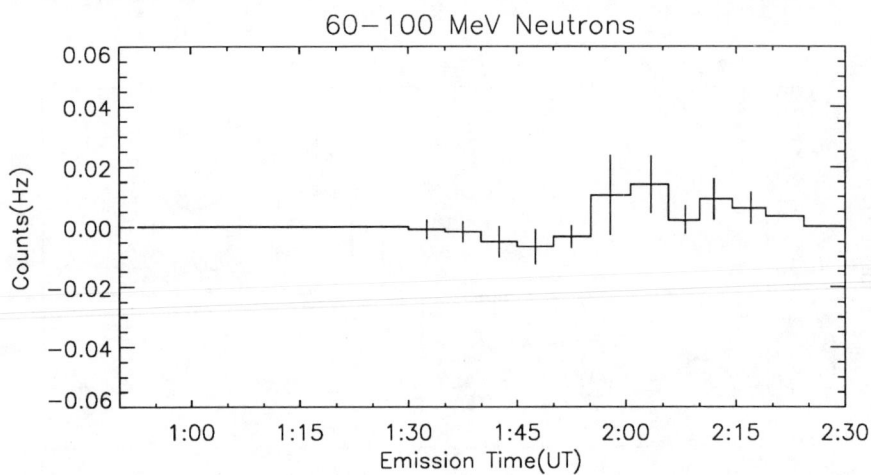

Fig. 4: Time history for neutrons in the energy range from 60 MeV to 100 MeV (see also Fig. 1).

SUMMARY

The spectrum of the 1991 June 11th flare shows a strong 2.223 MeV line that can be seen by COMPTEL for at least 4 hours after the impulsive phase.

Neutrons in the energy range from approximately 20 to 100 MeV can be detected during the flare orbit. Their calculated emission-time profile at the Sun matches well with the impulsive phase as defined by the γ-rays.

REFERENCES

X. M. Hua, R. E. Lingenfelter, *Solar Physics*, **107**, 351-383 (1987).

M. McConnell, K. Bennett, H. Bloemen, H. de Boer, M. Busetta, W. Collmar, A. Connors, R. Diehl, J.W. den Herder, W. Hermsen, L. Kuiper, G. G. Lichti, J. Lockwood, J. Macri, D. Morris, R. Much, G. Rank, J. Ryan, V. Schönfelder, G. Stacy, H. Steinle, A. W. Strong, B. N. Swanenburg, B. G. Taylor, M. Varendorff, C. de Vries, W. Webber, C. Winkler, *COMPTEL observations of solar flare gamma-rays* (COSPAR, 1992).

M. McConnell, K. Bennett, H. Bloemen, H. Debrunner, R. Diehl, D. Forrest, L. Hanlon, W. Hermsen, G. G. Lichti, J. Lockwood, M. Loomis, G. Rank, J. Ryan, V. Schönfelder, A. W. Strong, B. N. Swanenburg, M. Varendorff, C. Winkler, *An Overview of Solar Flare Results from COMPTEL*, (High Energy Physics of Solar Flares, 1993).

G. Rank, K. Bennett, R. Diehl, D. Forrest, L. Hanlon, G. G. Lichti, J. Macri, M. McConnell, J. Ryan, V. Schönfelder, B. N. Swanenburg, M. Varendorff, C. Winkler, *Observations of the 1991 June 11 solar flare with COMPTEL* (Compton Symposium, 1992).

J. Ryan, H. Aarts, K. Bennett, R. Byrd, C. de Vries, J. W. den Herder, A. Deerenberg, R. Diehl, G. Eymann, D. Forrest, C. Foster, W. Hermsen, G. G. Lichti, J. Lockwood, J. Macri, M McConnell, D. Morris, V. Schönfelder, G. Simpson, M. Snelling, H. Steinle, A. W. Strong, B. N. Swanenburg, T. Taddeucci, W. R. Webber, C. Winkler, *COMPTEL as a solar gamma ray and neutron detector* (Data Analysis in Astronomy IV, ed. V.D.Gesú et al., 1992).

SOME EVIDENCES OF PROLONGED PARTICLE ACCELERATION IN THE HIGH-ENERGY GAMMA-RAY FLARE OF JUNE 15, 1991

V.V.Akimov, N.G.Leikov
Space Research Institute, Moscow, 117810, Russia

A.V.Belov, I.M.Chertok
IZMIRAN, Troitsk, Moscow Region, 142092, Russia

V.G.Kurt
Insitute of Nuclear Physics, Moscow University,
Moscow, 119899, Russia

A.Magun
Institute of Applied Physics, University of Bern,
CH-3012 Bern, Switzerland

V.F.Melnikov
Radiophys. Research Inst.,N.-Novgorod, 603600, Russia

ABSTRACT

The γ-ray emission extending to energies greater than 1.5 GeV and lasting at least for two hours was observed from the 3B/X12 two-ribbon flare of June 15, 1991 with the telescope GAMMA-1 aboard the GAMMA satellite. A nuclear line emission was also registered in this flare by COMPTEL aboard the CGRO satellite. A comparison of the γ-ray flux time history with that of microwave emission and analysis of the data on solar cosmic ray fluxes provide an evidence of a long-term particle acceleration well after the impulsive phase of the flare. It is suggested that such an acceleration may be associated with a long post-eruptive energy release after a coronal mass ejection (CME).

INTRODUCTION

This work is devoted to the analysis of GAMMA-1[1] and partially of COMPTEL[2] γ-ray data for the June 15, 1991 flare in comparison with data on microwave bursts obtained at Bern and IZMIRAN as well as with world-network data on GLE and satellite particle measurements.

In order to illustrate the distinction of this event a short reminder about the well-known flare of June 3, 1982 would be appropriate. Fig. 1 shows the time profile of radio emission at 3 GHz in this flare. One can see that the burst consists of at least two main components separated by the time interval of about 15 min. The first component is an impulsive 10-min burst with two peaks and rather sharp decay. There is also the second, or delayed, component with a relatively weak but well visible enhancement. It should be emphasized that for this event the SMM gamma-ray observations[3] as well as numerous models and simulations[4-9] belong, in fact, to the two peaks of the impulsive component and its decay phase, but do not touch on the secondary delayed post-burst enhancement.

It is remarkable that the June 15, 1991 flare has very similar time profile of radio emission, although in this event the both impulsive and delayed components are more intense (Fig. 2). It happened so that observations by GAMMA-1 in the energy range >30MeV and by COMPTEL in the range 1-10 MeV were carried out just during the delayed component well after the impulsive phase. High sensitivity of these telescopes allowed to register for the first time γ-emission at this stage of a solar flare.

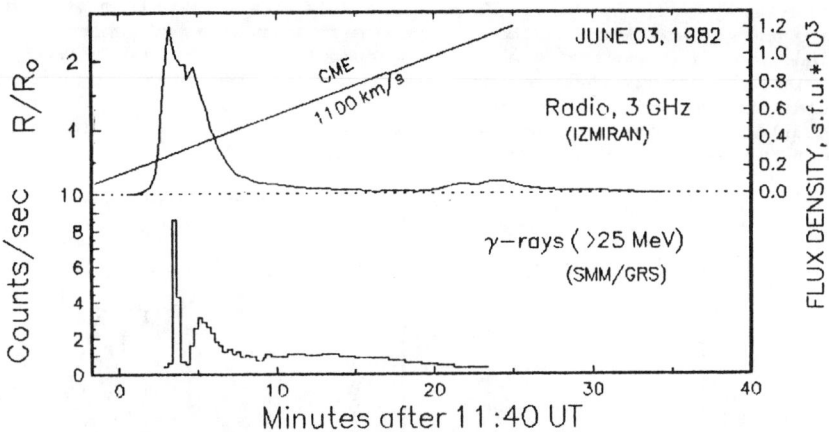

Fig. 1. Microwave and γ-ray time profiles of the June 3, 1982 flare and altitude-time trajectory of the CME.

The main purposes of the present consideration are to analyze the experimental data from the point of view of the origin of energetic particles and to discuss what model (long-term trapping or prolonged particle acceleration) is compatible with the late stage of the flare.

OBSERVATIONS AND DISCUSSION

The GAMMA-1 telescope being in shadow could not register the impulsive phase and was switched on in the solar attitude at 08:37:22 UT just at the time of the maximum of the delayed radio component (Fig. 2). During the next orbit observation (after 10:08 UT) the high energy γ-ray flux was still well above the background.

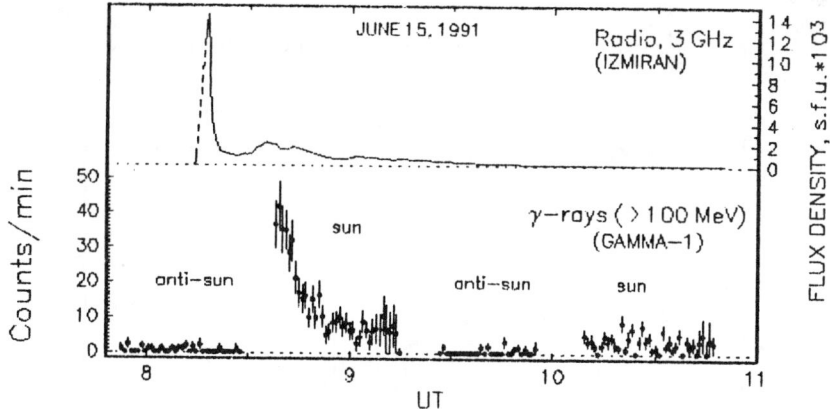

Fig. 2. Microwave and γ-ray time profiles of the June 15, 1991 flare

The γ-ray spectrum in the first orbit derived with the use of maximum likelihood method is shown in Fig. 3a[10]. A shape of the spectrum with a maximum at 70 - 100 MeV indicates that most of photons were born in neutral pion decay. A curve in Fig. 3b presents this photon spectrum folded with the energy dependence of the telescope effective area and energy response function. It demonstrates a good agreement with the registered counts (circles). The open squares show the counts in the second orbit. Scarce statistics does not allow to reliably restore a photon spectrum but general similarity of the two orbits data indicates that the photon spectrum has not changed significantly in 1.5 hour.

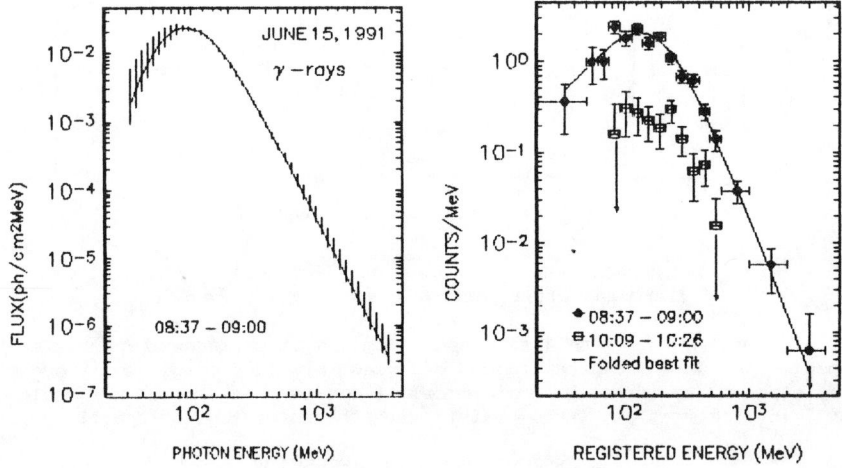

Fig. 3.
a) The γ-ray energy spectrum in the first orbit
b) Energy distributions of counts in the first and second orbits

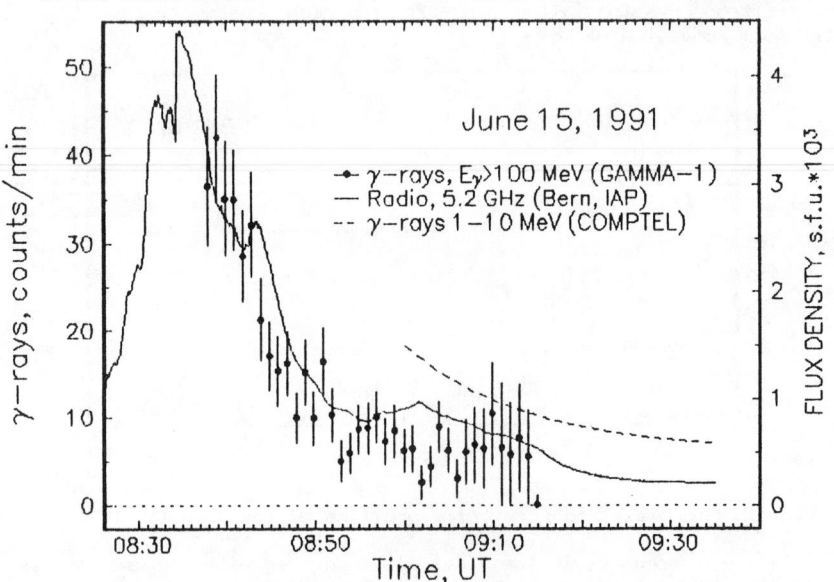

Fig. 4. Enlarged time profiles of the delayed component

There are two possible models to explain the presence of high-energy ions in the flare volume during the post eruptive phase: a long duration acceleration with fast precipitation, or a trapping of the impulsively injected particles with slow precipitation. As a rule processes of particle acceleration up to energies 1-10 GeV were attributed either to the impulsive flare phases or to the shock wave propagation through the corona, while prolonged gamma and neutron emissions were explained by different trapping models (sometimes with an additional acceleration)[4-9]. However the discovery of the delayed component in the radio and hard X-ray emissions forced to apply for the explanation of this phenomenon to the prolonged electron acceleration models based on the slow magnetodynamic process of the magnetic field restoration after a coronal transient eruption[11-14]. A total combination of data available for this flare can help to choose between these two models.

Enlarged picture of the microwave delayed component superimposed to the high energy gamma radiation time history is shown in Fig. 4. The microwave profile exhibits many sub-peaks with the typical rise and decay time ~100 s. This variations could be explained either by sharp changes of the electron flux or by fast variations of conditions in the emission region (magnetic field, matter density). The latter would show itself in changes of the microwave emission spectrum: an increase of the spectral maximum frequency or a steepening of the frequency spectrum at low frequencies. However, the comparison of microwave spectra for different moments depicted in Fig. 5 allows us to conclude that none of these changes took place. During the time of the gamma-ray observations, the spectral maximum frequency remains at a region of 5.2 GHz, and some relative increase of the intensity at low frequencies takes place rather than its decrease. So we have to choose the former explanation which means that a continuous acceleration and fast precipitation of electrons takes place during the delayed component of the microwave burst.

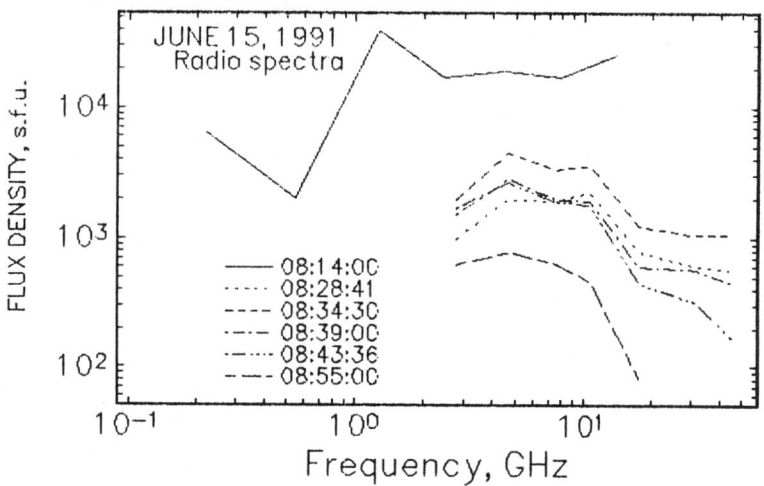

Fig. 5. The microwave spectra at different moments of the delayed component.

From Fig. 4 one can see that definite similarity is observed between the general trends (and perhaps some details) of the microwave and high energy γ-emissions. Moreover, shown by a dashed line time profile of the

γ-ray line emission registered by COMPTEL[2] agrees well with that of the microwave. To explain this similarity in terms of the trapping model one would have to suppose that just by chance conditions in the trapping regions for ions of tens and many hundred MeV were adjusted in some very special way. As it seems to be completely improbable we have to conclude that these ions responsible for production of the nuclear line emission and high energy γ-rays being accelerated in the same acceleration process as the electrons responsible for the microwave emission and had quite short precipitation time.

Some additional arguments in favor of the long-term acceleration come from the data on charge particle fluxes in this flare. Fig. 6 shows the onset and maximum times of the charged particle fluxes registered on satellites and by ground based neutron monitors versus the particle velocity[15,16]. One can see that protons with energies of several hundred MeV arrived with the normal velocity dispersion as if they escaped at the impulsive stage of the flare, while protons with energies above 500 MeV (as well as relativistic electrons) reached the Earth 10-15 min later ($08^h42^m \pm 2^m$ UT) than could be expected from the standard propagation model. Another argument is based on the time behavior of high energy proton spectra derived from the neutron monitor network data. Analysis shows that the spectra were not softening during at least 50 min after the onset. It can be interpreted as an indication of long, lasting for tens of minutes, emission of protons from the Sun.

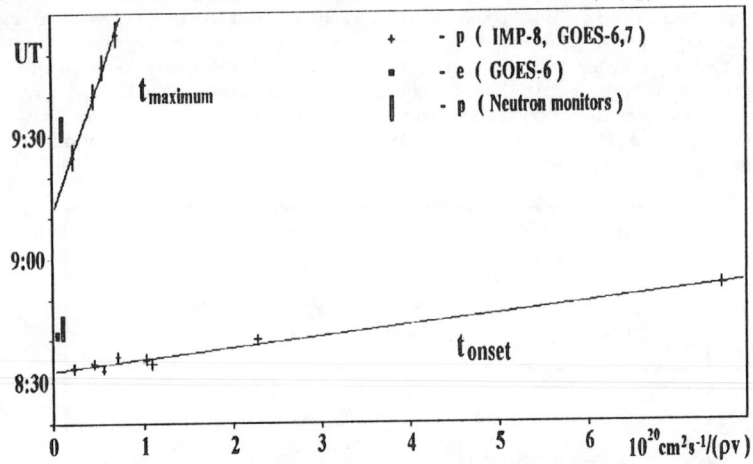

Fig. 6. Onset and maximum times of solar particle fluxes at 1 AU versus the effective parameter (ρV)
(ρ-Larmor radius for H=5nT, V-particle velocity)

It is worth to mention also that according to[2] production of neutrons at the Sun began just at the end of the impulsive phase and lasted for almost one hour in this flare.

CONCLUSION

The comparison of the electromagnetic emissions in the wide energy range and the analysis of the data on the charge particle fluxes lead to the conclusion that a long-term charge particle acceleration up to high energies took place at the delayed phase of the June 15, 1991 solar flare.

The shock wave acceleration appears to be not suitable for the explanation of the delayed component of the γ-ray and microwave emissions because at that time the shock wave is too far in the corona[17]. For example, from Figure 1 it is evident that for the June 3, 1982 flare by the time of delayed microwave component, an altitude of the CME front, whose location is close to that of the shock wave, was approximately 2.5 Ro above the photosphere[18]. The analysis of the dynamic spectrum of the type II burst for the June 15, 1991 flare gives a like estimation of the shock wave altitude.

It is reasonable to suppose that a prolonged energy release following the CME may be a source of particle acceleration at the late stage of the flare responsible for both the observed γ-ray and microwave delayed components. During this post-eruptive phase the magnetic field above the active region, strongly disturbed by the CME eruption, relaxes to its initial state through the magnetic field lines reconnection in a vertical current sheet accompanied by effective particle acceleration, generation of delayed microwave and X-ray components, formation of the post-flare loop system, and other remarkable phenomena[11-14]. Estimations[19] show that this process may result in a direct electric field acceleration of particles with power-law spectrum, in particular, of protons up to 20 Gev. Analysis of radio data reveals similar non-monotonous microwave bursts with analogous delayed components in other flares with the long-term high-energy γ-ray emission, such as the April 24, 1984, December 16, 1988, March 6, 1989, and June 11, 1991 events.

However, it is possible that at the final stage of a flare, the long-term acceleration is followed by the trapping of the accelerated particles which provides the long gradual decay of the microwave and γ-ray emissions lasting for hours as, for example, in this flare.

We are grateful to Dr. T.P.Armstrong for providing the IMP-8 data. I.M.Ch. acknowledges a support from the Soros Foundation Grant awarded by the American Physical Society.

REFERENCES

1. V.V.Akimov et al., 22nd ICRC, 3, 73 (1991).
2. J.M.Ryan et al., Proc. COMPTON Symposium St. Louis), in press (1993).
3. D.J.Forrest et al., 19th ICRC, 4, 146 (1985).
4. R.Ramaty et al., Ap. J. 316, L100 (1987).
5. G.E.Kocharov et al., Ioffe Phys.-Tech.Inst., Leningrad, Pr.1258 (1988).
6. V.G.Gueglenko et al., Solar Phys. 125, 91 (1990).
7. J.M.Ryan and M.A.Lee, Ap. J. 368, 316 (1991).
8. N.Z.Mandzavidze and R.Ramaty, Ap. J. 389, 739 (1992).
9. N.Z.Mandzavidze and R.Ramaty, Ap. J. (Letters), in press (1992).
10. N.G.Leikov et al., Astron. and Aphys. Suppl. 97, 345 (1993).
11. R.A.Kopp and G.W.Pneuman, Solar Phys. 50, 85 (1976).
12. E.W.Cliver, Solar Phys. 84, 347 (1983).
13. E.W.Cliver et al., Ap. J. 305, 920 (1986).
14. K.Kai et al., Solar Phys. 105, 383 (1986).
15. Solar-Geophysical Data, NOAA, Boulder (1991).
16. T.P.Armstrong, private communication (1993).
17. N.R.Sheeley, Jr., et al., J. Geophys. Res. 90, 163, (1985).
18. S.W.Kahler, Solar Phys. 90, 133 (1984).
19. Martens P.C.H., Ap. J. (Letters) 330, L131 (1988).

"EXTENDED PHASE" OF SOLAR FLARES OBSERVED BY *SMM*

P. P. Dunphy and E. L. Chupp
Physics Department and Space Science Center
Institute for the Study of Earth, Oceans and Space
University of New Hampshire
Durham, New Hampshire 03824, USA

ABSTRACT

Throughout the 10-year operating history of the *SMM* Gamma-Ray Spectrometer (GRS), only a handful of large solar flares have exhibited measurable emission of neutral pions (through pion-decay γ-rays) or high-energy neutrons (detected at the Earth). For 4 of the flares in which neutral pion-decay γ-rays have been detected by GRS, most of the neutral pions appear to have been produced after the "main" impulsive phase as determined from hard X-rays and low-energy γ-rays. The time history of the γ-ray emission above 10 MeV during this "extended" phase is also strikingly similar from flare to flare. Similar time histories have also been seen by EGRET and Comptel on CGRO. This may mean that the acceleration of protons to high energy, or, alternatively, the precipitation of trapped high-energy particles, in an "extended" or "delayed" phase may be a common feature of flares with significant production of pions and high-energy neutrons. The extended phase emission can be characterized by exponential "decay" times of a few minutes. The relaxation time can apparently be much longer for flares strong enough to have sufficient counting statistics and for sufficient observing time.

SMM/GRS OBSERVATIONS

The flare observations reported here were made with the Solar Maximum Mission Gamma-Ray Spectrometer (*SMM*/GRS), which operated between 1980 and 1989. The primary GRS data were from an array of 7 NaI(Tl) scintillation detectors, each 7.6 cm in diameter by 7.6 cm thick. These detectors produced high-resolution (476 channel) spectra covering the energy range from 0.3 to 9 MeV with an accumulation time resolution of 16 s. The GRS also had a high-energy mode (HEM) consisting of a matrix of counts made up of energy-loss bins from interactions in two detector layers: the NaI(Tl) scintillator array and/or a 7.5 cm thick CsI(Tl) rear shield. The HEM covered an energy loss range of 10 to 100 MeV in 4 channels for each layer with a time resolution of 2 s. The GRS and its high-energy mode has been described in detail by Forrest et al. [1]

An important property of the HEM is its ability to discriminate, at least on a statistical basis, between γ-ray and neutron fluxes incident on the detector.[2,3] Basically, the incident γ-ray and neutron fluxes are determined when the fluxes are deconvolved from the count matrix. The separation depends on the fact that multiple events (*i.e.* coincident interactions in both the NaI and CsI layers) are produced almost exclusively by γ-rays, while neutrons produce mainly singles events (*i.e.* interactions in the NaI or CsI, but not both). This characteristic of the HEM has been quantified by Monte Carlo calculations[4] and accelerator tests of a layered detector.[5]

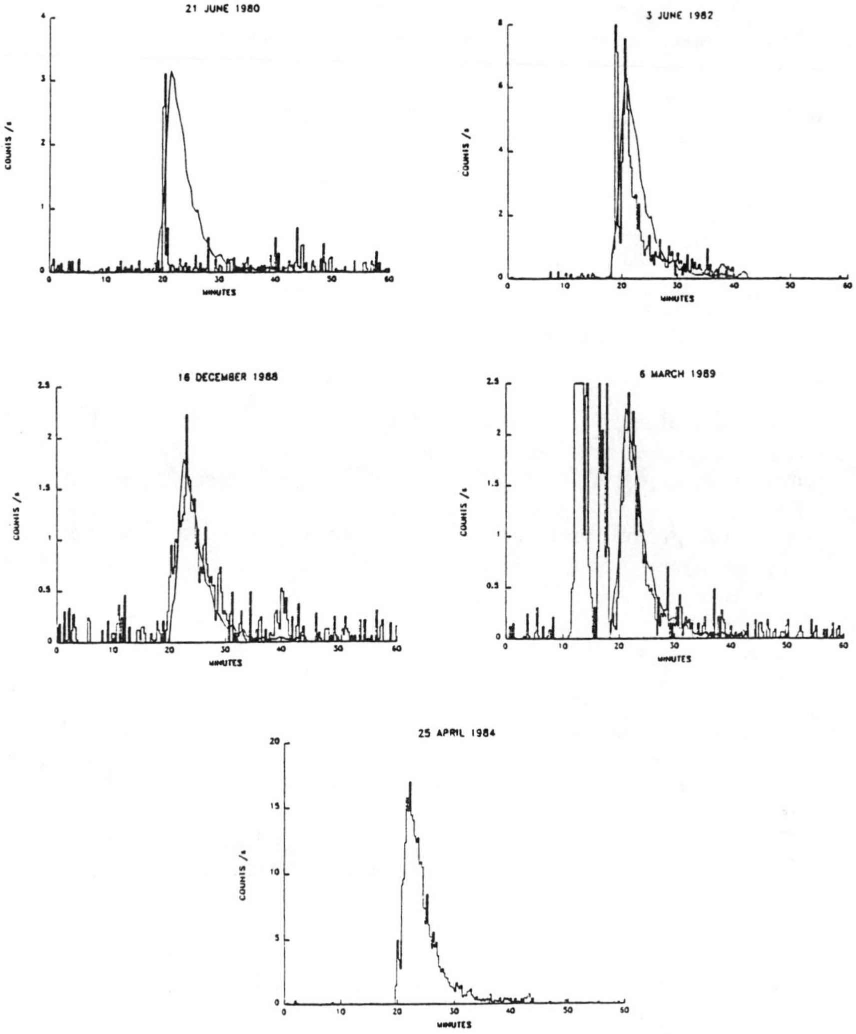

Figure 1. Time histories for 5 flares in the GRS HEM "multiples" channel for energy loss > 25 MeV. This channel is sensitive to high–energy γ–rays. The vertical axes are scaled and the time axis aligned to emphasize the similarity of a time–extended feature in some of the flares. A scaled and smoothed envelope of the 1984 April 24 flare is overlayed on the data from the other flares. All but the flare of 1980 June 21 show strong evidence of pion production.

Table 1
Parameters for bursts of "extended" high-energy emission (>25 MeV).

Flare Date	GOES/H_α	Position	c_1 ($[16\ s]^{-1}$)	τ_1 (min)	c_2 ($[16\ s]^{-1}$)	τ_2 (min)	χ^2/ν	ν
82/6/3	X8/2B	S09E71	108.1±8.2	1.15±0.14	17.1±3.9	11.7±3.0	0.96	68
84/4/25	X13/3B	S12E43	321.2±7.6	3.23±0.07	2.2±1.2	≥ 10	1.28	76
88/12/16	X4.7/1B	N27E33	30.5±2.7	3.34±0.30	-	-	0.75	53
89/3/6	X15/3B	N35E69	36.7±3.3	2.66±0.27	-	-	0.65	41

ν is the number of degrees of freedom for the fit.

Results of GRS observations of high-energy (> 10 MeV) photons and neutrons have been reported previously for the flares of 1980 June 21,[3,6] 1982 June 3,[3,7] 1988 December 16,[8] and 1989 March 6.[9,10] All of these flares have produced high-energy neutrons with sufficient intensity to be seen by the GRS HEM. In addition, ground level neutron monitors responded to flare neutrons > 1 GeV on 1982 June 3 [11] and protons from the decay of flare neutrons were also observed from this flare. [12]

Figure 1 shows the background-subtracted "light curves" from the GRS HEM multiple events for these flares. Since the multiple events are insensitive to neutrons, these plots give the time history of flare photons (> 25 MeV). Previous studies[3,8,9] have shown that the bulk of the pion production during these flares takes place after the x-ray and γ-ray onset during a "delayed" or "extended" phase. Although the light curves at the onset can be quite different (compare December 16 and March 6), the similarity among the light curves during the time intervals of enhanced pion production is striking. Figure 1 also shows the time history for the flare of 1984 April 24/25. A preliminary analysis of the GRS HEM data for this flare shows an extremely hard spectrum, consistent with pion decay, over a period of ~ 25 minutes. This flare is also known to have produced neutrons in the energy range 20 to 200 Mev[12] and possibly > 400 MeV.[13] In what follows, we assume that the 1984 April 24 HEM multiples light curve is dominated by pion-decay γ-rays.

TIME HISTORIES OF THE "EXTENDED PHASE"

Motivated by the analysis of the 1991 June 11 flare by Kanbach et al.,[14] where the > 30 MeV γ-rays showed an exponential decrease with time, and by the similarity among the time histories of the GRS high-energy flares, we have fit the time histories with an exponential decay model. The model is of the form
$$C(t) = c_1 e^{-(t-t_0)/\tau_1} + c_2 e^{-(t-t_0)/\tau_2}.$$

The results of the fitting are listed in Table 1. In the case of the flares of 1988 December 16 and 1989 March 6, a single exponential was sufficient to fit the decreasing part of the γ-ray light curve. For the 1982 June 3 flare, a sum of two exponentials was needed for an acceptable fit to the data. For the 1984 April 24

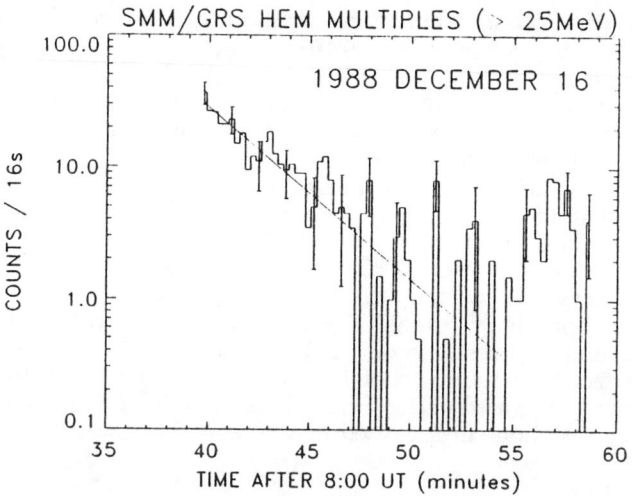

Figure 2. Time history of GRS HEM multiples events for the decreasing part of the 1988 December 16 flare. A single exponential is used to fit this part of the light curve. Selected error bars, based on counting statistics, are shown.

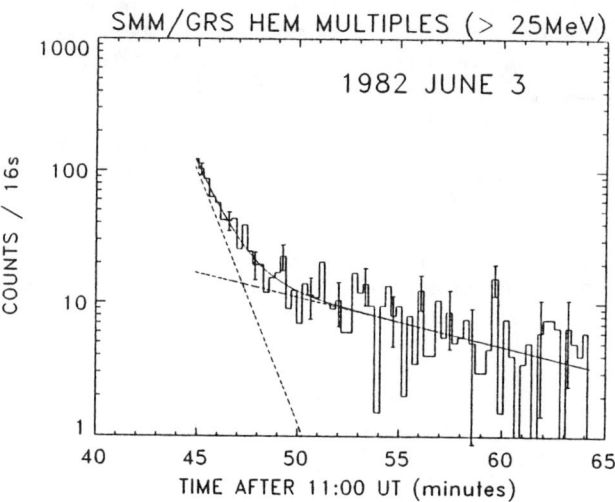

Figure 3. Time history of GRS HEM multiples events for the decreasing part of the 1982 June 3 flare. Two exponentials are used to fit this part of the light curve. The two components are shown by dashed lines, the sum of the components by a solid line. Selected error bars, based on counting statistics, are shown.

flare, the addition of a second exponential significantly improves the fit (from a probability of exceeding χ^2 of 0.01 to 0.10); however, the decay time of the second exponential is poorly determined because of limited counting statistics late in the flare and because the data was terminated at satellite nightfall. In all cases, the rise time is ~ 2 minutes to reach maximum intensity. Figures 2 and 3 show the fits for two of the flares.

It should be noted that subsequent bursts of γ-rays can have the effect of lengthening the apparent decay time of an earlier burst. Only in the case of the 1988 December 16 flare, however, do we clearly see the effect of a later burst. But here the burst (which occurs near minute 56 in Figure 2) comes late enough to be excluded from the data used in the fit. Since the exponential fits give formally acceptable χ^2 (probability of exceeding $\chi^2 \geq 0.1$ in all cases), it appears that secondary bursts do not make significant contributions to the light curves. The fact that the flares of 1988 December 16 and 1989 March 6 do not show the more extended decay times of the 1982 June 3 and 1984 April 24 flares could be a selection effect caused by the relatively lower flux from the former flares.

CONCLUSIONS

This comparison of 4 large flares observed by the SMM/GRS shows that pion–decay photons were produced in a time–extended phase. This phase can be characterized by an exponential decay or relaxation time. This time is at least of the order of several minutes. The largest flares, with the best counting statistics and sufficiently long observation times, have a component that can last more than 10 minutes and perhaps much longer. This behavior is similar to the very long duration emission from the 1991 June 11 flare reported by Kanbach et al.[14] In that case, the sensitivity of the EGRET detectors allowed a significant flux to be observed up to 8 hours after flare onset. Ryan et al.[15] have observed time-extended production (~ 10 minutes) of solar flare neutrons from the flare of 1991 June 9. This neutron production was taken as evidence for significant hardening of the proton spectrum over this period. Time–extended emission of γ–rays from pion decay has also been reported by Akimov et al.[16,17] for the flare of 1991 June 15. Finally, high–energy γ–ray emission detected by EGRET's NaI spectrometer from a number of flares during 1991 June has been described by Schneid et al.[18] These events also exhibited long–duration emission with a component consistent with pion decay.

There can be significant production of pions during the initial burst, for example on 1982 June 3.[3] However, from the above observations there is clear, if only circumstantial, evidence for an association between copious pion production and a time–extended phase of flares. From the SMM data alone, in all 4 flares for which there is strong evidence for pion–decay γ-rays, the pion production takes place mainly during a burst over a time scale of minutes or more. Since pion production is the result of interactions of protons with kinetic energies > 300 MeV, it follows that the acceleration of protons to these energies (or at least their transport to interaction regions) takes place over this time scale.

The decay–like profile which seems to be characteristic of these events probably reflects a process or morphology common to the flares. One likely possibilty is particle trapping and pitch angle scattering in flare loops (e.g. Mandzhavidze and Ramaty[19]). This would explain the decay of the radiation, with the decay time depending on the loop's size, shape, and location, and the amount of plasma turbulence in it. Another question is whether significant proton acceleration to > 300 MeV takes place during this extended phase.[20,21,22] This could be addressed by looking for changes in the proton spectral shape during the event using ratios of γ–ray line fluxes and the neutral–pion–decay γ–ray flux.[23] Given the high sensitivity of the Compton GRO detectors, it is likely that a much larger sample of flares with a detectable extended phase will become available for the relevant studies.

ACKNOWLEDGMENTS

This work was supported in part by NASA grant NAGW-2755. We also acknowledge the support of the Institute for the Study of Earth, Oceans and Space at UNH.

REFERENCES

1. Forrest, D.J., et al., Solar Phys. **65**, 15 (1980).
2. Chupp, E.L., et al., Proc. 19th Int. Cosmic Ray Conf. **4**, 126 (1985).
3. Forrest, D.J., et al., Proc. 19th Int. Cosmic Ray Conf. **4**, 146 (1985).
4. Cooper, J.F., et al., Proc. 19th Int. Cosmic Ray Conf. **5**, 474 (1985).
5. Dunphy, P.P., et al., Experimental Astron. **2**, 233 (1992).
6. Chupp, E.L., et al., Ap. J. (Letters) **263**, L95 (1982).
7. Chupp, E.L., et al., Ap. J. **318**, 913 (1987).
8. Dunphy, P.P., and Chupp, E.L., in Particle Acceleration in Cosmic Plasmas, ed. G. Zank and T.K. Gaisser (AIP, N.Y., 1992), p. 253.
9. Dunphy, P.P., and Chupp, E.L., Proc. 22nd Int. Cosmic Ray Conf. **3**, 65 (1991).
10. Rieger, E., and Marschhäuser, H., Proc. of MAX '91 Workshop # 3, ed. R.M. Winglee and A.L. Kiplinger (U. Colorado, Boulder CO, 1990), p. 68.
11. Debrunner, H., Flückiger, E., Chupp, E.L., and Forrest, D.J., Proc. 18th Int. Cosmic Ray Conf. **4**, 75 (1983).
12. Evenson, P., Kroeger, R., Meyer, P., and Reames, D., Ap. J. (Supplement) **73**, 273 (1990).
13. Smart, D.F., Shea, M.A., Flückiger, E.O., Debrunner, H., and Humble, J.E., Ap. J. (Supplement) **73**, 269 (1990).
14. Kanbach, G., et al., A & AS, in press (1993).
15. Ryan, J., et al., presented at IAU Colloquium #142, College Park, MD, Jan. 11–15 (1993).
16. Akimov, V.V., et al., Proc. 22nd Int. Cosmic Ray Conf. **3**, 73 (1991).
17. Akimov, V.V., et al., these proceedings (1993).
18. Schneid, E.J., et al., these proceedings (1993).
19. Mandzhavidze, N., and Ramaty, R., Ap. J. (Letters) **396**, L111 (1992).
20. Debrunner, H., Flückiger, E.O., and Lockwood, J.A., Ap. J. (Supplement) **73**, 259 (1990).
21. Ryan, J.M., and Lee, M.A., Ap. J. **368**, 316 (1991).
22. Forrest, D.J., these proceedings (1993).
23. Murphy, R.J., and Ramaty, R., Adv. Space Res. **4** (7), 127 (1985).

NUMERICAL MODELING OF PARTICLE TRANSPORT AND ACCELERATION OF SOLAR FLARE PARTICLES IN A CORONAL LOOP

E. Bennett, M.A. Lee and J.M. Ryan
Space Science Center
University of New Hampshire, Durham, NH 03824

ABSTRACT

We examine the behavior of energetic particles in a flaring solar coronal loop under the assumption that a uniform turbulent MHD wave field causes the particles to diffuse in momentum and real space. The analytical formulation of these processes, presented in a previously published work[1], is restricted by the choice of a simple energy and spatial dependence of the diffusion coefficients and the collisional energy loss term. Although good agreement is evident between the predicted and the observed time profiles when the model is applied to the 1982 June 3 flare, we have constructed a numerical model which will allow us relax the constraints inherent in the analytical version. Progress in applying this numerical model is presented.

INTRODUCTION

The emission of hard X-rays and γ-rays during a solar flare is caused by the precipitation of energetic protons and electrons onto the chromosphere, or the lower corona. These energetic particles, originally accelerated in the higher coronal regions, are influenced by the turbulent MHD wave field in the flaring solar loop. This turbulence causes the particles to diffuse in space, as they pitch-angle scatter in the MHD wave field. In the case of extended duration events, where the particles experience slow spatial diffusion, the MHD turbulence can actually supply energy to some particles via the process of second-order stochastic acceleration. During an impulsive event, the particles are transported directly to the footpoints of the loop and undergo minimal diffusion in momentum space.

We describe time-dependent particle transport and acceleration in a coronal loop based on the interaction of energetic particles with a turbulent wave field. Time profiles for the radiation emission are derived by calculating the precipitation rate of the parent particles for various injection positions and coronal densities. The spatial extent of the loop is assumed to be large, allowing the spatial diffusion time scale to be long enough for significant second-order stochastic acceleration to take place.

In order to gain insight into the primary mechanisms of particle acceleration in the complex solar flare environment, the geometry and the microphysics of the loop transport are simplified. The coronal loop is viewed as a one-dimensional structure connected at both ends to the chromosphere. It contains a uniform distribution of thermal plasma and a uniform turbulent wave field. Energy loss occurs via Coulomb scattering of the energetic particles with ambient electrons, while particle loss occurs via diffusion from the loop ends into the lower corona. The particles are otherwise confined to the coronal loop. In order to obtain a simplified transport, an appropriate collisional loss term is chosen and the spatial diffusion coefficient is taken to be independent of energy. A delta-function injection of particles in space, momentum, and time is assumed to obtain the time dependent distribution of particles in phase space, from which the particle precipitation rate is derived. We consider no specific origin for the MHD wave field.

Due to the simplifying assumptions, the analytic solution reviewed here is ultimately unable to provide optimized fits to the intensity-time profiles of available data. A more realistic scenario, in which the spatial diffusion coefficient, the momentum diffusion coefficient, and the collisional loss term have more general spatial and momentum dependence, is currently being investigated by integrating the diffusion equation numerically.

PRECIPITATION AND EXTENDED ACCELERATION

Energetic particles with mean free paths small compared with the loop length L can be efficiently accelerated within the loop via the process of second-order stochastic acceleration, while slowly diffusing in coordinate space. A two-dimensional diffusion equation, given for example by Schlickeiser[2], may be used to describe the time-dependent evolution of the particle distribution function in both momentum space and real space

$$\frac{\partial f}{\partial t} = p^{-2}\frac{\partial}{\partial p}\left\{p^2\left[D(p)\frac{\partial f}{\partial p} - f\dot{p}_c\right]\right\} + \frac{\partial}{\partial x}\left(\kappa\frac{\partial f}{\partial x}\right) + Q(x,p,t), \quad (1)$$

where x is position along the loop, $f(x, p, t)$ is the omnidirectional distribution function of the energetic particles, $D(p)$ is the diffusion coefficient in momentum space, $\dot{p}_c$ is the continuous collisional loss term, κ is the constant spatial diffusion coefficient, and $Q(x, p, t)$ is the injection rate. We take the impulsive injection rate to be

$$Q \equiv \left(\frac{N_o}{4\pi p_o^2}\right)\delta(x-x')\delta(t)\delta(p-p_o),$$

where N_0 is the total number of particles per unit cross section of the loop injected into the system. With the substitutions $\rho \equiv \ln(p/p_o)$ and $y = (x/L)$, and assuming that $\dot{p}_c = -\alpha p$ (where $\alpha \geq 0$) and $D = Bp^2$, the diffusion equation becomes

$$\frac{\partial f}{\partial t} = B\frac{\partial^2 f}{\partial \rho^2} + C\frac{\partial f}{\partial \rho} + Hf + \frac{\kappa}{L^2}\frac{\partial^2 f}{\partial y^2} + Q, \quad (2)$$

where $B = V_A^2/(9\kappa)$, $C=3B + \alpha$, $H = 3\alpha$, and V_A is the Alfvén speed. The energy dependence of $D(p)$, chosen for this calculation, produces an enhanced acceleration of higher energy particles. However, this effect is partially counteracted by collisional energy loss described by $\dot{p}_c$. Equation (2) is solved by using a Fourier transform in ρ and an eigenfunction expansion in x. The expressions for the omnidirectional distribution function and the flux of precipitating particles (given by the sum of $|\kappa\nabla_x f|$ evaluated at each loop footpoint), are presented in Ryan and Lee[1]. The peak of the particle distribution at any given time occurs at $\rho = \rho_{max}$, defined by $c\tau + \rho_{max}=0$, (where c (>0) is related to the sum of the ratios of the spatial-diffusion time scale to the acceleration time scale and the collisional loss time scale). As time increases, ρ_{max} drifts from zero toward larger negative values and acceleration is evident in the high-energy tail of the broadening Gaussian distribution.

We model the behavior of the 1982 June 3 γ-Ray flare, which is described in detail by Forrest et al.[3,4] and Chupp et al.[5]. For simplicity, the collisional losses due to the ambient matter within the loop are neglected. The product of the diffusion and the acceleration time scales ($t_d \cdot t_{acc}$) is constant. This suggests that a lower limit on the length of the loop, or an upper limit on κ, will establish an upper limit on t_{acc} (since $t_d = L^2/(\pi^2 \kappa)$) and may indicate inefficient acceleration. Therefore, in order to model accurately the temporal and spectral development of particle precipitation, the spatial scale of the system must be explicitly included along with the plasma parameters.

APPLICATION AND DISCUSSION OF THE MODEL

A possible model describing the sequence of events observed in the 1982 June 3 γ-ray flare (detected by the *Gamma Ray Spectrometer* on SMM) is presented by Ryan and Lee[1]. An unspecified initial proton acceleration is assumed to take place somewhere within a coronal loop attached at both ends to the denser regions of the solar atmosphere. These accelerated protons must have energies greater than 20 MeV in order to account for the impulsive phase nuclear line emission. As the protons diffuse within the turbulent environment of the coronal loop toward its footpoints, their precipitation into the chromosphere is seen as the impulsive-phase γ-ray flash. The protons which do not immediately precipitate into the chromosphere diffuse within the loop in both real and momentum space. Some of these protons eventually reach the π meson production threshold (about 300 MeV for p-p collisions) and become visible through pion related emission to form the gradual phase.

We require that the temporal behavior of the γ-ray flux observed during the 1982 June 3 γ-ray flare matches the particle precipitation time profile predicted by the model. The rise time of the impulsive phase 20 MeV proton precipitation, required to produce the γ-ray flux in the nuclear line part of the spectrum, must be less than 20 seconds. In addition, the peak of the 300 MeV proton precipitation responsible for the π meson production must occur more than one minute later than the peak of the impulsive phase. Finally, it is necessary that the overall acceleration time scale be on the order of 5-10 minutes. Protons continue to be accelerated to higher energies and precipitate for approximately 20 minutes. The free parameters are the injection position within the loop, the ratio of the spatial diffusion time scale to the acceleration time scale, and the global Alfvén speed. In order for the diffusion approximation to be valid, it is necessary that the proton mean-free path be much shorter than the loop length, and the loop size be consistent with typical loop dimensions.

The expression for the observable flux of precipitating particles (given by Ryan and Lee[1]) is evaluated and plotted in Figure 1. The dimensionless injection position is chosen to be 0.25. The spatial-diffusion time scale t_d is determined by requiring that the peaks of the 20 and 300 MeV fluxes be separated by 100 seconds (t_d=460 s). Then, the acceleration time scale t_{acc} follows from an assumed ratio t_d/t_{acc}, taken to be 1 in this case. The loop length is chosen to be 10^5 km. The spatial diffusion coefficient is calculated from $\kappa = L^2/(\pi^2 t_d)$ to be $\kappa = 2 \times 10^{16}$ cm^2 s^{-1}. The mean-free path is given by $\kappa = \lambda v/3$ (where v is the proton velocity), and has the value $\lambda = 110$ km. The acceleration time scale in an Alfvénic wave field is related to the Alfvén speed by the expression $t_{acc} = 9\kappa/V_A^2$ (Schlickeiser[2]), resulting in V_A= 210 km/s. Note that the inferred value of the Alfvén speed V_A is lower than the commonly accepted value of 1000 km s^{-1}. This discrepancy arises because the spatial diffusion coefficient, κ, is assummed to be independent of energy.

We can compute the spectrum of precipitating protons in both the impulsive and the gradual phases of the event by integrating the spectrum given by the expression for the flux of precipitating particles (Ryan and Lee[1]). The integration is carried out for the first 70 seconds of the event, as well as over the remainder of the event, using the parameters given above. The resulting spectrum of protons/MeV is shown in Ryan and Lee[1], where the integration over the impulsive phase can be fitted by the exponential spectrum of Forman, Ramaty & Zweibel[6] with $\alpha T=0.075$.

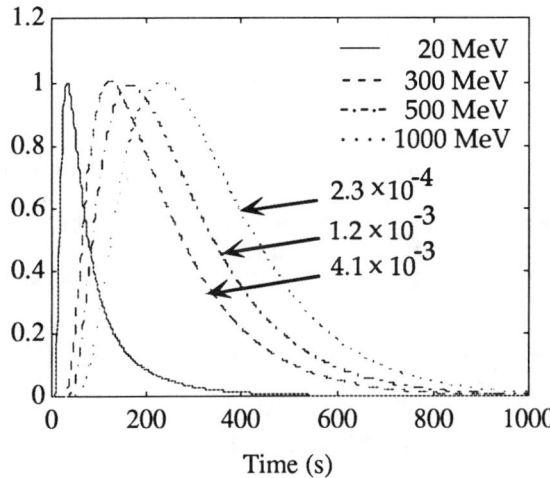

Fig. 1. - Dimensionless precipitation rates for protons of 20 (injection), 300, 500, and 1000 MeV, modeling the temporal behavior of the γ-ray flux observed in the 1982 June 3 flare. The ratio of the diffusion timescale to the acceleration time scale is assumed to be 1. Higher energy curves are normalized to that of the 20 MeV protons with the scale factors indicated.

The remainder of the event after the first 70 seconds may be fitted by a power law. Thus, the proton spectrum naturally evolves from a soft exponential function to a power law if the effects of spatial diffusion in the acceleration process are explicitly included.

The model presented here is limited by the assumed energy dependence of the collisional loss term, and of the momentum and spatial diffusion coefficients. The last assumption results in a somewhat optimistic efficiency of the acceleration process. Thus, no detailed fitting of the 1982 June 3 data was attempted to this model. However, when spatial diffusion is explicitly included with second-order stochastic acceleration in a reasonable flare geometry, many of the energetic particle features formerly ascribed to distinct multiple acceleration phases can be reproduced.

WORK IN PROGRESS

More complicated cases, in which the spatial diffusion coefficient, the momentum diffusion coefficient, and the collisional loss term take on more realistic spatial and momentum dependence, may be treated by integrating the diffusion equation (1) numerically.

We choose the Alternate-Direction-Implicit method (von Rosenberg[7]) to integrate the diffusion equation in two phase-space dimensions and one time dimension, using the analytical solution evaluated at small time as the initial condition (Ryan and Lee[1]). In order to test the algorithm for accuracy, the particle distribution function obtained numerically is compared to the analytical solution.

The alternate-direction implicit method involves the use of two separate finite difference equations to represent a partial differential equation in two phase-space dimensions and time. For the first finite difference equation, all the derivatives with respect to one space variable are written at a new time level, t_{n+1}. The derivatives with respect the other variable are still expressed at the old time level, t_n. In this case, equation (2) may be rewritten as

$$\frac{f_{x,\rho}^{n+1} - f_{x,\rho}^n}{\Delta t} = \frac{\kappa}{L^2}\left[\frac{f_{x+1,\rho}^{n+1} - 2f_{x,\rho}^{n+1} + f_{x-1,\rho}^{n+1}}{(\Delta x)^2}\right]$$
$$+ B\left[\frac{f_{x,\rho+1}^n - 2f_{x,\rho}^n + f_{x,\rho-1}^n}{(\Delta \rho)^2}\right] + C\left[\frac{f_{x,\rho+1}^n - f_{x,\rho-1}^n}{2(\Delta \rho)}\right] + Hf_{x,\rho}^n \quad (3)$$

Equation (3) is implicit in the x direction and explicit in the ρ direction, and the coefficients of the distribution function at the unknown time form a tridiagonal coefficient matrix. We proceed to solve for f at t_{n+1}. If equation (3) were used continuously, the method would be unstable except for restricted values of the ratio $(\Delta t)/(\Delta \rho)^2$, since only the x derivative is advanced in time. Instead, an equation which is explicit in the x direction and implicit in the ρ direction is used *with the same time interval Δt*

$$\frac{f_{x,\rho}^{n+2} - f_{x,\rho}^{n+1}}{\Delta t} = \frac{\kappa}{L^2}\left[\frac{f_{x+1,\rho}^{n+1} - 2f_{x,\rho}^{n+1} + f_{x-1,\rho}^{n+1}}{(\Delta x)^2}\right]$$
$$+ B\left[\frac{f_{x,\rho+1}^{n+2} - 2f_{x,\rho}^{n+2} + f_{x,\rho-1}^{n+2}}{(\Delta \rho)^2}\right] + C\left[\frac{f_{x,\rho+1}^{n+2} - f_{x,\rho-1}^{n+2}}{2(\Delta \rho)}\right] + Hf_{x,\rho}^{n+1} \quad (4)$$

Once again, the coefficients of f at the time interval t_{n+2} form a tridiagonal matrix, and can be solved. The appropriate boundary conditions can be incorporated directly into the coefficients.

The ADI method has been shown to be stable for any ratio of the time increment to the space increment, as long as *the same time increment* is used for the successive applications of equations (3) and (4). The time increment can be increased after any double time step for the solution of problems which approach steady-state conditions. However, the initial choice for the time step must be sufficiently small to obtain numerically accurate results. The finite differencing scheme is second-order correct in x, ρ, and t.

The algorithm was tested using the physical parameters listed in the previous section. The analytical solution, evaluated at 10 seconds, was taken as the initial condition for the numerical integration. After 900 time steps of 0.001 second each, the numerical and the analytical solutions agreed to within 0.3%. This accuracy may be improved by taking a smaller time step and a finer x-ρ mesh.

CONCLUSION

The simple analytical model presented here has been shown to reproduce the general features of the intensity-time profiles and the spectral evolution of the high-energy gradual phase of particle precipitation observed in the 1982 June 3 solar flare. In solving the diffusion equation describing the time-dependent evolution of the particles, the spatial and the momentum dependence of the diffusion coefficients, and of the collisional loss term have been chosen to facilitate the calculation. A numerical solution of the same diffusion equation has been compared to the analytical results with good agreement. A more general numerical solution, where the functional dependence of the diffusion coefficients and the collisional loss term has been relaxed, is currently being investigated.

REFERENCES

1. J. M. Ryan, M. A. Lee, Astrophys. J. **368**, 316 (1991).
2. R. Schlickeiser, in Cosmic Radiation in Contemporary Astrophysics, edited by M. M. Shapiro (Dordrecht:: Reidel, 1986), p. 27.
3. D. J. Forrest, W. T. Vestrand, E. L. Chupp, E. Reiger, J. Copper, G. H. Share, Proceedings of the 19th International Cosmic Ray Conference (La Jolla, 1985), Vol. 1, p. 146.
4. D. J. Forrest, W. T. Vestrand, E. L. Chupp, E. Reiger, J. Copper, G. H. Share, Advances in Space Research 6, No. 6, 115 (1986).
5. E. L. Chupp *et al.*, Proceedings of the 19th International Cosmic Ray Conference (La Jolla, 1985), Vol. 1, p. 126.
6. M. A., Forman, R. Ramaty, E. G. Zweibel, in Physics of the Sun, edited by P. A. Sturrock *et al.* (Dordrecht:: Reidel, 1986),Vol II, p. 249.
7. D. U. von Rosenberg, Methods for the Numerical Solution of Partial Differential Equations (Am. Elsevier Publishing Company, Inc., New York, 1969),p. 87.

ACKNOWLEDGMENTS

The authors would like to thank Dr. Dawn Meredith and Albert B. Bennett III for their advice on the numerical computations.

DIRECTIVITY OF GAMMA-RAYS FROM NEUTRAL PI-MESONS DECAY

D. Heristchi and R. Boyer

Observatoire de PARIS-DASOP, URA 326
F-92195, Meudon Principal Cedex, FRANCE

ABSTRACT

An isobaric model is used to determine the spectrum of the neutral pi-mesons produced in proton-proton interaction. The angular distribution of isobars in the center-of-momentum system is taken to be isotropic or of the form $A + B cos^2 \theta^*$. It is shown that the total meson spectrum is approximately the same, even if outgoing isobars run in the same initial direction of the colliding proton. However, the double differential spectrum of mesons is considerably dependent on the isobar distribution. The spectrum of gamma-rays from neutral pion decay is calculated. Here also, the double differential spectrum is dependent on the isobar distribution.

INTRODUCTION

The γ−rays from π° decay certainly play a major role in the high-energy radiation field in astrophysics, the π° being produced by the high energy proton-proton (p-p) interaction, or more exactly from nucleon-nucleon interaction. The main characteristics of the high energy γ−rays generated this way are that i)the threshold of the p-p interaction yielding a π° is low ($\approx 280 MeV$), and ii) π° decays into two γ−rays of high energy. These characteristics are amplified under astrophysical condition where protons and other nuclei are present covering a large energy range.

This high energy emission mechanism has often been refered to in the discussion of γ−rays from the interstellar medium. A brief review of these works is given in Dermer[1] (hereafter paper 1). It is to be noted that the total cross section of π° production from p-p interaction is well known. The difficulty arises when we need to known the differential cross section with respect to the pion energy.

The isobaric model of Stecker[2] used in paper 1 is based on the theory of Lindenbaum and Sternheimer[3]. This theory has it that the pion production in p-p interaction, at least in the low-energies, is mediated by isobar that subsequently decays into a nucleon and a pion. If we go by this theory we see that the problem of an interaction giving three outgoing particles is replaced by two successive interactions with two outgoing particles.

This requires that we study two and three successive interactions for π° and γ−ray production. Nevertheless, the isobaric model differs considerably from a simple two-particle interaction. Here the rest mass of the isobar is variable. Then, all possibilities must be taken into account in the computation of the cross section and of the maximum energy of the pions in the Center-of-Momentum system (CMS) or in the Laboratory System (LS).

From the study of paper 1, it appears clearly that the isobar model of Stecker[2] give a reasonably good fit with the accelerator experimental data. Using this hypothesis, the differential cross section, with respect to the pion's energy, can be obtained[1,4]. This cross section can be used when the particles have an isotropic distribution (Interstellar Medium) or when the total differential spectrum is needed.

In this paper we study the differential cross section of pions and γ-rays from the p-p interaction taking in consideration the isobar distribution. We also give the double differential cross section (with respect to the energy and the direction) of these particles.

MAXIMUM ENERGY OF PIONS

In the CMS, the maximum energy of pions is the same in all directions. In the general case when multiple particles are produced, the maximum energy is obtained when the particle considered moves in one direction and all the others move together in the opposite direction. The kinematics of the interaction are then easy to analyze by considering that there are two particles: the one considered, and the one having the rest mass of all the others.

In the case of the maximum energy of π^0 in the CMS, we have:

$$E^*_{\pi^0} = \frac{W^{*2}_T + m^2_{\pi^0} - 4m^2_p}{2W^*_T} - m_{\pi^0} \tag{1}$$

Throughout this paper we use the following notation for the physical parameters: m for the rest mass, ξ the total energy, E the kinetic energy, γ the Lorentz factor, P the momentum, and β the velocity of a particle. The subscript of p, Δ, π^0 and γ refers to the incident proton, the isobar, the pion and the γ-ray, respectively. All these parameters in the CMS of the p-p interaction have the superscript *. The total energy of the system is W_T in the LS and W^*_T in the CMS. The velocity and the Lorentz factor of the CMS in the LS are β_c and γ_c.

The maximum energy in the CMS, as a function of the kinetic energy of the incident proton, is given in Figure 1. We recall that this energy is the same in any direction in the CMS.

The maximum energy of pions in the LS, as a function of the angle, is obtained by considering two outgoing particles: one pion and two protons together. The conservation law of 4-momenta leads to a complicated formula. This maximum energy can also be calculated by Lorentz transformation of the same value in the CMS. Figure 1 shows the maximum energy as a function of proton energy and for $\theta_{\pi^0} = 0$.

MAXIMUM AND MINIMUM GAMMA-RAY ENERGY

The decay of pions with maximum energy give the maximum and minimum energy of γ-rays, forward and backward. In the pion rest frame each γ-ray has an energy of $m_{\pi^0}/2$.

In the CMS, the maximum and minimum γ-rays energies are obtained by the Lorentz transformation from the pion rest frame:

$$E^*_\gamma(max) = \frac{\xi^{*M}_{\pi^0} + P^{*M}_{\pi^0}}{2} = \frac{m_{\pi^0}}{2}\left(\frac{1 + \beta^{*M}_{\pi^0}}{1 - \beta^{*M}_{\pi^0}}\right)^{1/2} \tag{2}$$

$$E^*_\gamma(min) = \frac{\xi^{*M}_{\pi^0} - P^{*M}_{\pi^0}}{2} = \frac{m_{\pi^0}}{2}\left(\frac{1 - \beta^{*M}_{\pi^0}}{1 + \beta^{*M}_{\pi^0}}\right)^{1/2} \tag{3}$$

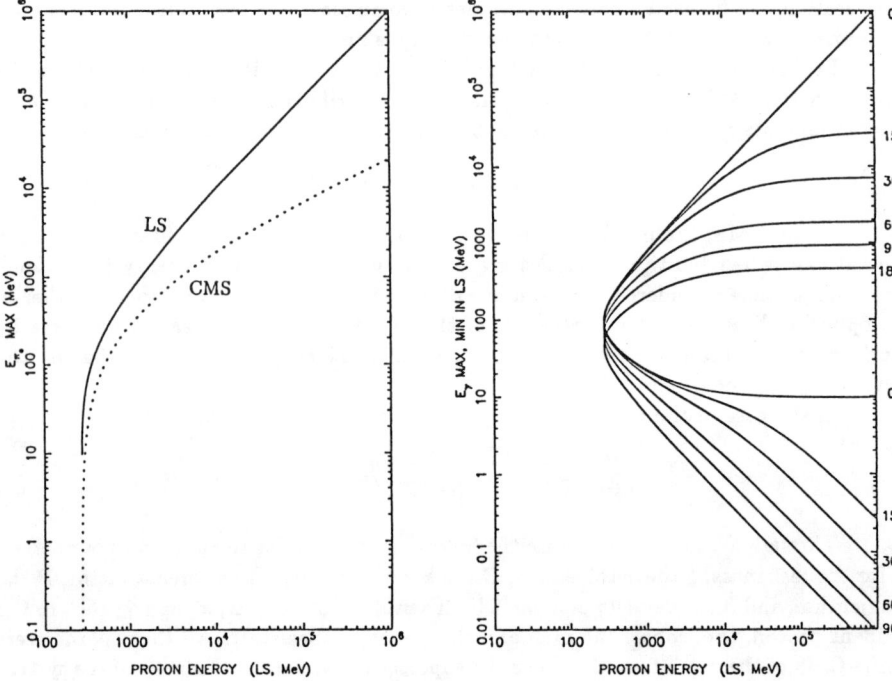

Fig. 2- Maximum and minimum energy of γ-rays in the LS

Fig. 1- Maximum energy of pions in the CMS and the LS

We see from these equations that the maximum and minimum energies are equal to $m_{\pi^0}/2$ multiplied or divided by the same value.

The maximum and minimum pion energy in the LS is obtained by the Lorentz transformation of the same parameters in the CMS:

$$E_\gamma(max, min) = \frac{E_\gamma^*(max, min)}{\gamma_c(1 - \beta_c \cos\theta_\gamma)} \qquad (4)$$

$E_\gamma(max)$ and $E_\gamma(min)$ are given in Figure 2 for selected angles.

PION PRODUCTION

The cross section of pion production in the CMS is obtained by an integration on the isobar mass. In fact, for a fixed mass of the isobar, its energy and its velocity are determined in the CMS, and when it decays, the pion has a flat energy spectrum in the CMS of p-p interaction between two limits that can easily be calculated. We find exactly the same energy distribution as in paper 1. The angular distribution of isobars in the last system is without effect on this energy distribution.

In the LS, we determine the cross section by considering a normalized angular distribution $f_\Delta(\theta^*)d\Omega^*$ for isobar in the CMS. We have:

$$\frac{d\sigma}{dE_{\pi^\circ}} = \int_{m_p+m_{\pi^\circ}}^{W^*-m_p} B(m_\Delta) \frac{\pi m_\Delta}{\beta_c \gamma_c P'_{\pi^\circ} P^*_\Delta} dm_\Delta \int_c^d \frac{f(\theta^*)}{P_\Delta} dE_\Delta \quad (5)$$

where B is the normalized distribution of isobar mass m_Δ (see paper 1); P'_{π° the pion momentum in the rest frame of the isobar; P^*_Δ the isobar momentum in the CMS; d the smaller of following two isobar energies: the highest isobar energy (with a fixed mass) in the LS, and the lowest value of it needed for pion production; and c the greatest of the two following quantities: the minimum isobar available energy and the minimum isobar energy needed for the creation of a pion with the energy considered.

The right-hand side integral of (5) is evaluated analytically in the distributions considered (isotropic or even polynomial of cosine[5]). The resulting cross section is to be compared with the experimental values[6].

The pion double differential (with respect to the energy and angle) cross section in the CMS can be calculated. We consider a direction in the last system with zenith angle $\theta^*_{\pi^\circ}$ measured from the direction of the colliding proton (z-axis). The cross section for pion production with an energy E_{π° in this direction within a solid angle element $d\Omega^*_{\pi^\circ}$ is given by:

$$\frac{d\sigma}{dE^*_{\pi^\circ} d\Omega^*_{\pi^\circ}} = \int_{m_p+m_{\pi^\circ}}^{W^*-m_p} B(m_\Delta) \frac{m_\Delta}{4\pi P^*_\Delta P'_{\pi^\circ}} dm_\Delta \int_0^{2\pi} f(\theta^*_\Delta) d\eta^*_\Delta \quad (6)$$

This formula is obtained by changing the origin of the coordinate system to the direction considered. In this new system the direction of the isobar is specified by ϑ^*_Δ and η^*_Δ. The angle θ^*_Δ is related to η^*_Δ, from trihedral formed by the z-axis, the isobar, and the pion:

$$\cos\theta^*_\Delta = \cos\theta^*_{\pi^\circ} \cos\vartheta^* + \sin\theta^*_{\pi^\circ} \sin\vartheta^* \cos\eta^*_\Delta \quad (7)$$

The right-hand side integration of (6) can be performed analytically, and gives the pion distribution in the CMS. We find that it is in the form of an even polynomial of $\cos\theta_{\pi^\circ}$ if the isobar have the same distribution form.

The double differential cross section of pions in the LS is obtained by the transformation of the same value in CMS.

This cross section is calculated to illustrat the method that will be used to determine the γ-ray cross section.

GAMMA-RAY PRODUCTION

The γ-ray spectrum can be calculated easily using the energy spectrum of the π° (paper 1). We give here a direct method for these calculations in the CMS and LS. The energy spectrum of γ-rays extend between a maximum and a minimum as already mentioned, and it is strictly symmetrical on a logarithmic scale to the axis of $m_{\pi^\circ}/2 \approx 67.49$ MeV, independently of the pion production model.

The γ-ray production in the CMS can be written as:

$$\frac{d\sigma}{dE_\gamma} = \int_{m_p+m_{\pi^\circ}}^{W^*-m_p} B(m_\Delta) \frac{m_\Delta}{2P^*_{\pi^\circ} P'_\Delta} dm_\Delta \int_c^d \frac{1}{P_{\pi^\circ}} dE_{\pi^\circ} \quad (8)$$

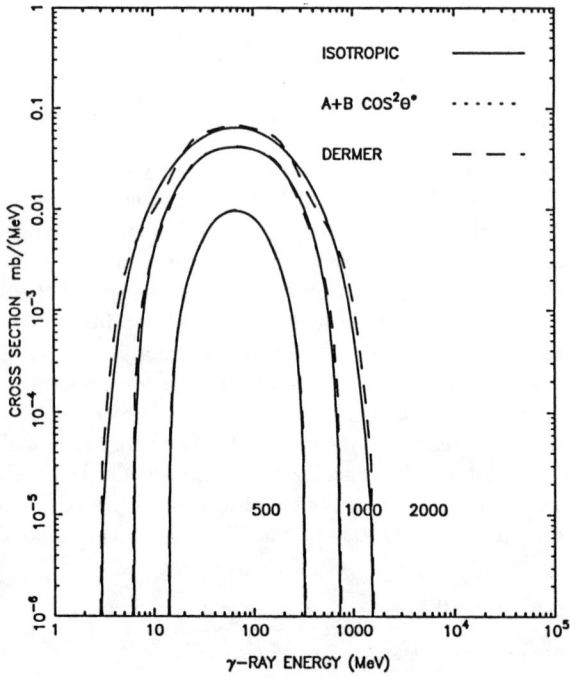

Fig. 3- Cross section of γ-rays in the LS for 500, 1000 and 2000 MeV

where c and d are chosen as in the case of equation (5). The maximum and minimum pion energy needed for E_γ production is given by:

$$\xi_{m_{\pi^0}} = E_\gamma + \frac{m_{\pi^0}^2}{4E_\gamma} \tag{9}$$

The cross section in LS is given as:

$$\frac{d\sigma}{dE_\gamma} = \int_{m_p+m_{\pi^0}}^{W^*-m_p} B(m_\Delta) dm_\Delta \int_a^b \frac{f(\theta^*)}{P_\Delta} dE_\Delta \int_c^d \frac{dE_{\pi^0}}{P_{\pi^0}} \tag{10}$$

where a, b, c, and d are chosen as for (5). The integral on the pion energy is equal to $log_e(\xi - P)]_c^d$, and then we have two numerical integral to compute. Figure 3 gives three examples of γ-ray cross sections. All cross sections are symmetrical about the $m_{\pi^0}/2$ axis.

The differential cross section relative to the γ-ray energy, as studied above, is useful only in the case of an isotropic proton distribution. We give here a double differential (with respect to γ-rays energy and direction) cross section.

The method used here is similar to the case of pions (6). We have four successive integrations for γ-ray production in the CMS.

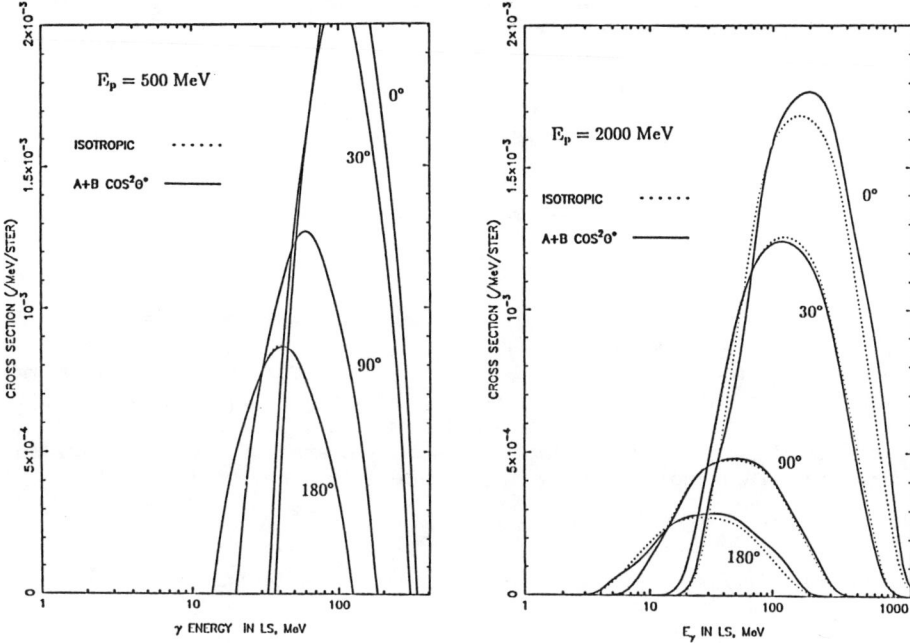

Fig. 4- Two examples of the normalized γ−ray production cross section in the LS.

In the LS, this cross section is calculated by the Lorentz transformation of the radiation from the CMS. Two examples of the cross section obtained are given in Figure 4. The characteristics of the directivity are clearly shown in this Figure.

REFERENCES

1. C. D. Dermer, Astr. Ap. **157**, 233 (1986).
2. F. W. Stecker, Ap. Space Sci. **6**, 377 (1970).
3. S. J. Lindenbaum and R. M. Sternheimer, Phys. Rev. **105**, 187 (1957).
4. R. J. Murphy, C. D. Dermer and R. Ramaty, Ap. J. (Suppl.) **63**, 721, (1987).
5. X. M. Hua and R. E. Lingenfelter, Solar Phys. **107**, 351 (1987).
6. D. W. Bugg et al. Phys. Rev. **133**, 1017 (1964).

THE GAMMA-1 DATA ON THE MARCH 26, 1991 SOLAR FLARE

V.V.Akimov, N.G.Leikov
Space Research Institute, Moscow, 117810, Russia

V.G.Kurt
Insitute of Nuclear Physics, Moscow University,
Moscow, 119899, Russia

I.M.Chertok
IZMIRAN, Troitsk, Moscow Region, 142092, Russia

ABSTRACT

Two different components of γ-emission with energies from 30 MeV to several hundred MeV were observed in the solar flare with total duration of 10 minutes. The first, impulsive one, originates as a bremsstrahlung of electrons accelerated at high plasma density ($n > 10^{14}$ cm^{-3}). The second, delayed component, is associated with a pion production of high energy ions accelerated at low densities in the corona during the stage of magnetic field restoration.

INTRODUCTION

The 3B/X4.7, white light, two-ribbon solar flare of March 26, 1991 (coordinates: S28 W23), observed in high energy γ-rays by the telescope GAMMA-1[1], fell fairly well into a class of so-called electron dominated events[2] due to its extreme impulsivity and pure electron bremsstrahlung spectrum[3]. These distinctions were established for the first 1.5 min of the flare comprising the main bulk of registered photons. We present here the more detailed time profile of the flare main peak and discuss possible source of the observed rapidly variable γ-emission.

Recent analysis of the data collected at a late stage of the flare revealed existence of a hard delayed component which become distinguishable only at energies greater than 100 MeV. We associate this delayed γ-emission with nuclear interactions of high energy ions accelerated at the stage of magnetic field restoration. Some common features allow to consider the March 26 event as a typical example of high energy γ-ray solar flare that owing to its shortness for the first time has been seen from the very beginning to the very end.

1. EXPERIMENTAL RESULTS AND DISCUSSION

1.1. Impulsive component

Figure 1 shows a time profile of the flare impulsive phase in γ-rays above ~30 MeV derived with the best resolution permitted by counts statistics. The starting point corresponds to a moment of the first significant excess above background. The clearly seen two sub-peaks structure exhibits the rise time of 1-2 sec and shortest decay time of 0.5-1 sec. The data available allowed to perform spectral maximum likelihood analysis of the impulsive component up

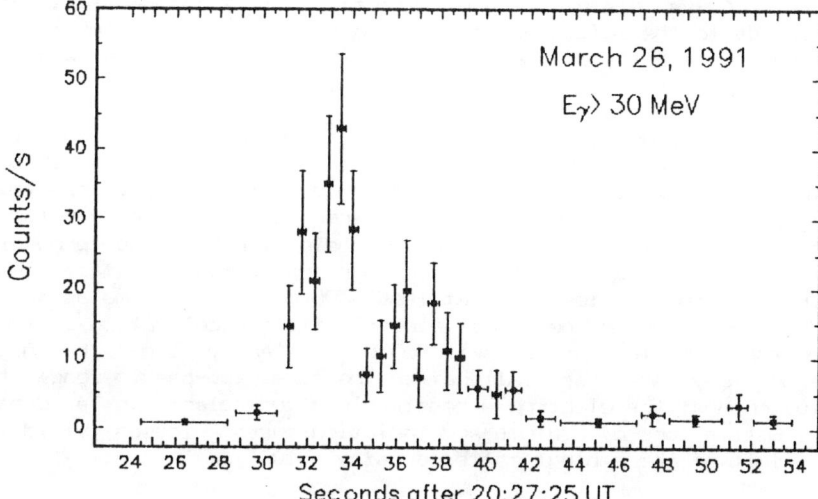

Fig. 1. The detailed time profile of the flare impulsive phase in γ-rays above the instrumental threshold of GAMMA-1.

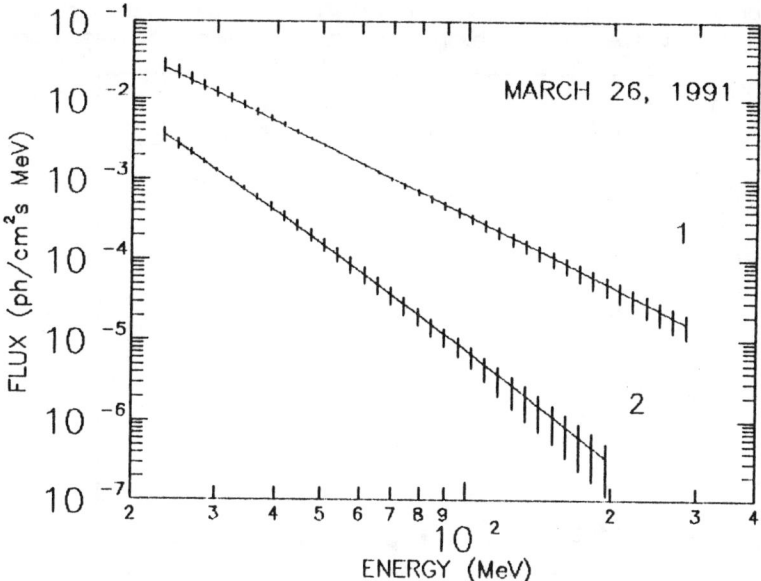

Fig. 2. Differential γ-ray energy spectra of the impulsive component (curve 1, 20:27:56-20:28:07 UT) and trailer (curve 2, 20:28:07-20:29:20 UT). Shaded area contains 68% of the bootstrap test solutions.

to 300 MeV and of gradually decreasing γ-emission during the subsequent 73 sec up to 200 MeV. Excellent agreement with a simple power-law spectrum model was found within both intervals, but the spectral indexes obtained were different: -3.0±0.2 and -4.1±0.4 that corresponds to the softening of γ-emission with time. Figure 2^3 illustrates stability of the maximum likelihood solution when a flexible trial spectrum (logarithmic parabola) was applied along with a multiple bootstrap sampling.

The pure power-law spectra of γ-emission during the impulsive and gradual phases unambiguously indicate an electron bremsstrahlung origin of the emission. It is worth to note that apparent spectrum steepening after the main peak is rather typical for impulsive events, as it follows from observations below 1 MeV[4], and probably reflects regime of acceleration. In our opinion, the fact that electrons reach energies up to at least 300 MeV in the time as short as 1 sec testifies to the direct electric field acceleration[5]. If we assume that "accelerator" is switched off instantly then the high emission decay rate, especially after the first sub-peak, leads to conclusion that the electron bremsstrahlung took place in a dense media. Estimations of Coulomb and radiation energy losses yield a lower limit of the ambient matter density $n=10^{14}$ cm^{-3}.

1.2. Delayed component

Figure 3 shoes in a scale of minutes time profiles of the flare for γ-rays below and above 100 MeV. The impulsive component is smoothed due to the larger channel width and comprises less counts

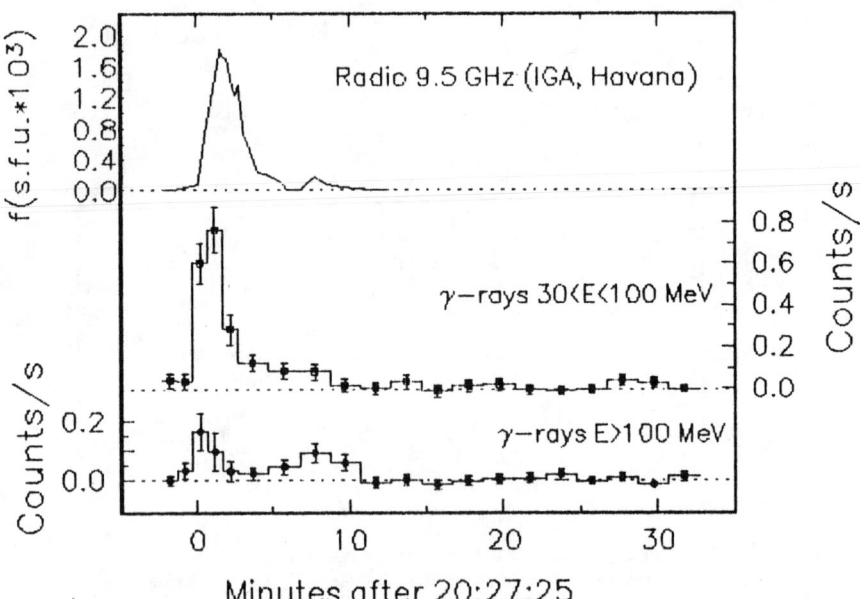

Fig. 3. Microwave and γ-ray time profiles (March 26, 1991).

than in Figure 1 because of the dead time data loss in the GAMMA-1 slow parallel registration mode. This mode was used here since it has an advantage of technically more accurate energy determination, while the dead time of 0.5 sec after each trigger is not important at the flare late stage which is essential for us now. Both γ-ray time profiles are dead time corrected and background subtracted.

The time profile for γ-rays with energies above 100 MeV exhibits a delayed component about 8 minutes after the flare onset. In view of a scarce counts statistics we resort to a comparative spectral analysis of the excess in terms of spectral ratio R=N(100 MeV)/N(<100 MeV). For delayed component we have R=1.1, while for impulsive one R=0.2. This difference is significant at the confidence level of 4σ. So hard spectrum of the delayed component can be attributed to a decay of neutral pions born in nuclear interactions of high energy ions accelerated to many hundreds MeV at the late stage of the flare.

Radio profile at 9.5 GHz (Figure 3) indicates that, in fact, additional energy release took place just at the time of the delayed γ-ray component. Analogy with the June 15, 1991 flare suggests itself. That flare had very similar radio profile, and intense γ-emission with a well-shaped pionic spectrum was observed at the time of delayed radio component[6]. Acceleration of ions was associated there with magnetic field relaxation high in the corona after the coronal mass ejection (CME). For the March 26 event we have no evidence of CME, in particular the type II radio emission was not observed. However it is known that some CME even with a high ascent speed are not followed by the type II radio emission[7].

In general, it would be reasonable to suppose that any strong enough magnetic field deformation at the flare initial stage is then followed by the prolonged stage of restoration with acceleration of charged particles in the induced electrical field. We consider the fact that high energy ions appear just at this stage to be a result of acceleration at higher altitudes and so at lower plasma density than at the time of impulsive component.

In comparison with the June 15 event this flare is scaled down in amplitude and duration that may be connected with absence of CME and therefore with less energy accumulated by the magnetic field.

We are grateful to O.Alvarez Pomares and E.del Pozo Garsia (Institute of Geophysics and Astronomy, Havana, Cuba) for providing microwave time profiles.

REFERENCES

1. V.V.Akimov et al., 22nd ICRC, 3, 73 (1991).
2. E.Rieger, H.Marschhauser, MAX '91, Workshop #3, 68 (1990).
3. N.G.Leikov et al., Astron. and Ap. Suppl. 97, 345 (1993).
4. B.Dennis, Solar Phys. 100, 465 (1985)
5. H.Alfven and P.Carlqvist, Solar Phys. 1, 220 (1967)
6. V.V.Akimov et al., this issue.
7. N.R.Sheeley et al., Ap. J. 279, 839 (1984)

STOCHASTIC FERMI ACCELERATION AND SOLAR COSMIC RAYS

M. A. Lee
University of New Hampshire, Durham, NH 03824

ABSTRACT

A brief history of stochastic Fermi acceleration is presented together with a review of the astrophysical sites of its application. A derivation of the transport equation appropriate to the process is outlined, and its relation to the transport equation for quasilinear energy diffusion is discussed. Finally stochastic Fermi acceleration is assessed as a mechanism for the acceleration of solar cosmic rays.

1. INTRODUCTION

The particle acceleration mechanism known as stochastic Fermi acceleration was originally suggested and formulated by Fermi[1] to account for the energization of cosmic rays in interstellar space. Fermi proposed that cosmic rays scatter from *clouds* of enhanced or turbulent magnetic field by being mirrored or channeled by curved magnetic flux tubes, so that a particle leaves the cloud with a guiding center velocity different from the one with which it entered the cloud. The scattering is elastic in the frame of the cloud. However, if the cloud has a velocity **V**, then the relativistic particle changes its energy by an amount $\Delta E = O(Vc^{-1}E)$. The cosmic ray gains energy following head-on-collisions and loses energy following overtaking collisions. Since the collision rate is proportional to the relative speed of the particle and cloud, head-on collisions are more probable by an amount $O(V/c)$. Thus, averaged over many collisions,

$$dE/dt \sim (V/c)^2 \, E \, c/\lambda, \tag{1}$$

where λ is the particle scattering mean free path.

Cosmic rays may also be effectively destroyed by a spallation collision with an interstellar gas nucleus. The characteristic loss time T depends on the collision cross section which is approximately independent of E at typical cosmic ray energies. Neglecting cosmic ray escape from the Galaxy, Fermi wrote down the transport equation

$$\partial N/\partial t + \partial/\partial E\,[(dE/dt)N] + N/T = Q(E), \tag{2}$$

where Q(E) is the source, or injection term. Assuming a stationary distribution, injection at low energies, and energization described by equation (1), he derived the power-law distribution

$$N(E) \propto E^{-1-\lambda c/(TV^2)} \qquad (3)$$

Since a power-law distribution was perhaps *the* distinguishing observational characteristic of cosmic rays at that time, prediction of a power law was a major achievement of the theory.

In fact stochastic Fermi acceleration of cosmic rays by turbulent cloud motion in the interstellar medium is no longer considered a viable acceleration mechanism. The required cloud motions give extreme viscous losses[2]. Also, the observed ratio of secondary cosmic ray species (produced in spallation collisions) to primary species decreases with increasing energy, indicating that average cosmic ray lifetime decreases with energy[3]. This finding is inconsistent with an acceleration mechanism which operates continually[4].

However, stochastic Fermi acceleration has been applied at many other sites. It was recognized in the 1970's that the observed secondary to primary ratio does not rule out stochastic Fermi acceleration in or near compact objects, where secondary production is low and acceleration is rapid. Scott and Chevalier[5] and Chevalier et al.[6,7] proposed that cosmic rays are accelerated by Fermi acceleration in supernovae remnants (such as Cas-A) due to randomly moving magnetic *knots*, which are sufficiently vigorous to overcome energy loss due to the expansion of the remnant. Based on these ideas, Cowsik[8] obtained excellent agreement with the observed radio spectrum of Cas-A assuming the radiation is due to synchrotron emission from the electrons which are injected impulsively and subjected to acceleration, expansion energy loss, and a constant probability of loss from the remnant.

More recently, stochastic Fermi acceleration has been invoked to accelerate the cosmic ray anomalous component in the outer heliosphere[9,10], and to accelerate interstellar and cometary pickup ions in the solar wind[11-16]. Finally, it has been a popular mechanism for the acceleration of solar cosmic rays; we shall address this application in Section 3.

Stochastic, or second-order (in V/v where v is particle speed), Fermi acceleration is related to first-order Fermi acceleration. Implicit in early work on Fermi acceleration[1,17], was the idea that the acceleration rate can be proportional to V/v, and therefore much more efficient, if the collisions with clouds are predominantly head-on. This can occur with special configurations of clouds[1], oppositely directed hydromagnetic pulses[17], or astrophysical accretion flows[18-20]. However, the most effective configuration for the operation of first-order Fermi acceleration is a shock wave at which the compression dictates head-on collisions for particles traversing the shock and scattering on magnetic irregularities upstream and downstream[21].

In view of the more rapid acceleration timescales, the common occurrence of shocks, the potentially large energy content of shocks, and their high acceleration efficiency, shock acceleration is generally accepted to be the primary acceleration mechanism for galactic cosmic rays[22], the cosmic ray anomalous component[23], and a host of energetic ion populations in the heliosphere[24].

In spite of its dramatic early promise, favored status for stochastic Fermi acceleration is generally reserved for the modest energy gains of pickup ions, and possibly acceleration of solar cosmic rays in solar flares.

2. AN APPROPRIATE TRANSPORT EQUATION

The transport equation (2) clearly neglects the stochastic feature of stochastic Fermi acceleration; particles systematically increase their energy according to equation (1). Davis[25] first formally recognized that the process must be described by a transport equation with diffusive second-order derivatives in energy. The first rigorous and correct derivation was provided by Parker and Tidman[26]. They considered elastic scattering of particles described by the distribution function $f(\mathbf{p},t)$ from massive spheres described by the distribution $T(\mathbf{v}_1,t)$. The time evolution of the spatially homogeneous average distribution $f(\mathbf{p},t)$ is governed by the Boltzmann equation

$$\partial f/\partial t = \int d\sigma \int d^3 \mathbf{v}_1 \, \bar{v} \, T(\mathbf{v}_1) \, [f(\mathbf{p}') - f(\mathbf{p})], \tag{4}$$

where we generalize the work of Parker and Tidman[26] to include relativistic particles. Here we assume that the spheres are infinitely massive and $(v_1/c)^2 \ll 1$, $d\sigma$ is an element of cross section, $\mathbf{p}$ and $\mathbf{p}'$ are the pre-collision and post-collision momenta of the particle, $\bar{v}$ is the relative speed of the particle and sphere, and $\int d^3 \mathbf{v} \, T(\mathbf{v}) = N$, the number density of the spheres.

Assuming isotropy, $f(\mathbf{p}) = f(p)$ and $T(\mathbf{v}) = T(v)$, and expanding $f(\mathbf{p}')$ in a Taylor series about $f(\mathbf{p})$ we obtain

$$\partial f/\partial t = p^{-2} \partial/\partial p \, [p^2 D(p) \, \partial f/\partial p], \tag{5}$$

where

$$D(p) = 1/3 \, \pi \, a^2 \, V^2 N \, p^2 \, v^{-1} = 1/3 \, V^2 \, \lambda^{-1} \, p^2 \, v^{-1}. \tag{6}$$

The sphere radius is given by "a", $V^2 = N^{-1} \int d^3 \mathbf{v} \, v^2 \, T(v)$, and as before λ is the scattering mean free path.

Equation (5) is the only form of the energy diffusion equation consistent with particle conservation and the *Born approximation*, which is based on the assumption $v_1 \ll v$ and allows $f(p')-f(p)$ to be expanded in a Taylor series. Equation (5) may be transformed with E as an independent variable. With $D \equiv \alpha p^2 v^{-1}$ and $F dE = 4\pi p^2 f \, dp$ we obtain

$$\alpha^{-1} \partial F/\partial t = \partial^2/\partial E^2 (p^2 vF) - \partial/\partial E (4pF). \tag{7}$$

In the relativistic limit (E = pc) we obtain

$$c\alpha^{-1} \partial F/\partial t = E^2 \partial^2 F/\partial E^2 - 2F. \tag{8}$$

It is interesting to note that the appropriate transport equation for stochastic Fermi acceleration has caused confusion in the past. Davis[25], Scott and Chevalier[5] and Chevalier et al.[6,7], all used versions which differed from equation (8) in the coefficient of F and/or with inclusion of a term of the form $b\partial/\partial E(EF)$, where b is a constant. However, equations (5) and (6) are in agreement with the equations derived by Tverskoi[27] and Kulsrud and Ferrari[28] from other points of view.

Illustrative and useful solutions of equation (5) may be derived. Here we repeat those derived by Ramaty and myself, described by Ramaty[29]. With a source term of the form $\delta(t) \delta(p-p_o)$ included on the right hand side we derive the solution for impulsive monoenergetic injection in the nonrelativistic and relativistic limits, respectively:

$$f = p_o(p\alpha m t)^{-1} I_2[2(p_o p)^{1/2} (\alpha m t)^{-1}] \exp[-(p_o + p)(\alpha m t)^{-1}], \tag{9}$$

$$f = p_o^2 p^{-3} (4\pi \alpha c^{-1} t)^{-1/2} \exp[-(\ln p/p_o - 3\alpha c^{-1} t)^2 (4\alpha c^{-1} t)^{-1}], \tag{10}$$

where as above $\alpha = 1/3 \, V^2/\lambda$, m is the particle mass, and $I_2(x)$ is the standard modified Bessel function of the first kind. Solution (9) was derived by Tverskoi[27] to describe the energy spectral evolution of solar cosmic rays. A generalized version of solution (10) was utilized by Cowsik[8] to describe the transport of electrons in the Cas-A supernova remnant.

A stationary solution of equation (5), including a constant characteristic loss time T and monoenergetic injection, satisfies

$$0 = p^{-2} \partial/\partial p [p^2 D(p) \partial f/\partial p] - T^{-1} f + \delta(p - p_o). \tag{11}$$

Solutions in the nonrelativistic and relativistic limits are, respectively:

$$f(p > p_o) = 2 p_o (\alpha m p)^{-1} I_2 [2 p_o^{1/2} (\alpha m T)^{-1/2}] K_2 [2 p^{1/2} (\alpha m T)^{-1/2}], \tag{12}$$

$$f(p > p_o) = c p_o^2 p^{-3} \alpha^{-1} (9 + 4 c\alpha^{-1} T^{-1})^{-1/2} (p/p_o)^{-\Gamma}, \tag{13}$$

where $\Gamma = 1/2\,(9 + 4\,c\alpha^{-1}\,T^{-1})^{1/2} - 3/2$ and $K_2(x)$ is the modified Bessel function of the second kind. Solution (12) was shown by Ramaty[29] to give excellent fits to solar cosmic ray energy spectra. Indeed the Bessel function parametrization of solar cosmic ray spectra has become virtually universal[30]. Solution (13) is the correct solution for the case considered by Fermi[1], described by equation (3). We note that although the spectra are both power law, they are identical only in the limit $c\lambda(TV^2)^{-1} \leq 1$. Since the observed cosmic ray spectrum requires $c\lambda(TV^2)^{-1} \sim 1.7$ in equation (3), solution (13) rather than equation (3) is appropriate. Solutions to more general versions of equation (11) involving energy-dependent escape (described by $T^{-1}f$) and collisional losses, have been given by Schlickeiser[31].

Equation (5) may also be derived by considering the quasilinear evolution of particles in the presence of a broad spectrum of waves[32-34]. We consider the spatially homogeneous special case in which the waves are transverse hydromagnetic waves propagating either parallel, $I_+(k)$, or antiparallel, $I_-(k)$, to $\hat{e}_z$ with $\mathbf{B} = B\hat{e}_z$. If the Alfvén speed V_A is small compared with energetic ion speeds, then the ions, presumed to be gyrotropic initially, scatter in pitch-angle to isotropy in the average wave frame[16]. On a subsequent slower timescale the omnidirectional distribution function of the particles satisfies equation (5) with[16]

$$D(p) = \pi q^2\, V_A^2\, c^{-2}\, v^{-1} \int_{-1}^{1} d\mu\, (1-\mu^2)\, |\mu|^{-1}\, I_+(\Omega v^{-1}\mu^{-1})$$
$$\cdot I_-(\Omega v^{-1}\mu^{-1})[I_+(\Omega v^{-1}\mu^{-1}) + I_-(\Omega v^{-1}\mu^{-1})]^{-1}, \qquad (14)$$

where q is ion charge, $\Omega\;(= qB\,m^{-1}\,\gamma^{-1}\,c^{-1})$ is the ion cyclotron frequency, γ is the Lorentz factor, μ is the cosine of ion pitch angle, and the wave intensities are normalized as $\langle|\delta\mathbf{B}|^2\rangle = \int_{-\infty}^{\infty} dk\,[I_+(k) + I_-(k)]$. The argument of the wave intensities, $\Omega v^{-1}\mu^{-1}$, arises from the cyclotron resonance condition under the assumption that $V_A \ll v$.

In contrast with the original view of stochastic Fermi acceleration involving nonresonant interactions with the scattering magnetic field (mirroring or channeling by a curved flux tube), equation (14) involves pitch-angle scattering of ions by a resonant set of waves. As a result $D(p)$ is related to the wavenumber dependence of the resonant waves and is not restricted to being proportional to p^2/v, as is the case in equation (6). Indeed, energization by a spectrum of waves is not generally called stochastic Fermi acceleration because it can involve fluctuating electric fields parallel to the magnetic field (which cannot be removed by Lorentz transformation). Also, Fermi acceleration was originally reserved for nonresonant scattering by magnetic fluctuations. However, in the case of parallel transverse wave propagation discussed above, the quasilinear energy diffusion shares the essential feature of stochastic Fermi acceleration that the particles gain energy by being scattered elastically by scattering centers which move relative to each other with no divergence.

A net divergence of scattering centers would lead to first order Fermi acceleration, or adiabatic deceleration. In the case of quasilinear energy diffusion described by equation (14), the scattering centers are the waves. Note that in equation (14) $D = 0$ if all waves propagate in one direction. In the wave frame there is then no electric field and no acceleration, and therefore in no other frame can there be energy diffusion. Thus, in common usage, energization by a spectrum of transverse waves is often called stochastic Fermi acceleration. We shall follow this usage in the following section. It is interesting to note in equation (14) that if $I_{\pm}(k) \propto k^{-2}$, then $D(p) \propto p^2/v$ – the dependence for nonresonant scattering.

3. STOCHASTIC FERMI ACCELERATION OF SOLAR COSMIC RAYS

Stochastic Fermi acceleration has also been proposed as a mechanism for accelerating ions and electrons to high energies during or following solar flares[27,35,36]. It is easy to imagine the generation of substantial magnetic turbulence by the relaxation of an unstable magnetic field configuration in the corona, which is thought to initiate a flare. The turbulence would naturally accelerate ions and electrons. Since Fermi acceleration basically changes a particle's velocity, acceleration is generally much more rapid for ions. This was a positive feature of Fermi acceleration as the origin of cosmic rays, which are dominated by protons[37]. Ramaty et al.[38] pointed out that the same feature favors Fermi acceleration (as opposed to, say, direct acceleration in an electric field) as the origin of the solar flare particles observed in space, which tend to exhibit a large proton to electron ratio at a given energy. Ramaty et al.[38] also showed that the Bessel function energy spectrum described by equation (12) could fit the energy spectra of particles observed in space extremely well with reasonable values for the escape time, the scattering velocities, and the scattering mean free path.

With the development and immediate success of the theory of diffusive shock acceleration, however, attention turned to the possibility that coronal shocks accelerate solar cosmic rays[39]. Moreton waves and metric type II bursts were known to be signatures of coronal shocks, and it was natural to assume that these shocks accelerated solar cosmic rays via first-order Fermi acceleration. Lee and Ryan[40] calculated the time profiles and spectra of particles accelerated by a coronal/interplanetary shock via a very simplified model. However, a shock origin of solar cosmic rays had obvious deficiencies - not all solar cosmic ray events show obvious shock association. The gamma ray spectrometer on SMM revealed that ion acceleration to 10's of MeV in flares can occur in seconds, but it is difficult to understand how shocks can both be generated and then accelerate so rapidly. Furthermore, shocks are not expected to be very strong in the corona since the Alfvén speed is very high.

The current view appears to be the following[41,42]: Flare events generally divide into two classes, *impulsive* and *gradual* [43]. The large gradual events tend to exhibit clear shock associations and a coronal/interplanetary shock probably accelerates at least those solar cosmic rays observed in interplanetary space. The very impulsive events are still a mystery, but may be accelerated by direct electric fields at the flare site. However, the more gradual phase of impulsive events (or the particles confined near the sun in gradual events) may very well be accelerated by stochastic Fermi acceleration.

A popular scenario is the following[44]: a pair of coronal magnetic loops are squeezed together by photospheric motions; they reconnect, and a sequence of post-flare loops is formed. Reconnection can produce a downward jet which shocks in the denser plasma below[45]. The reconnection process and/or shock can produce turbulence and "seed" particles in the flare loops, which then undergo stochastic Fermi acceleration within the loops. The ions eventually leak out of the loops at the footpoints, precipitate into the photosphere, and produce via nuclear collisions the gamma rays and neutrons observable in space . There are potential problems and limitations with this scenario: excitation of turbulence by the reconnection jet shock may be limited; waves may escape the loop; wave power in the cyclotron resonant frequency range must be replenished as the ions are accelerated. A positive feature of this view is that the ions escaping the loop into the photosphere may excite additional turbulence.

The promise that stochastic Fermi acceleration operates in flares has generated much additional work on the process. Miller et al.[46] generalized the energy spectrum predicted by stochastic Fermi acceleration by bridging the nonrelativistic and relativistic regimes. Smith and Brecht[47], Miller et al.[48], and Bogdan et al.[16] have begun to study the difficult problem of wave excitation by the ions, mode coupling of the waves, and wave damping due to ion acceleration. In summary, prospects for a successful theory of the acceleration of some of the solar cosmic rays by stochastic Fermi acceleration appear to be promising. This time-honored mechanism appears to have finally found a home!

ACKNOWLEDGEMENTS

The author wishes to acknowledge the extensive help of Elaine Bennett in preparing the typescript and the patience of the Editors. This work was supported, in part, by NSF Grant ATM-9215279 and NASA Space Physics Theory Program Grant NAG5-1479.

REFERENCES

1. E. Fermi, Phys. Rev. 75, 1169 (1949).
2. E.N. Parker, Phys. Rev. 99, 241 (1955).
3. J. Ormes and P. Freier, Astrophys. J. 222, 471 (1978).
4. R. Cowsik, Astrophys. J. 241, 1195 (1980).
5. J.S.Scott and R.A. Chevalier, Astrophys. J. 197, L5 (1975).
6. R.A. Chevalier, J.W. Robertson, and J.S. Scott Astrophys. J. 207, 450 (1976).
7. R.A. Chevalier, W.R. Oegerle, and J.S. Scott, Astrophys. J. 222, 527 (1978).
8. R. Cowsik, Astrophys. J. 227, 856 (1979).
9. L.A. Fisk, J. Geophys. Res. 81, 4633 (1976).
10. B. Klecker, J. Geophys. Res. 82, 5287 (1977).
11. W.-H. Ip, and W.I. Axford, Planet. Space Sci., 34 1061 (1986).
12. P.A. Isenberg, J. Geophys. Res. 92, 1067 (1987a).
13. P.A. Isenberg, J. Geophys. Res. 92, 8795 (1987b).
14. T.I. Gombosi, K. Lorencz, and J.R. Jokipii, J. Geophys. Res. 94, 15,011 (1989).
15. D. D. Barbosa, Astrophys. J. 341, 493 (1989).
16. T.J. Bogdan, M.A. Lee, and P. Schneider, J. Geophys. Res. 96, 161 (1991).
17. E.N. Parker, Phys. Rev. 109, 1328 (1958).
18. R. Cowsik and M.A. Lee, Proc. R. Soc. Lond. A. 383, 409 (1982).
19. G.M. Webb and T.J. Bogdan, Astrophys. J. 320, 683 (1987).
20. P. Schneider and T.J. Bogdan, Astrophys. J. 347, 496 (1989).
21. W.I. Axford, E. Leer, and G. Skadron, 15th Int. Cosmic Ray Conf. (Plovdin, 1977), Vol. 2, p. 273.
22. R.D. Blandford and D. Eichler, Phys. Rept. 154, 1 (1987).
23. J.R. Jokipii, J. Geophys. Res. 91, 2929 (1986).
24. M.A. Lee, in Particle Acceleration in Cosmic Plasmas, edited by G. Zank and J. T. Gaisser (New York: AIP, 1992), p.27.
25. L. Davis, Phys. Rev. 101, 351 (1956).
26. E.N. Parker and D.A. Tidman, Phys. Rev. 111, 1206 (1958).
27. B.A.Tverskoi, Soviet Phys. JETP 25, 317 (1967).
28. R.M.N. Kulsrud and A. Ferrari, Astrophys. Space Sci. 12, 302 (1971).
29. R.Ramaty, in Particle Acceleration Mechanisms in Astrophysics, edited by J. Arons, C. McKee, and C. Max, (New York: AIP, 1979), p.135.
30. M.A. Forman, R. Ramaty, and E.G. Zweibel, in Physics of the Sun, edited by P.A. Sturrock et al., (Dordrecht: Reidel, 1986), p. 249.
31. R. Schlickeiser, in Cosmic Radiation in Contemporary Astrophysics, edited by M.M. Shapiro (Dordrecht: Reidel, 1986), p.27.
32. C.F. Kennel and F. Engelman, Phys. Fluids 9, 2377 (1966).
33. I. Lerche, Phys. Fluids 11, 1720 (1968).
34. D.E. Hall and P.A. Sturrock, Phys. Fluids 10, 1593 (1967).
35. D.J. Mullan, Astron. Astrophys., 52, 305, (1976).
36. A.A. Korchak, Sol Phys. 56, 223 (1978).

37. P. Morrison, A. Olbert, B. Rossi, Phys. Ref. $\underline{94}$, 440 (1954).
38. R. Ramaty et al., in Solar Flares (Colorado Assoc. Univ. Press: Boulder, Colorado 1980), pp. 117-185.
39. A. Achterberg and C.A. Norman, Astr. Ap. $\underline{89}$, 353 (1980).
40. M.A. Lee and J.M. Ryan, Astrophys. J. $\underline{303}$, 829 (1986).
41. M.A. Lee, 22nd Internat. Cosmic Ray Conf. $\underline{5}$ (Dublin: Ireland, 1991), p. 293.
42. B. Klecker, E. Cliver, S. Kahler, and H. Cane, EOS 71, 1102 (1990).
43. D.V. Reames, Astrophys. J. $\underline{358}$, L63 (1990).
44. J.M. Ryan and M.A. Lee, Astrophys. J. $\underline{368}$, 316 (1991).
45. T.G. Forbes, J.M. Malherbe, and E.R. Priest, Solar Phys. $\underline{120}$, 285 (1989).
46. J.A. Miller, N. Guessoum, and R. Ramaty, Astrophys. J. $\underline{361}$, 701 (1990).
47. D.F. Smith and S.H. Brecht, Astrophys. J. $\underline{373}$, 289 (1991).
48. J.A. Miller, R. Ramaty, and N. Guessoum, 21st Int. Cosmic Ray Conf. (Adelaide, 1990), Vol. 5, p. 36.

GAMMA RAYS FROM AN "OVER-THE-LIMB" FLARE

W. Thomas Vestrand and David J. Forrest
Physics Department and Space Science Center
University of New Hampshire
Durham, NH 03824

ABSTRACT

We present observations taken with the SMM Gamma-Ray Spectrometer of a giant flare that occurred on 1989 September 29 at a position located approximately 10° beyond the solar limb. The spectral measurements show a rich gamma-ray spectrum with a remarkably strong neutron capture line at 2.22 MeV. We briefly examine several possible interpretations of the measurements. We conclude that the best interpretation is that the gamma-ray emission had a spatially extended component which subtended more than 30° on the solar surface in addition to the spatially compact component that normally dominates the total flare fluence. We suggest that the "back diffusion" of solar energetic particles resident on open field lines may have generated the extended component.

INTRODUCTION

Existing experiments do not have the spatial resolution needed to image flare gamma-ray emission. However, the observed limb darkening of the neutron capture line at 2.22 MeV suggests that the bulk of that emission originates in a relatively small, bright, kernel located below the Hα flare site[1,2]. On the other hand, interplanetary measurements of Solar Energetic Particles (SEP) indicate that flare accelerated particles can access magnetic field lines whose footpoints have heliolongitudes that are 30° or more from the flare position[3]. The question therefore arises: Is there a component of spatially extended gamma-ray emission from flares? The occultation of the intense compact component during "over-the-limb" flares provides an opportunity to search for a spatially extended component.

OBSERVATIONS OF THE 1989 SEPTEMBER 29 FLARE

On 1989 September 29 a flare that was rich in high-energy phenomena erupted on the Sun. The flare was first detected at 10:47 UT by soft x-ray detectors on a GOES satellite which subsequently recorded a peak (X9.8) at 11:33 UT [4]. Intense high-energy emission was measured by the Gamma-Ray Spectrometer (GRS) aboard the Solar Maximum Mission (SMM) satellite when the instrument was switched back on at 11:33 UT after a spacecraft South Atlantic Anomaly passage[5]. The GRS flare measurements show a complex spectrum with an electron bremsstrahlung continuum, a positron annihilation line, prompt nuclear emission, high-energy emission extending to energies >50 MeV, and a strong neutron capture line at 2.22 MeV (see figure 1). Detection of the x-ray and gamma-ray emission was followed by the onset of the largest cosmic-ray ground level event (GLE) observed by neutron monitors since 1956 [6] and the first clear detection of a solar cosmic ray event by an underground muon telescope [6].

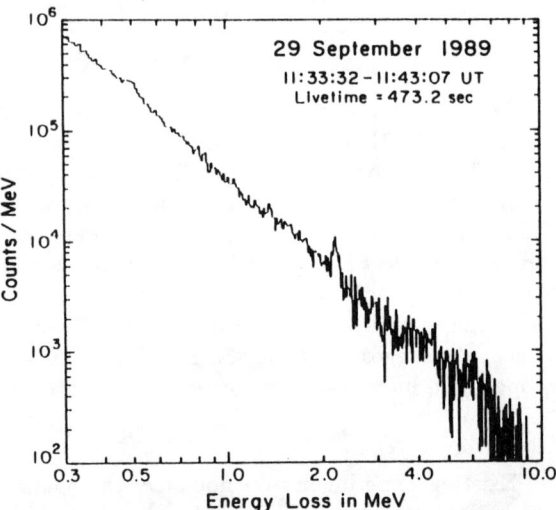

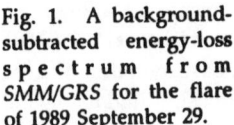
Fig. 1. A background-subtracted energy-loss spectrum from SMM/GRS for the flare of 1989 September 29.

A number of other interesting phenomena were associated with this powerful flare. Several interplanetary spacecraft measured SEPs with energies >30 MeV[7]. Images taken by the Coronagraph/Polarimeter on SMM showed evidence for a Coronal Mass Ejection (CME) with an inferred velocity which is consistent with the formation of a coronal/interplanetary shock[8]. A metric type II radio burst that is indicative of a coronal shock was also detected. The relative timing of the CME, metric radio bursts, and the inferred injection profile for the 20 GeV protons responsible for the ground level event suggest that a CME driven shock was the source of the SEP event[7].

The Hα observations for this period do not show any flare on the visible solar hemisphere that is likely to be associated with such a large ground level event. However, solar flare patrol observations [3] showed a "behind the limb" flare-like brightening indicative of major activity in NOAA region 5698 that was located beyond the southwestern limb. To estimate the flare position we studied the positions of AR5698 flares measured by three or more observatories as recorded in the Hα flare list of the NOAA Solar Geophysical Data comprehensive reports[4]. Extrapolation of the linear least squares fit to that data (figure 2) yields the position for a AR5698 flare at 11:33 UT on 29 September 1989 of W97.9° ± 5.3° and S25.6° ± 2.0°. This places the nominal flare position well beyond the limb at a heliocentric angle of $\Theta = 100.0° \pm 4.7°$.

The relative strength of the neutron capture line at 2.22 MeV in the gamma-ray spectrum is remarkable for a flare located at such a large heliocentric angle.

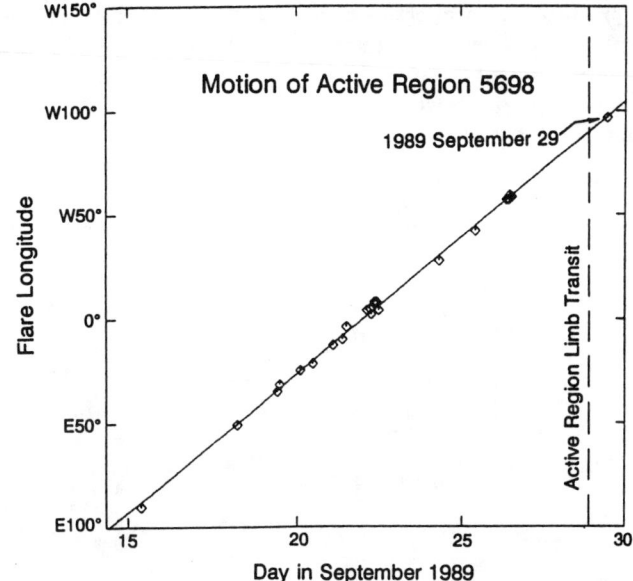

Fig. 2. The measured positions for Hα flares as active region 5698 traversed the solar disk.

Since the GRS instrument was switched off to protect it from SAA particles during the intense early phase of the flare, deviation of the diagnostic fluence ratio $\Re$, which is given by the ratio of the delayed neutron capture line fluence to the prompt nuclear de-excitation fluence in the 4-7 MeV band, requires modeling of the flux time histories. Such a temporal study[5] yields a fluence ratio $\Re \simeq 0.2$ which is less than one naively derives from the integrated spectrum, but that is still much larger than one would predict for an over-the-limb event. We have plotted in figure 3 the fluence ratio as a function of heliocentric angle predicted[1] for "point-like" flares with a range of ion spectral shapes. The ratio derived for 1989 September 29 is clearly much higher than predicted.

INTERPRETATION OF THE GAMMA RAY MEASUREMENTS

A potential explanation for the anomalous fluence ratio is that the flare has been attributed to the wrong active region. Unfortunately, the only Hα flare detected during the gamma-ray emission interval was a disk subflare at position S16E44 that seems an unlikely source of the measured gamma-ray emission. First, the gamma-ray time history would be odd for a GRS flare because it persisted for at least ten minutes after the end of the Hα emission. Second, the fluence ratio of $\Re \simeq 0.2$ is smaller than for any other disk event yet measured; requiring a very soft ion spectrum. The presence of >10 MeV emission and a hard bremsstrahlung spectrum ($s \simeq 2.5$) would also be unusual for a disk event with such a soft ion spectrum. Furthermore, that relatively small eastern hemisphere flare cannot explain the GOES x-ray time profile nor, since it was poorly connected, could it produce the properties of the GLE observed by the neutron monitors. This explanation therefore requires the unlikely temporal coincidence of an unrelated eastern hemisphere disk flare having unprecedented gamma-ray properties with a flare beyond the western limb that produced the largest GLE detected in more than three decades.

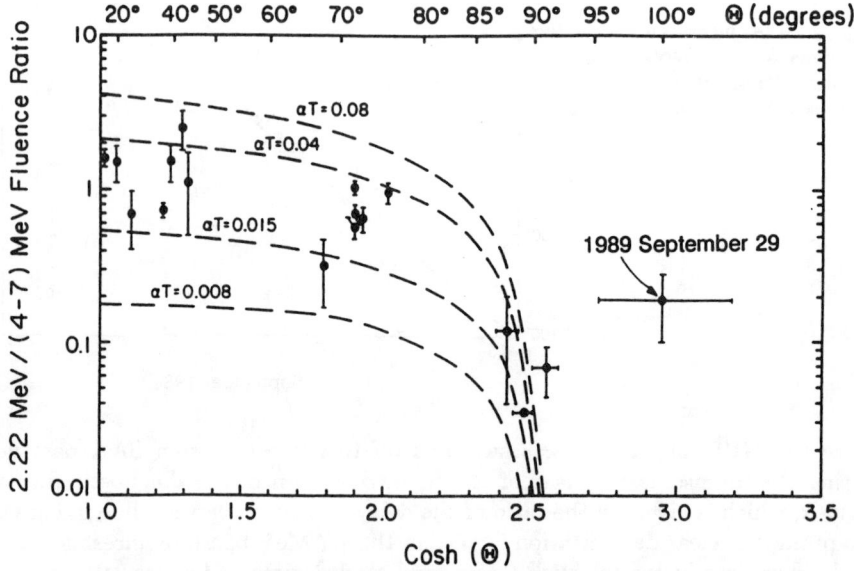

Fig. 3. The fluence ratios measured for a number of flares plotted versus the heliocentric angle of the flare. The dashed curves show the predictions for Bessel function ion energy spectra with a horizontal fan beam angular distribution[1].

If, as seems more likely, the gamma-rays are associated with the giant "beyond-the-limb" flare then we must add some new elements to the standard picture of flare gamma-ray production. The standard picture[9] assumes that the prompt nuclear emission originates in a relatively compact chromospheric region located at the Hα position and that the neutron capture emission originates in a compact photospheric patch situated below the prompt emission region. However, the gamma-rays observed during the 1989 September 29 flare could not have originated in this manner because the column depth along the line of sight to the Hα region was so large that photons generated in either the chromosphere or the photosphere could not reach the earth. An attractive idea is that the gamma-rays were produced in a coronal region—perhaps at the top of one of the large flare loops. In fact, it has been suggested[10] that recent observations of another over-the-limb flare by instruments on the GRANAT spacecraft are best explained by a coronal emission region. An important difference between the two data sets is the presence of a strong neutron capture line feature at 2.22 MeV in the 1989 September 29 spectra. The difficulty of trapping and ultimately capturing neutrons at the top of a coronal loop would make it a very inefficient source of neutron capture emission.

A coronal loop model for 1989 September 29 gamma-rays therefore needs an additional element to explain the strong neutron capture line at 2.22 MeV. One possibility[2] is that feature was a ^{32}S de-excitation line that can be mistaken for neutron capture line emission by moderate resolution spectrometers like SMM/GRS. To explain the dominance of the 2.2 MeV feature, that hypothesis would require an enhancement of the ^{32}S abundance by more than an order of magnitude over all other elements that typically generate lines in flare gamma-ray spectra. While one cannot rule out such a possibility, the spectra of other limb events do not show such a dramatic ^{32}S enhancement. Furthermore, the only known fractionation mechanism[11] which could generate such an enhancement would also enhance the ^{20}Ne abundance whose line at 1.634 MeV is not visible in the 29 September 1989 spectrum. An alternate possibility is that the feature is generated by the capture of albedo neutrons that manage to find their way to the visible hemisphere of the Sun. For example, low energy neutrons that are gravitationally trapped can follow ballistic trajectories that return them far from the flare site. This mechanism is quite inefficient since typically only $\sim 10^{-3}$ of the total neutron capture emission is generated by gravitationally trapped neutrons. One could also invoke the presence of a neutron reflecting layer located high in the corona.

The explanation we favor for the 1989 September 29 gamma-ray measurements is that the emission was generated low in the solar atmosphere, but that in addition to the compact interaction region — which normally dominates the emission — there was an interaction region that was quite extended in heliolongitude. In this picture the fraction of the extended region that extends into the visible hemisphere generates the observed emission. The measured fluence ratio $\Re$ is then given by a convolution of the limb darkening curve as a function of the viewing angle with the spatial distribution of the bombarding particles. If we assume the particles uniformly bombard an axially symmetric region that is centered at the nominal flare position, then the observed fluence ratio of $\Re = 0.19 \pm 0.09$ requires an emission patch with an angular diameter of at least 30° [5]. If such a component were present in all flares then consistency with the observed limb darkening constrains the diffuse fluence to a value that is typically less than 1/10 of the fluence from the compact component.

The extended emission region could be powered by bombarding particles that diffuse from a compact acceleration region at the flare site, or by particles accelerated over a large range of heliolongitudes by a coronal shock[5,12]. We find the latter possibility particularly attractive for the 1989 September 29 because of the evidence that a CME driven coronal/interplanetary shock was associated with the flare[7]. Here particles on open field lines that are accelerated by the shock can either back-diffuse to bombard the solar atmosphere or escape into the interplanetary medium to be observed as SEPs. Simple estimates[7] show that less than a third of the particles accelerated at the shock would have to precipitate into the solar atmosphere to power the observed nuclear emission.

ACKNOWLEDGMENTS

This work was supported by NASA grants NAG 5-1561 and NAGW-3538.

REFERENCES

1. X.-M., Hua and R.E. Lingenfelter, Solar Phys., **107**, 351 (1987).
2. R. Ramaty, N. Mandzhavidze, B. Kozlovsky, and J. Skibo, Adv. Space Res. (COSPAR), in press (1993).
3. D.V. Reames, Ap. J. Supp., **73**, 235 (1990).
4. NOAA Solar Data Comprehensive Reports, no. 547, part II (1990).
5. W.T. Vestrand and D.J. Forrest, Ap. J. Letters, **409**, L69 (1993).
6. D.B. Swinson and M.A. Shea, Geophys. Research Letters, **17**, no.8, 1073 (1990).
7. E.W. Cliver, Kahler, S.W. and W.T. Vestrand, Proc. 23th Internat. Cosmic Ray Conf., **3**, 91 (1993).
8. J.T. Burkepile and O.C. St. Cyr, NCAR/TN-389+STR, (NCAR:Boulder, CO), (1990).
9. E.L. Chupp, Ann. Rev. Astron. Ap., **22**, 359 (1984).
10. G. Trottet and N. Vilmer, private communication (1993).
11. A. Shemi, MNRAS, **251**, 221 (1991).
12. R. Ramaty, Murphy, R.J. and Dermer, C.D., Ap. J. Letters,**316**, 100 (1987).

YOHKOH OBSERVATIONS/HIGH-ENERGY ELECTRON PHENOMENA

THE YOHKOH CONTEXT FOR HIGH-ENERGY PARTICLES IN SOLAR FLARES

Hugh S. Hudson
Institute for Astronomy, University of Hawaii 96822

ABSTRACT

Yohkoh, a satellite dedicated to high-energy observations of solar flares, began observations in September, 1991. It carries (i) a soft X-ray telescope with arcsecond resolution and excellent temporal sampling; (ii) a hard X-ray imager making the first images above 30 keV; (iii) a sensitive Bragg crystal spectrometer for soft X-ray emission lines, and (iv) a set of proportional and scintillation counters. The flare observations confirm the central role of impulsive-phase electron acceleration in causing "evaporation" and white-light flare emission. SXT has found impulsive soft X-ray time profiles at the footpoints. It also shows compact bright structures apparently at the tops of flaring loops during the gradual phase. Large flares may show cusp-shaped structures that strongly resemble the usual picture of coronal magnetic reconnection, but otherwise do not match the details of the classical flare scenario. The data taken as a whole suggest that large-scale magnetic reconnection in the solar corona does not drive flare energy release, but rather is driven by the flare; the reconnection may have an important role in flare triggering.

INTRODUCTION: YOHKOH

Our physical interpretation of solar high-energy observations of the Sun, thus far mainly non-imaging at hard X-ray and γ-ray energies, depends intimately upon our knowledge of the ambient conditions at the sources of the hard radiations. The corona represents a difficult observational problem at optical wavelengths because of the proximity of the bright photosphere, in spite of the invention of the coronagraph many decades ago. Soft X-radiation represents the natural product of the most prominent coronal plasma, which has temperatures on the order of $(1-5) \times 10^6$ K. Until the present time, our knowledge of coronal structure and conditions based upon soft ($\sim$1 keV) X-ray observations has been limited to *Skylab* (ca. 1973–1974), plus limited sounding-rocket observations mainly using grazing-incidence telescopes with obsolete technology.

The *Yohkoh* satellite, launched at the end of August 1991 and almost fully functional at the time of writing of this paper (April 1993), has filled this observational gap by use of a modern soft X-ray telescope (SXT) with "superpolish" mirror technology and a CCD readout[1]. The CCD has 1024 × 1024 square pixels 2.46 arc s across. These advantages let SXT observe with large image dynamic range (because of the low-scattering optics and the CCD linearity) and with high time resolution. In addition to the SXT observations, *Yohkoh* carries a novel hard X-ray imager sensitive to flare radiations in the 15–100 keV range, plus spectrometers and photometers. The results are quite remarkable[2] and are beginning to inform us about many aspects of coronal physics, especially the physics of solar flares, the main target of *Yohkoh*. Initial results from *Yohkoh* have been published in a special issue of the *Publications* of the Astronomical Society of Japan.

Figure 1 demonstrates the sensitivity of the SXT observations by showing soft X-ray emission to be detectable some 20 arc min (1.1×10^{10} cm) above the solar limb in a set of medium-length exposures made on 8 May 1992. These clear low-scattering views of the direct X-ray emission of the hot corona may rank with Lyot's introduction of the coronagraph as a tool for general coronal investigation.

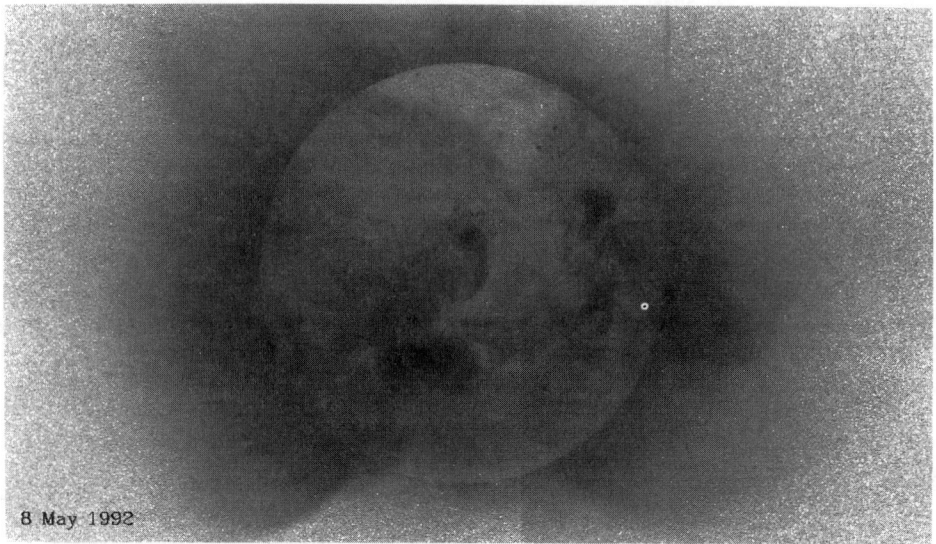

Fig. 1. Composite image of the soft X-ray corona obtained 8 May 1992. This and other images shown later are data from the Soft X-ray Telescope (SXT) on board *Yohkoh*. This telescope uses grazing-incidence "superpolish" optics, broad-band spectral filters sensitive in the 1-3 keV range, and a 1024 × 1024-pixel CCD readout. At the time of the Waterville workshop, it had returned almost 10^6 images of the corona, the quiet Sun, and flares. The images presented here are negatives.

This paper reviews the early results from *Yohkoh*, first in the area of "quiet" coronal structure, and then for flares specifically. As implied in the title, the major objective here is to present these results in a manner that will help to place non-imaging observations, for example from *Compton/GRO*, in the context of the solar phenomena that give rise to the hard radiations.

CORONAL STRUCTURES

Quasi-Static Structures

The large-scale structures visible in Figure 1 include the well-known active regions and coronal holes, plus a not-very-well-investigated "general corona" apparently consisting of plasma trapped in closed magnetic fields. Here "closed" means re-entrant into the photosphere relatively near to the exit point. The general corona has a more amorphous appearance than the active-region corona, which consists of discrete bright loops. Finally, the "outer" corona, extending in Figure 1 to distances greater than 10^{11} cm above the solar surface, appears to fall off in brightness with a power-law dependence upon radial distance. This presumably represents the region in which the solar wind is formed on open field lines that extend outwards into the heliosphere. None of these large-scale structures have been analyzed thoroughly yet in the *Yohkoh* data, largely because their low intensity and diffuse character make it necessary to have exact knowledge of telescope properties such as point-response function, vignetting, spectral calibration, and scattering. Information in these areas is now quite comprehensive, and quantitative results are in progress.

The loops seen mainly at lower altitudes and in active regions represent regions of high gas pressure, which can vary by perhaps as many as four decades across the loop boundaries. The magnetic pressure presumably does not vary as rapidly with position because of the observed persistence of the structures, which must thus have a nearly magnetostatic or force-free character. What causes a particular loop (or perhaps more appropriately, 'channel') to have strong energy input and a resulting high pressure represents an important sub-question to the general problem of coronal heating. The *Yohkoh* data, in comparison with ground-based vector magnetograph data, show that the relationship between X-ray brightness and the distribution of vertical electric currents in an active region is not simple[3].

The compact active-region bright loops appear to arise in sunspot penumbrae, rather than in the stronger fields of the umbrae themselves[4]. Although analysis is at a preliminary level, it appears that the main heating of these loops is continuous, rather than episodic[5], consistent with the finding that "microflares" cannot support coronal heating without having a physical nature different from that of ordinary flares[6].

Dynamic Structures

The SXT data show a great deal of variability in the solar corona, seemingly on all spatial or temporal scales. The types of variability range from jets and other transient ejecta, which can occur on small spatial scales, to huge developments in large volumes of the polar coronal regions[7]. This variability is best appreciated from a movie representation of the SXT data, which are hard to represent in a simple manuscript. The SXT movie consists of composite images assembled from multiple exposure times to increase the dynamic range, typically several frames per 96-minute orbit of *Yohkoh*. The sampling restriction acts like a "slow-pass filter", but nevertheless many events resembling slow coronal mass ejections have been observed.

Figure 2 shows a remarkable jet event of 7 Dec. 1991, in which a blob of plasma rose out of a compact active region, moved along a high-lying magnetic loop (about 8×10^{10} cm)[8]. at a velocity of some 1000 km/sec through the corona, and then re-entered the photosphere and apparently causing a secondary soft X-ray brightening. Such remote brightenings have previously been inferred from Hα and meter-wave observations, but with soft X-ray observations we can trace out the entire structure in which the motions occur. The X-ray plasmas appear to be well-collimated in jets of this type, unlike the geometry that might be expected from field line diverging in a potential-field configuration from photospheric sources alone[9]. A movie representation may give the appearance of whip-like motions of some of the jet trajectories[10], and there are many interesting unanswered questions regarding the physics of phenomena of these types — in the event shown, for example, why is the subsidiary brightening at the N end of the long trajectory so energetic and so confined?

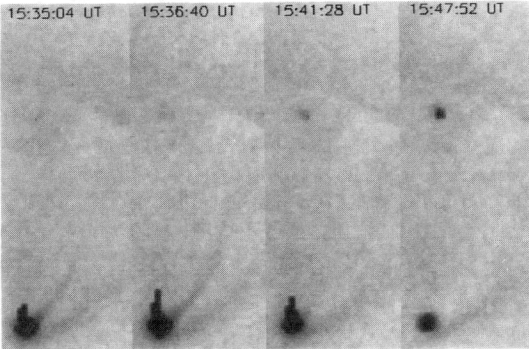

Fig. 2. Sequence of soft X-ray images showing the eruption of a jet from a flare-like brightening[8]. The ejected plasma followed a well-collimated trajectory, at a speed of approximately 1000 km/s, and produced a secondary brightening at the impact point near the N pole.

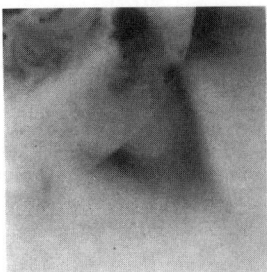

Fig. 3. X-ray helmet streamer formed after the eruption of a filament and a coronal mass ejection seen in white light[11].

On larger scales, SXT shows well–developed helmet–streamer structures, both in active regions (see the discussion of the 21 Feb. 1992 flare below) and in the aftermath of filament eruptions outside active regions[11], as shown in Figure 3. Finally, on the largest size scales observable with the SXT field of view of about 42 arc min (some 2×10^6 km), huge diffuse brightenings occur, often in the vicinity of the polar coronal holes[7]. In all of these dynamical effects, there is a general tendency to make interpretations in terms of magnetic reconnection, especially in view of the extremely suggestive helmet–streamer geometry.

The active–region loops in general appear to ignore magnetostatic constraints on intermediate time scales, and limb observations show[12] a general tendency for expansion at velocities of tens of km/sec. We conjecture that these active–region expanding loops may lead to solar wind formation outside the context of the classical Parker mechanism, *i.e.* in which the flow is perpendicular to the field lines; such a mechanism, if established, could help to explain the slow–speed streams of the solar wind.

FLARES

The *Yohkoh* results on flares are probably of greatest interest to the high–energy community. During a flare, *Yohkoh* switches to a high–time–resolution observing mode, with an SXT image cycle time through several filters typically on the order of 10–12 sec. The use of multiple filters allows the data to be used diagnostically, *i.e.* to characterize the temperature distribution in the soft X–ray sources. At the same time the flare mode enables observations by the *Yohkoh* hard X–ray imager, HXT, which observes in four energy bands (∼15–100 keV) with an array of 64 individual scintillation counters each viewing the Sun through a different image–modulating grid collimator.

In general the *Yohkoh* results confirm earlier results on flare morphology, with the important new dimensions of high time resolution in soft X–radiation. In hard X–radiation, the HXT instrument[13] provides great sensitivity, approximately equal to that of the HXRBS non–imaging instrument on *Solar Maximum Mission* . Its response at high energies definitely extends to the non–thermal bremsstrahlung domain, and it has better angular resolution than instruments on *Solar Maximum Mission* and *Hinotori*. The best–studied of the *Yohkoh* flares to date is doubtless the 3B/X1.5 event of 15 Nov. 1991[14−18], partly because of the excellent ground–based observations from Mees Observatory. This event exhibited most of the interesting features of a major flare, with the exception of a long–duration post–flare loop system. The papers cited will give a comprehensive view of the *Yohkoh* capabilities for flare observation in general.

The following sections discuss different aspects of the flare observations.

The 'Thick Target' Paradigm

The most significant finding thus far from the *Yohkoh* flare observations has been a rather complete confirmation of the 'thick target' paradigm for the energetic parts of flare development[19,20]. In this model, internal stresses in coronal magnetic–field structures relax, accelerating 10—100 keV electrons. These electrons fly to the chromospheric footpoints of the field lines, energizing the plasma there and causing evaporation. This results in high gas pressure in the corona and the observed soft X-radiation.

Several key new *Yohkoh* observations support this general picture:

- The non-thermal hard X-ray sources closely match the footpoints of magnetic loops defined by the soft X-rays[14];
- SXT commonly observes *impulsive* soft X-radiation from the footpoints of flaring loops, closely synchronized with non-thermal bremsstrahlung[21];
- White-light continuum, from the lower solar atmosphere, matches the impulsive phase quite well[22,16];
- The Neupert relationship between hard and soft X-rays[23] holds generally, even for "slow" LDE events[24].

These findings all support the ideas of chromospheric evaporation that were developed during the analysis of data from the *Solar Maximum Mission* and *Hinotori* satellites. These ideas were first clearly described by Neupert[23] and the main photometric signature has been termed the "Neupert Effect"[25]. Spectroscopically, the coincidence of blue–shifted X-ray emission lines with the impulsive phase established the upflow of chromospheric material into coronal trapping regions[26].

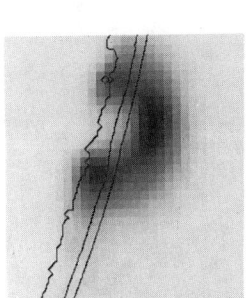
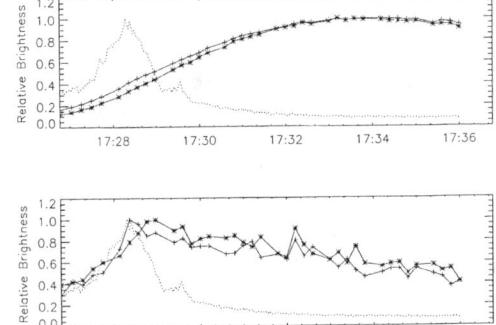

Fig. 4. (Left) Limb flare of 13 January 1992, early in its development, showing bright footpoint regions. Image dimension is 1.2×10^4 km. (Right) Light curves of soft and hard X-rays. The upper panel shows the SXT light curves from the whole flare and from the loop top, overlaid on hard X-ray photometric data from HXT (22.7–32.7 keV). The lower panel shows light curves from the footpoints, overlaid on the same HXT data. The comparison shows that the energization of the loop system begins with the footpoints.

Figure 4 shows the existence of impulsive soft X-ray emission at the footpoints of flaring loops. This phenomenon occurs commonly, and the soft X-ray footpoint emission appears to have a relatively cool thermal spectrum, consistent with the expectation for particle-driven evaporation. This finding essentially completes the observational picture of chromospheric evaporation, in the sense that SXT directly images the dense chromospheric plasma in the process of its upflow. The remarkable appearance of compact bright points at the tops of loops[27] remains unexplained in this picture, however, and this remains one of the interesting puzzles of the SXT flare observations.

The energetic importance of non-thermal electrons has been clearly recognized for powerful impulsive flares[28], as confirmed by the observations of impulsive soft X-ray footpoints just discussed. How about more gradual events? Among the most-cited results from the *Skylab* observations[29] has been the idea of post-impulsive-phase energy release, found specifically in large two-ribbon flares[30]. Is this also mediated by particle acceleration? What role is played by the filament eruption and coronal mass ejection, in flares they are associated with?

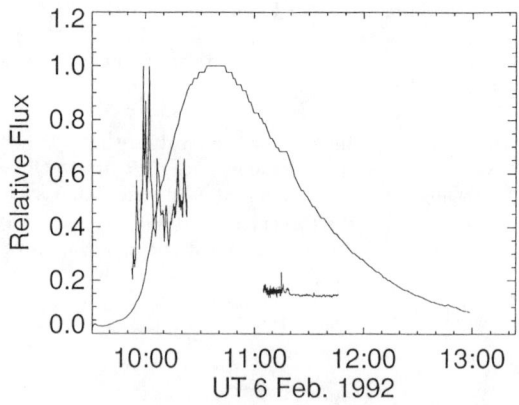

Fig. 5. Light curves of a "slow LDE" flare (6 February 1992), comparing the GOES soft X-ray flux (linear scale) with the hard X-ray photometric data from HXT. The hard X-ray burst lasted for more than 30 minutes, at a low flux level as expected from the Neupert effect. During the orbital gap in the *Yohkoh* data the large-area BATSE detector on board *Compton/GRO* continued to observe hard X-radiation[24] during the remainder of the rise phase of the event.

The comprehensive and sensitive observations of soft X-rays from *Yohkoh*, and of hard X-rays from the BATSE instrument on *Compton/GRO*, now make it possible to look more deeply for non-thermal effects in gradual events[24]. This has been done for a sample of "slow LDE events", one of which is shown in Figure 5. These events were chosen to have slow rise times, long durations, and no apparent impulsive-phase effects, in an effort to isolate as strongly as possible any intrinsically thermal energy release. As Figure 5 shows, non-thermal hard X-radiation does occur in these slow and apparently thermal events. We found that all of the slow LDE events selected had powerful acceleration of non-thermal electrons, just as in the more impulsive events. The fluxes are lower because of the longer time scales involved, consistent with the integral relationship of the Neupert effect. As Figure 5 shows, the hard X-ray emission coincides closely in time with the rise phase of the event in soft X-radiation, as expected.

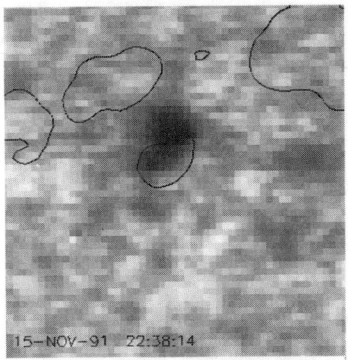

Fig. 6. White–light flare of 15 November 1991, as observed by the aspect camera of the *Yohkoh* SXT experiment[16]. Image dimension is 1.2×10^4 km. This negative image represents the difference between an image at the peak of the hard X–ray emission and a pre–flare image. The sunspots thus do not appear in the gray scale, but they have been drawn back in with the contours. All major flares observed to date with *Yohkoh* have been shown white–light continuum emission.

Finally, with respect to the thick–target paradigm, the SXT contributions to the observation of white–light flares are most important. The data come from the aspect sensor of SXT, a simple telescope with two–inch aperture normally viewing through a 30Å bandpass centered at 4308Å. We find all major flares observed by *Yohkoh* to be accompanied by emission in this passband. The emissions have a strong correlation with the impulsive phase and in particular with the hard X–ray burst. This is consistent with the thick–target paradigm[31] and implies that particle acceleration to the highest energies occurs during the epoch of major energy release in a flare, because high–energy particles have sufficient range to be able to penetrate to the height of formation of the continuum radiation.

Coronal Magnetic Structures and Reconnection

The SXT soft X–ray images give us our first systematic look, with reasonably high resolution, at the coronal structures of solar flares. Because the highly–ionized material of the hot coronal plasmas is tied to the magnetic field lines, these data promise a major step forward in our understanding of the role of magnetic reconnection in the generation of solar activity. In this section we discuss large flares with an eye towards understanding their geometrical properties. Each appears to consist of an arcade of large coronal loops, as expected from the *Skylab* data, and we see them in all orientations according to their locations on the disk. Often a single bright loop will dominate the appearance of the arcade in soft X–rays. The soft X–ray perspective on the dynamical development of these flare structures contains much of *Yohkoh*'s contribution to understanding macroscopic reconnection processes in solar flares.

The flare structure of the remarkable *Yohkoh* limb flare of 21 February 1992 is a good starting point for discussion of events of this type. This flare[32], seen in Figure 7, displays a helmet–streamer configuration that closely resembles flare models involving coronal neutral sheet formation[33] that we will refer to below as the "classical" picture of flare development, in which initially closed field lines open prior to the flare, and the flare energy release then originates in the reconnection of the open field lines. Cusped structures such as that seen in this flare commonly occur in the SXT images, although not always identifiably with a flare (see also Figure 3).

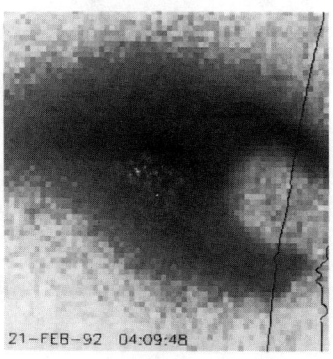

Fig. 7. Limb flare of 21 February 1992 as observed by the *Yohkoh* soft X–ray telescope[33]. This event, discussed in detail below, also was a "slow LDE" and exhibited long–enduring particle acceleration throughout the rise phase[24].

The geometry of this flare strikingly suggests magnetic reconnection in a coronal neutral sheet as a source of flare energy, but the other half of the classical picture is not obvious in the SXT data: there is no apparent tendency for the opening of magnetic field lines prior to the reconnection thought to drive the flare. In fact this active region provides a good example of the general expansion of active–region coronal flux tubes noted in the SXT data[12], with a relatively clear example at about 01:40 UT (some two hours before the flare) at speeds of about 30 km/sec. This opening could of course occur on relatively long time scales, involving magnetic structures with low gas pressure, and thereby escape easy detection by SXT. In the 21 February event, however, there is clearly no restructuring of the corona during the flare brightening itself, since the pre–flare configuration was well observed and had essentially the same geometry. Essentially the previously existing structure brightened in place, expanding gradually as is the pattern for Hα loop prominence systems. In particular, there is no evidence for inflow of the type required to drive energy release by reconnection in the classical model. The projected geometry of the event gives the clear appearance of closed field lines above the flaring loop, rather than open ones. These findings disagree substantially with the expectations from the classical model.

There is a speculative way in which we can rationalize the clear evidence for large–scale reconnection in a coronal neutral sheet, with the equally clear disagreements between the *Yohkoh* data and with the standard model of solar flares. This would be to hypothesize that the flare energy release actually takes place predominantly in closed magnetic field structures. The perturbations associated with the flare energy release then drive macroscopic reconnection, rather than the other way around. The key test of this hypothesis lies in analyzing the energetic relationships of the different components of solar flares, thereby tracking the energy released to its immediate origins and also revealing the mechanisms of its transport. This kind of quantitative analysis is still in a preliminary state in the *Yohkoh* data base, but much progress can be anticipated.

Finally, I would like to point out the attractiveness of using high–energy particles to study the connectivity of flaring plasma structures. The common occurrence of foot–point brightenings and the other paraphernalia of the thick–target paradigm for solar flare development clearly indicate that high–energy particles are usually present at the time of flare energy release. At exactly this time the plasma is undergoing some convulsion, whose details we really don't understand yet. The geometry of the field is central to our understanding, and an extension of the concept of *conjugacy* from magnetospheric physics to solar physics will be helpful. Essentially the high–energy particles can serve as a natural tool for discerning the connectivity of the magnetic field, via the matching of the footpoints of a given field line by use of simultaneity or correlation between the variations at the two locations. This tool is actually physically more correct than any technique based upon the extrapolation of photospheric field observations, or even from more direct coronal magnetic field measurements if they were possible. The gyroradii of the particles are far smaller than the angular resolutions of the magnetographs, for example. To make good use of observations of conjugacy, we need detailed modeling of the relevant observations of the type pioneered by Newkirk[34]. Such observations and calculations probably should start with slowly–developing large events, such as these "slow LDE" events, because of limitations of angular resolution.

CONCLUSIONS

The high–energy particles responsible for solar high–energy (hard X–ray and γ–ray emissions, as well as for non–thermal radio emissions, reside in the solar corona. Here they are accelerated, propagate, and radiate. With the *Yohkoh* data we now have soft X–ray observations that can show the coronal structures of the corona clearly, even on the disk. It is a natural prediction that we will be able to put these data together in a manner in such a manner as to be able to learn a great deal about acceleration, propagation, and radiation of high–energy particles. At present, unfortunately, little of this comparative work has been done in detail. In the future we expect a rich harvest of understanding from comparisons of SXT data with microwave and longer–wavelength radio data in particular. We are also waiting for the recurrence epoch of the historically great flares of August 1972 or April 1984 — these times translate to 1994 and 1995, respectively, in terms of a mean solar cycle duration. At this epoch we might hope to obtain simultaneous *Yohkoh* and *Compton/GRO* observations of γ–ray flares of similar significance.

The *Yohkoh* data by themselves, however, are already making important contributions to our understanding of flares and other dynamic structures in the corona, as are the *Compton/GRO* data. The most striking of these contributions, described in the preceding sections, confirms the relationship of high–energy particles and flare energetics that is implied in the Neupert effect and the general thick–target picture of flare evolution[19,20]. At the same time the soft X–ray observations from SXT clearly show helmet–streamer structures presumably associated with neutral sheets and magnetic reconnection. I have argued above that these structures probably do not supply the main energy of a flare, because of their passivity, but are instead driven by the flare. The difficult point in this argument is the well–known fact that coronal mass ejections are often associated with flares and may have large energy content[35], whereas they at least begin early in the flare process (as do filament eruptions, when they occur), well before the main energy release of the impulsive phase. The outstanding problem of *Yohkoh* flare research, therefore, seems to be to reconcile these different discoveries into a single self–consistent picture of what we normally think of as distinct problems: flare energy storage, triggering, energy release, and particle acceleration. On the observational side, the outstanding problem in interpreting the SXT images is the three–dimensionality of the source, and the corresponding difficulty of inferring the true geometry of the complicated optically-thin phenomena that we observe. As mentioned above, the concept of foot–point conjugacy introduced earlier may be helpful in clarifying the geometry and the physics dictated by it.

Acknowledgements. NASA supported this work under contract NAS 8–37334. The *Yohkoh* satellite is a project of the Institute of Space and Astronautical Sciences of Japan. I am grateful to all of my Japan, U.S., and U.K. colleagues for their participation in this endeavor, and especially to the Project Manager for ISAS, Prof. Y. Ogawara, who made it all work so well. I also thank Lidia Van Driel-Gesztelyi for critical comments on the manuscript.

REFERENCES

1. Tsuneta, S., Acton, L, Bruner, M., Lemen, J., Brown, W., Caravalho, R., Catura, R., Freeland, S., Jurcevich, B. Morrison, M., Ogawara, Y., Hirayama, T., and Owens, J., *Solar Physics* **136**, 37 (1991).
2. Acton, L., S. Tsuneta, Y. Ogawara, R. Bentley, M. Bruner, R. Canfield, L. Culhane, G. Doschek, E. Hiei, T. Hirayama, H. Hudson, T. Kosugi, J. Lang, J. Lemen, J. Nishimura, K. Makishima, Y. Uchida, T. Watanabe, *Science* **258**, 618 (1992)
3. Metcalf, T.W., Canfield, R. C., Hudson, H. S., Mickey, D. L., Martens, P. C. H., Tsuneta, S., in preparation (1993).
4. Sams, B. J., III, Golub, L., and Weiss, N. O., *Astrophys. J.* **399**, 317 (1992).
5. Tsuneta, S., and Kano, R., in preparation (1993).
6. Hudson, H. S., *Solar Physics* **133**, 357 (1991).
7. Tsuneta, S., Takahashi, T., Acton, L. W., Bruner, M. E., Harvey, K. L., and Ogawara, Y., *Publ. Astr. Soc. Japan* **44**, L211 (1992).
8. Strong, K. T., Harvey, K., Hirayama, T., Nitta, N., Shimizu, T., and Tsuneta, S., *Publ. Astr. Soc. Japan* **44**, L161 (1992).
9. Shibata, K., Y. Ishido, L. W. Acton, K. T. Strong, T. Hirayama, Y. Uchida, A. H. McAllister, R. Matsumoto, S. Tsuneta, T. Shimizu, H. Hara, T. Sakurai, K. Ichimoto, Y. Nishino, and Y. Ogawara, *Publ. Astr. Soc. Japan* **44**, L173 (1992).
10. Klimchuk, J. A., Lemen, J. R., Feldman, U., Tsuneta, S., and Uchida, Y., *Publ. Astr. Soc. Japan* **44**, L181 (1992).
11. Hiei, E., Hundhausen, A., and Sime, D., in preparation (1993).
12. Uchida, Y., McAllister, A., Strong, K.T., Ogawara, Y., Shimizu, T., Matsumoto, R., and Hudson, H.S., *Publ. Astr. Soc. Japan* **44**, L155 (1992).
13. Kosugi, T., Makishima, K., Murakami, T., Sakao, T., Dotani, T., Inda, M., Kai, K., Masuda, S., Nakajima, H., Ogawara, Y., Sawa, M., and Shibasaki, K., *Solar Physics* **136**, 17 (1991).
14. Sakao, T., T. Kosugi, S. Masuda, M. Inda, K. Makishima, R. C. Canfield, H. S. Hudson, T. R. Metcalf, J.-P. Wülser, L. W. Acton, and Y. Ogawara, *Publ. Astr. Soc. Japan* **44**, L83–L88 (1992).
15. Canfield, R. C., Hudson, H. S., Leka, K. D., Acton, L. W., Strong, K. T., Kosugi, T., Sakao, T., Tsuneta, S., Culhane, J. L., Phillips, A., and Fludra, A., *Publ. Astr. Soc. Japan* **44**, L111 (1992).
16. Hudson, H. S., Acton, L. W., Hirayama, T., and Uchida, Y., *Publ. Astr. Soc. Japan* **44**, L77 (1992).
17. Anwar, B., Acton, L. W., Hudson, H. S., McClymont, A. N., Makita, M., and Tsuneta, S., *Solar Physics* (to be published, 1993).
18. Wuelser, J.-P., Canfield, R. C., Acton, L. W., Culhane, J. L., Phillips, A., Fludra, A., Sakao, T., Masuda, S., Kosugi, T., and Tsuneta, S., *Astrophys. J.*(submitted, 1993).
19. Brown, J. C., *Solar Physics* **18**, 489 (1971).
20. Hudson, H. S., *Solar Physics* **24**, 414 (1972).

21. Hudson, H.S., Strong, K.T., Dennis, B.R., Zarro, D.M., Kosugi, T., Inda, M., and Sakao, T., submitted to *Nature* (1993).
22. Neidig, D., and Kane, S. R., *Solar Physics* (to be published, 1993).
23. Neupert, W., *Astrophys. J. Letters* **153**, L59 (1968).
24. Hudson, H. S., L. W. Acton, A. S. Sterling, S. Tsuneta, J. Fishman, C. Meegan, W. Paciesas, and R. Wilson, in preparation (1993).
25. Dennis, B.R., and Zarro, D.M., *Solar Physics*, to be published.
26. Antonucci, E., Gabriel, A.H., Acton, L.W., Culhane, J.L, Doyle, J.G., Leibacher, J.W., Machado, M.E., Orwig, L.E., and Rapley, C.G., *Solar Physics* **78**, 107 (1983).
27. Acton, L. W., Feldman, U., Bruner, M. E., Doschek, G. A., Hirayama, T., Hudson, H. S., Lemen, J. R., Ogawara, Y., Strong, K. T., and Tsuneta, S., *Publ. Astr. Soc. Japan* **44**, L71 (1992).
28. Lin, R. P., and Hudson, H. S., *Solar Physics* **50**, 153 (1976).
29. Sturrock, P.A. (ed.) *Solar Flares (A Monograph from Skylab Solar Workshop II)* Colorado Associated University Press (1980).
30. MacCombie, W. J., and Rust, D. M., *Solar Physics* **61**, 69 (1971).
31. Hudson, H. S., *Solar Physics* **24**, 414 (1972).
32. Tsuneta, S., Hara, H., Shimizu, T., Acton, L. W., Strong, K. T., Hudson, H. S., and Ogawara, Y., *Publ. Astr. Soc. Japan* **44**, L63 (1992).
33. Priest, E. R., *Phil. Trans. R. Soc. Lond. A* **336**, 363 (1991).
34. Newkirk, G., in R. Ramaty and R. G. Stone (eds.), *High Energy Phenomena on the Sun* (NASA SP-342), 453 (1973).
35. Kahler, S., *Ann. Revs. Astron. Astrophys.* **30**, 113 (1992).

ACCELERATION OF ELECTRONS IN SOLAR FLARES

Vahé Petrosian
Center for Space Science and Astrophysics
Stanford University, Stanford, CA 94305-4055

ABSTRACT

The significance of deviations of hard x-rays and gamma-ray continuum spectra from a simple power law is discussed in the framework of thick target nonthermal models. It is shown that in some flares such deviations can be explained neither by the radiative transfer effects nor by the effects of electron transport processes. Some of these deviations, therefore, must be attributed to the acceleration mechanism. The equations and various approximations involving stochastic acceleration by plasma turbulence is reviewed and it is shown that this acceleration process when combined with the expected collisional and synchrotron loss processes can account for these deviations.

INTRODUCTION

I would like to review recent progress in our understanding of the acceleration mechanism for electrons in solar flares based primarily on spectral observations of radiation during the impulsive phase of the flares. The radiations most directly associated with accelerated electrons are at microwave wavelengths, and at hard x-ray and gamma-ray energies (>20 keV). As is the case in most astrophysical studies involving nonthermal electrons, the initial observations were described by simple power law spectra. Because of the absence of an energy scale in such spectra, they shed little light on the acceleration process. However, with increasing spectral resolution and widening energy range of observations of flares, we are beginning to see various deviations from simple power laws. I will concentrate on these kinds of deviations and discuss their implications for the acceleration process.

I will begin with a brief review of the different processes and mechanisms that shape the radiation spectrum from accelerated electrons in flares.

RADIATION PROCESSES

There are only two primary mechanisms of radiation: *Bremsstrahlung* of accelerated electrons with the ambient ions gives rise to the continuum emission in the hard x-ray and gamma-ray range. In the gamma-ray range there is contribution from nuclear line excitation in the 1 to 7 MeV range, and from pion decay in the >50 MeV range. These radiations are produced by accelerated protons and will not be discussed here. The other mechanism is *synchrotron* which contributes to radiation at microwave frequencies up to $\simeq 100$ GHz. The emission of these radiations is well understood and they produce power law photon spectra from power law electron spectra.

RADIATION TRANSFER EFFECTS

Scattering and absorption of the emitted radiation becomes an important factor in shaping the observed spectra in some energy ranges. At low microwave frequencies (≤10 GHz) several absorption processes, including synchrotron self-absorption, can play significant roles depending on the ratio of the plasma frequency to gyro-frequency of the electrons which is related to the density of the plasma n and magnetic field B:

$$\zeta \equiv \omega_{pe}^2/\Omega_e^2 = \left(\frac{n}{10^9 \text{cm}^{-3}}\right)\left(\frac{100\text{G}}{B}\right)^2. \qquad (1)$$

The ambient plasma is optically thin at hard x-ray energies. A somewhat significant effect in this range is Compton back-scattering of photons directed into the photosphere. This could produce some enhancement of radiation in the 10 to 30 keV range, the amount of which depends on the pitch angle distribution of accelerated electrons[1,2]. On the other hand, very high energy photons (gamma-rays) are produced by higher energy electrons which can penetrate to lower depths in the solar photosphere. This depth depends on the pitch angle distribution of acclerated electrons and on the transport effects, such as trapping due to convergence of magnetic field of the flaring loop or pitch angle scattering by plasma waves. The total column depth an electron of energy E (in units of mc^2) traverses before a substantial loss of energy due to collision alone (for Coulomb logarithm $\ln\Lambda = 20$) is

$$N = 5 \times 10^{22} \text{ cm}^{-2} E^2/(1+E). \qquad (2)$$

For zero pitch angle electrons on a radial field line, this is the column of matter their emitted bremsstrahlung photons (of comparable energy) must cross to emerge out of the sun. Thus, for energies $E > 30$ (i.e. >15 MeV) the Thompson optical depth will exceed one and scattering or absorption of the photon must be included. Deviations from power law spectra are, therefore, expected at high energies.

PARTICLE TRANSPORT EFFECTS

The most important transport process is *Coulomb collision* of the accelerated electrons with the ambient plasma, which causes energy loss and pitch angle scattering with comparable rates. These effects clearly are dominant at x-ray energies and as a result give rise to nearly isotropic power law emission (so that back-scattering is not very important). At higher energies (>100 keV) pitch angle scattering diminishes and more of the photons are directed toward the sun. This, combined with the fact that bremsstrahlung radiation becomes more efficient, could give rise to deviation from power law spectra. At even higher energies and for larger magnetic fields *synchrotron losses* may become important and increase the energy loss rate. This can cause spectral steepening at higher energies. Magnetic *field convergence* can trap particles of all energies in the upper atmosphere altering the rate of energy loss and the radiation yield of the electrons.

Finally, pitch angle scattering due to *wave-particle interactions* can alter the spectrum by isotropising the electrons and the emitted photons even at higher energies. Unfortunately, there is less direct information on the density of plasma waves or turbulence as compared to plasma density, field strength and convergence. However, on theoretical grounds one can expect production of substantial amounts of turbulence during the flare process.

ACCELERATION MECHANISMS

Possible agents of acceleration of electrons are parallel (to magnetic field) electric fields, shocks and plasma turbulence[3]. Because the normal coronal Dricer field is small ($\sim 10^{-5}$V/cm), acceleration by *electric fields* requires anomalous resistivity. Acceleration to energy E would require enhancement of resistivity by a factor η over a region of size L such that $\eta L > 10^{13}$ cm($E/200$). It is therefore unlikely that electric field can accelerate particles to hundreds of MeV required by recent observations, some of which show flare emission up to 2 GeV. Electric fields could be important in accelerating electrons up to 100 keV, but simple theoretical arguments[4] and laboratory experiments[5] seem to show that particle distributions arising from acceleration by electric fields are unstable and that their energy is quickly dissipated by generation of plasma turbulence.

Acceleration by *shocks* requires presence of a strong shock and turbulence, the latter for pitch angle scattering. This gives rise to a rapid first order Fermi acceleration. However, there is no evidence for the existance of a single dominant strong shock for acceleration of impulsive phase electrons. Acceleration by many smaller shocks have also been proposed[6] but this is very similar to stochastic acceleration by turbulence and leads to similar spectra.

Stochastic acceleration by *plasma turbulence* is another possibility which would be present even if the above two processes are in operation, because they either lead to generation (or require the presence) of plasma turbulence. Turbulence could also be generated during the reconnection process which is believed to be the source of the energy of flares. We shall describe some features of this process below.

In what follows, after a brief description of the model for impulsive phase, I will discuss some gamma-ray spectra and general results which lead to the conclusion that acceleration plays an important role in shaping these spectra. I will then describe the acceleration due to stochastic processes.

MODEL

The model for impulsive phase nonthermal emission of solar flares, which has become standard by now, is the loop model. This is a thick-target, nonthermal model with particle acceleration occuring at the top of the loop in the corona. These particles, as they travel down the loop along the field lines, give rise to the observed radiations. The characteristics of the emerging radiations depend on the energy and pitch angle distribution of the accelerated electrons and on the variation along the loop of magnetic field, plasma particle density, and wave energy density and spectrum $W(k)$. The first quantitative analysis of this model, including pitch angle scattering and field convergence, was done by Leach and Petrosian[7] (see also J. Leach's thesis[8]). Over the years, we have continued to improve this model by inclusion of the various processes described in the preceding four sections. Comparison of model predictions with observations allows us to set some limits on these physical parameters.

For example, the spatial structure of microwaves[9], spectral variation of hard x-ray over time scales less than a second[10], and the centre-to-limb variation of x-ray and gamma-ray emissions[11] indicate that the field convergence is small. If $N = \int_0^s nds$ is the column depth as measured from the top along the length s of the loop, then the above result implies that in the coronal part of the loop $(d\ln B/dN) \times N_{tr} < 5$ or $b \equiv B_{min}/B_{max} \leq 5$ throughout the loop: Here N_{tr} is the column depth to the transition region. Most flares show two hard x-ray emission spots, presumably coming from the two footpoints of the loop, which indicates that electrons emitting these x-rays must penetrate below the transition region. From equation (2) for a 20 keV electron (pitch angle cosine $\mu = \frac{1}{2}$) we can set the limit $N_{tr} \leq 4 \times 10^{19}$ cm^{-2}. Furthermore, from observed optically thin microwave and x-ray fluxes and the spectral index[12] we can estimate the integrated magnetic field $\int B^2 ds \simeq 1.2 \times 10^{15}$ G^{2}cm. From these we estimate $B^2/n \geq 3 \times 10^{-5}$ G^{2}cm^3, and that the frequency ratio in 00equation (1) $\zeta \leq 3$. This also gives Alfven velocity divided by speed of light, $\beta_A \geq 0.04$. These factors play important roles in the interaction between plasma waves and energetic particles and therefore on the acceleration of electrons.

SPECTRAL DEVIATIONS

X-ray and gamma-ray bremsstrahlung spectra of flares show considerable variation from flare to flare and during the evolution of long-lasting individual flares. At first glance a power law fit seems adequate for most spectra, but this generally fails for spectra observed with high resolution or long dynamic range. Deviations from power law are observed both at high and low energies.

At the low end some spectra seem to steepen below a break energy which may range from tens of keV to one MeV. This is evident in the spectra of individual flares[13] and statistically in a large sample[14]. It turns out that transport effects cannot produce such deviations for a power law spectrum of accelerated electrons. This may be an indication of the presence of another emission source (hot or superhot[13] thermal component) or a sign of deviation of energy spectrum of the *accelerated electrons* from a simple power law. It should be noted that similar deviations from power laws are observed directly in the spectrum of interplanetary electrons[15] associated with flares which can be explained more readily by the second of the above two interpretations.

At the high end (> 1 MeV) the observed electron bremsstrahlung spectra are contaminated by nuclear line emission and pion decay gamma-rays which are produced by high energy protons. As expected, there is considerable variation in the relative contribution of protons and electrons in this energy range. The nuclear line emission is very weak in some flares, so-called electron dominated flares, so that the high energy continuum can be almost fully attributed to electrons[16]. In two such impulsive bursts (13:57 and 13:59 UT, 6, March, 1989) preliminary indications are that the gamma-ray spectra seems to fall off rapidly at above 40 MeV. In Figure 1 we compare the observed spectra for the first impulsive peak with expected spectra from several models. The solid lines show the expected bremsstrahlung spectra from a power law electron energy distribution injected at the top of a magnetic loop. Here we[17] have included transport effects due to Coulomb collisions, field convergence and synchrotron losses. First we note that there is some steepening at low energies due to transport effects but not to the degree observed. This is similar to our earlier[11] analysis of SMM data from GRS. Second we note that for reasonable magnetic field strengths synchrotron losses steepen the spectrum at the high energy end as well but not as rapidly as observed. Synchrotron losses (during transport) can increase the absolute value of the bremsstrahlung spectral index approximately by two units. Finally we note that an unreasonably large magnetic field $B_0 = 5000$ G is required to explain the low flux in the last channel but the spectrum for this model shows a poor fit at lower energies.

The dashed lines in Figure 1 show the optical depth effects which also fail to account for the rapid spectral steepening[17]. Even though higher energy photons are produced at higher depths, the Compton scattering optical depth remains small because of the decrease with energy of the Klein-Nishina cross section. Pair production on the other hand, can result in a large optical depth, with approximately linear increase with energy for >50 MeV photons. This can steepen the spectrum, but the spectral index increases only by one unit which is not sufficiently fast to account for the observations.

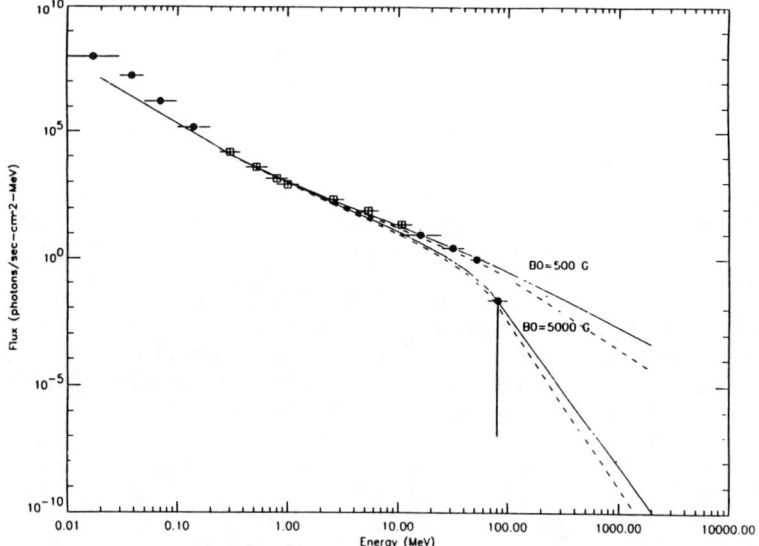

Figure 1. Spectrum of the first impulsive bump (13:57:12 to 13:58:50 UT) of the March 6, 1989 flare from Rieger and Marcshhäuser (1991). For clarity of presentation we have binned the data in the middle portion shown by squares. The solid lines show the whole loop spectra for a thick target model with a power law electron spectrum and a gaussian pitch angle distribution injected at the top of the loop with magnetic field B_0 and with a field convergence $b = 5$. The model parameters are chosen to obtain the best fit for points marked as squares. The dotted line shows the optical depth effects.

From this we conclude that the origin of the observed deviations from a simple power law may be characteristic of the acceleration process. We show below that in realistic models of acceleration of electrons by plasma waves, such deviations arise naturally.

STOCHASTIC ACCELERATION

Stochastic or second order Fermi acceleration, presumably by plasma turbulence, has been applied to the acceleration of protons in flares by various authors for more than a decade. In particular, Ramaty and his collaborators[18] have simplified application of this mechanism to determination of a single parameter, the so-called αT, which is a product of the acceleration rate α and escape time T. Similarly simple[19] and more complete models have been used for the study of acceleration of electrons by Schlikheiser and coworkers[20] and by Hamilton and Petrosian[21]. In what follows I will first describe the many approximations involved in obtaining some of the simplifications.

1) The most general equation dealing with transport of particles is the *Lionville or Beltzmann-Vlasov Equations*. The first useful simplification is obtained by *quasi-linear approximation* which leads to the *Fokker-Planck Equation* with energy loss rate $\dot{E}_L$ and the diffusion coefficients D_{ij} due to the interaction of electrons with plasma waves. In the presence of magnetic field only two phase space variables, say pitch angle cosine μ and momentum p (or energy E) are needed. The Fokker-Planck coefficients depend on the type and spectral and spatial distribution of plasma waves.

2) In the majority of situations it turns out that the pitch angle scattering rate, $D_{\mu\mu}$, is larger than the energy diffusion rate, D_{EE}/E^2, (proportional to α introduced above) approximately by $(\beta/\beta_A)^2$, when $\beta \leq 1$ is the accelerated particle velocity. As shown above for a flaring loop this factor could be as high as 10^3. Consequently, particle distributions will be nearly isotropic and we can use the *isotropic approximation* which leads to the *Diffusion-Convection Equation*[22,23]. This equation is described by three transport coefficients (say, d_{ss}, d_{sE} and d_{EE}) which are related to integrals over pitch angle of the Fokker-Planck coefficients.

3) The next, and not so easily justifiable simplification, comes from the assumption of *spatial homogeneity* (transport coefficients independent of spatial coordinates) and a finite size for acceleration region. This allows the equations to be integrated over the finite region so that the spatial diffusion can be approximated by a term describing the escape of the particles from this region which can be characterized by an *escape rate* or *escape time T*.

4) Another common assumption is neglect of the energy loss term $\dot{E}_L$. The expected losses (e.g. due to collisions or synchrotron emission) when considered are assumed to be occurring during the transport process in a separate region from the acceleration site. This clearly is not always justified.

5) And finally the last and even less justifiable assumption which leads to simple power law or Bessel function spectra characterized by the parameter αT is that the acceleration rate and the escape time or more generally their product $\alpha T \propto D_{\mu\mu} D_{EE}/E^2$ are independent of energy.

We have begun a systematic examination of the effects of the above assumptions. Before describing some of our progress in this area I would like to clarify a misunderstanding about acceleration by plasma turbulence. It is generally believed that plasma waves cannot accelerate electrons from the low energy background thermal plasma. This is based on the assertion that only electrons of velocities greater than $1836 v_A$ and $43 v_A$ can be accelerated by Alfven waves and Whistlers, respectively[26,27]. This is not always the case, even though it is true that it is difficult to accelerate lower energy than higher ones by certain plasma waves. In fact in a more detailed analysis[21] we show that lower energy electrons can be accelerated by high wave number Whistlers as we approach the electron cyclotron branches of the dispersion relation. Furthermore, in high magnetic field low density plasmas, where the ratio $\zeta \ll 1$, not only this possibility becomes more likely but other waves from other branches of the dispersion relation can also accelerate electrons from very low energies. Of course, whether or not these other waves or very short wavelength Whistlers can be excited by the energy release process is unknown.

Assuming that such agents of acceleration are present I now present some results arising from removing the restrictive assumptions of items 5 and 4 above.

Figure 2 shows some preliminary results from an on going work[24] where we have relaxed assumptions by including energy dependent *acceleration rate* and *escape time*. As evident both high and low energy steepenings can be obtained depending on the relative energy dependence of these terms.

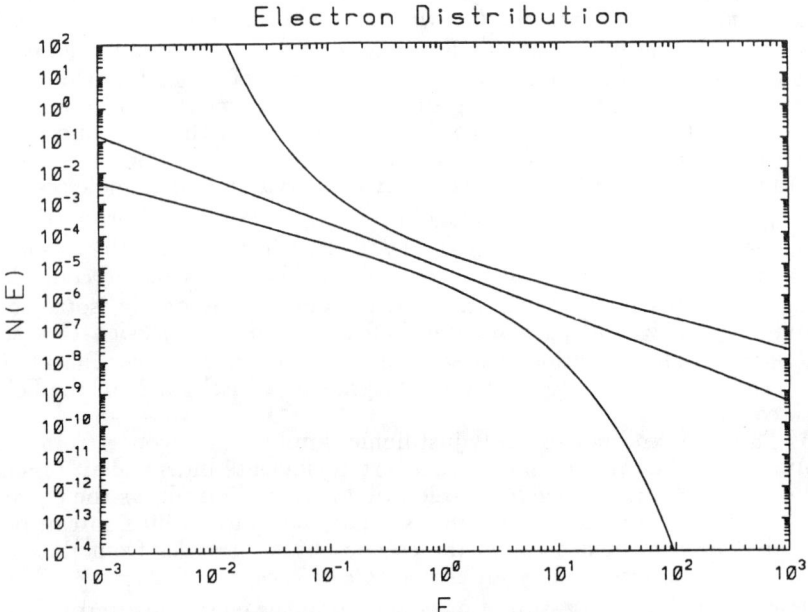

Figure 2. Steady state electron spectra obtained from stochastic acceleration of thermal electrons (see equation 9 in reference 21) with energy dependence diffusion coefficient $\alpha \propto D_{EE}/E^2 \propto E^{q-2}$ and escape time $T \propto E^s$. The three curves from top to bottom are for the hard sphere case ($q = 2$) and for positive, zero and negative values of the exponent s, respectively; relative normalization is arbitrary; $\dot{E}_L = 0$.

Next let us consider the effects of losses in the acceleration site versus those considered above during transport. In an earlier work we considered the effects of energy loss by Coulomb collisions[21] where we showed that for acceleration time scale of the order of one second, and for reasonable plasma densities, inclusion of this loss term leads to steepening of spectra at low energies. This can explain some of the low energy steepening of x-ray spectra mentioned above.

In our current work mentioned above[24] we have also shown that inclusion of *synchrotron losses* during the acceleration process can lead to an exponential cutoff of electron spectra at high energies. This could be the explanation for the steep cutoff in gamma-ray energies of the electron-dominated flares. Approximate analytic expressions describing effects of both these losses were also given by Beck et al[25]. Figure 3 shows the effects of these losses on the spectrum of the stochastically accelerated electrons.

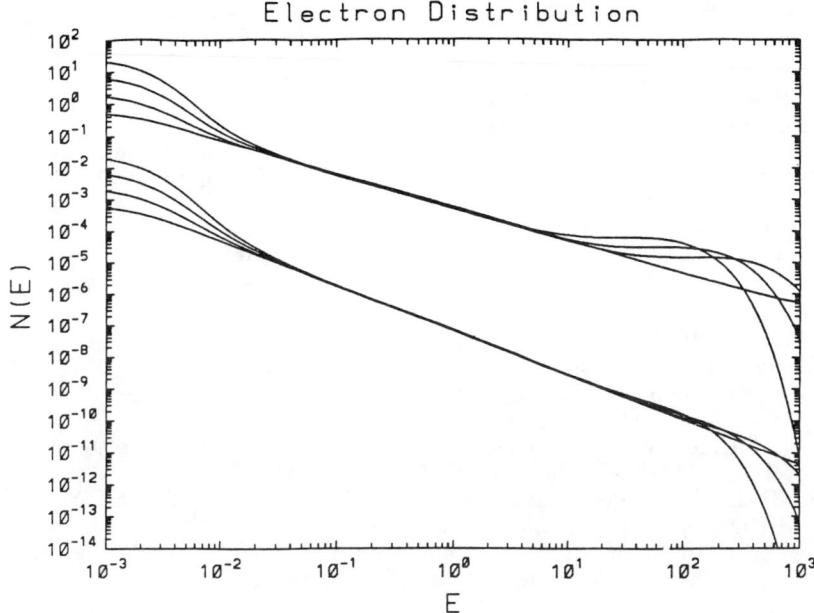

Figure 3. Steady state electron spectra obtained from stochastic acceleration of thermal electrons in the presence of Coulomb collisions and synchrotron losses and with a constant (independent of energy) escape time $T = 10s$ (top set) and $T = 1s$ (lower set). Note low energy steepening which increases with increasing plasma density (0,1,2,4 in some units) and high energy exponential cutoffs shifting to lower energies with increasing magnetic field (0,1,2,4 in some units).

SUMMARY

I have stressed here the importance of deviations from power law of x-ray and gamma-ray spectra during the impulsive phase of flares. Some of these deviations can only be produced in the acceleration region and during the acceleration process so that they provide important clues about the nature of the acceleration mechanism and the physical condition of the flaring plasma. We have shown that stochastic acceleration by plasma turbulence when combined with expected energy loss processes and more realistic assumptions can account easily for the rate and spectrum of the accelerated electrons. In the future we hope to examine the effects of the other assumptions as we climb up the ladder toward more realistic models.

Acknowledgment: Over the years, several graduate students' theses at Stanford have dealt with the details of the effects of all the processes described here in shaping the spectrum of the flare impulsive phase radiations. The students are S. Langer, J. Leach, J. McTiernan, E. Lu and R. Hamilton. I would like to thank them for their contributions. Special thanks go to Brian Park, a current graduate student, for production of Figure 2. This work is supported by NASA grant NAGW 1976 and NSF grant ATM 90-11628.

REFERENCES

1. S. H. Langer and V. Petrosian, ApJ 216, 666 (1977); see also S. H. Langer Thesis, Stanford University (1978).
2. T. Bai and R. Ramaty, ApJ 219, 705 (1978).
3. M. A. Forman, R. Ramaty & E. G. Zweibel, in Physics of the Sun, Vol. 2, eds. P. A. Sturrock, T. E. Holzer, D. M. Mihalas & R. K. Ulrich (Boston: Reidel, 1986) p. 249.
4. G. D. Holman, ApJ 293, 584 (1985).
5. J. P. Boris, J. M. Dawson & J. H. Orens, Phys. Rev. Lett. 25, 706 (1970).
6. V. Petrosian, in Dynamics of Solar Flares, eds. B. Schmieder and E. Priest. Observatoire de Paris, DASOP, 109 (1991).
7. J. Leach & V. Petrosian, ApJ 269, 715 (1983).
8. J. Leach, Ph.D. Thesis, Stanford University (1984).
9. V. Petrosian, ApJ Letters 255, L85 (1982).
10. E. T. Lu & V. Petrosian, ApJ 338, 1122 (1989).
11. J. M. McTiernan & V. Petrosian, ApJ 359, 541b (1990b); see also J. M. McTiernan Thesis, Stanford University (1990).
12. E. T. Lu & V. Petrosian, ApJ 354, 735 (1990); see also E. Lu Thesis, Stanford University (1990).
13. A good example of this was obtained by high spectral resolution observations of R. P. Lin, R. A. Schwartz, R. M. Pelling & K. C. Hurley, ApJ 251, L109 (1981).
14. W. T. Vestrand, D. J. Forrest, E. L. Chupp, E. Riger & G. H. Share, ApJ 322, 1010 (1987).
15. W. Dröge, P. Meyer, P. Evanson, & D. Moses, Sol. Phys. 21, 95 (1989).
16. E. Riger & H. Marschhäuser, in "MAX91/SMM Solar Flares, Observations and Theory" eds. R. M. Winglee and A. L. Kiplinger. Proc. of MAX91 Workshop No. 3 (Boulder, Colorado, 1991), p. 68; see also Marschhäuser et al in these proceedings.
17. V. Petrosian & J. M. McTiernan, submitted to ApJ, (1992).
18. See for example, R. J. Murphy & R. Ramaty, Adv. in Space Res. 4, No 7, 127 (1984).
19. J. A. Miller, N. Guessoum & R. Ramaty, ApJ 361, 701 (1990).
20. See for example, R. Schlickeiser, ApJ 336, 243 (1989) or J. Steinacker, W. Dröge & R. Schlikheiser, Sol. Phys. 115, 313 (1988).
21. R. J. Hamilton & V. Petrosian, ApJ 398, 350 (1992); see also R. Hamilton Thesis, Stanford University (1992).
22. L. J. Gleeson & W. I. Axford, ApJ 149, L115 (1967).
23. J. G. Kirk, R. Schlickeiser & P. Schneider, ApJ 328, 269 (1988).
24. B. T. Park & V. Petrosian, manuscript in preparation, (1993).
25. F. W. Beck, J. Steinacker & R. Schlickeiser, Sol. Phys. 129, 195 (1990).
26. D. B. Melrose, Instabilities in Space and Laboratory Plasmas (New York: Cambridge Univ. Press, 1986).
27. D. B. Melrose, ApJ 344, 973 (1986).
28. R. Dung & V. Petrosian, submitted to ApJ, (1993).

TEMPORAL EVOLUTION OF BREMSSTRAHLUNG-DOMINATED GAMMA-RAY SPECTRA OF SOLAR FLARES

H. Marschhäuser, E. Rieger and G. Kanbach
Max-Planck-Institut für extraterrestrische Physik
8046 Garching, Germany

ABSTRACT

SMM/GRS- observations of the initial impulsive phase of the solar flare on 6 March 1989 are well suited for the study of temporal and spectral variations of bremsstrahlung caused by relativistic electrons. The bemsstrahlung-dominated spectra of the impulsive bursts are unique among gamma-ray spectra due to the strong electronic continua above 1 MeV. By estimating the nuclear contributions from narrow de-excitation lines, we find that the bremsstrahlung can be roughly approximated by power laws with spectral indices as low as n=1.5. This continuum extends to higher energies showing an exponential cutoff at several tens of MeV. Between 0.3 and 0.8 MeV, the spectra are steeper and can be well described by power laws with indices between n=2.4 and n=3.0 The degree of spectral flattening, losing its significance at later bursts, seems not to be correlated with the intensity of the radiation < 1 MeV. Ions are accelerated throughout the whole flare, even well before the onset of the intense bursts. The typical line structure however is suppressed by the dominant continuum during the initial bursts.

INTRODUCTION

The study of gamma-ray line spectra of solar flares allow a unique insight into high energy processes at the site of flaring loops. The interactions of flare accelerated electrons and ions in the solar atmosphere produce a large variety of secondary particles such as excited nuclei, neutrons, positrons and pions. The resulting gamma-radiation below $\sim$ 8 MeV consists of a characteristic nuclear line spectrum superposed on a continuum of bremsstrahlung which extends up to the highest energy of the creating primary electrons. The decay of neutral pions produced in nuclear reactions also forms a characteristic spectrum peaked at about 70 MeV. Charged pions and directly measured solar neutrons can further complicate the interpretation of the emission > 10 MeV. Spectra of events with dominant nuclear lines [1,2], π°-emission [3] and bremsstrahlung from electrons [4] were reported. For details on the physical implications of the numerous solar flare observations the reader is referred to recent reviews [5,6].

To analyze the gamma-ray spectra one has to face two major problems: 1) the determination of the true photon distribution from the measured energy loss spectrum by means of a correct instrumental response model and 2) the unfolding of the complex superposition of the radiation components as discussed above. By using nuclear cross sections and by assuming different abundances for the target and the energetic particle beam [7,8], a forward-folding technique was applied successfully to the line spectrum of the flare on 27 April, 1981. It was shown that, even with the limited energy resolution of the SMM/GRS NaJ-scintillator, details in the elemental composition in target and beam can be discriminated.

To study such second order features, one needs to know the form of the underlying bremsstrahlung continuum to a high precision. As we will show in this paper, the dominant bremsstrahlung continua of the gamma-emission of the impulsive phase of the 6 March, 1989 flare can deviate strongly from simple power laws. Therefore, we analysed the spectra in a most model-independent way by applying a direct inversion technique.

ANALYSIS

The deconvolution of energy-loss spectra leads to matrix inversion problems widely known in all fields of astrophysics. Any applied technique must cope with the fundamental problem of numerical instabilities [9]. Most importantly, high frequent noise terms in count space must be kept in bay and should not appear in the recovered solution as spurious oscillations. The spectral resolution, given by the number and bin width of the data points in photon space, has to be traded for larger 1-σ uncertainity intervals. This leads to a reduction of the dimension in photon space compared to the oversampled 476 channel energy loss by at least a factor of 6. We found an inversion method which is based on the widely used singular value decomposition to be most useful [10,11]. For details on the applied iterative inversion see [12].

To include the SMM/GRS high energy matrix (HEM) data above 10 MeV, we calculate photon values by use of monte carlo calculations of [13]. The response model for the entire energy range up to 100 MeV is therefore not uniform. However work on such a model which differs slightly from the off-diagonal elements of the used UNH response (vers. 3.1) is under way [14].

Life-time measurements of some of the intense spectra reach values which are even below 50%. Such low values have a potential for gain change and pulse pileup. Although we did not find any sign for the latter, gain shifts were caused by a significant high voltage increase affecting slightly the structure of the nuclear lines during the later bursts (phases I3 and I4, see below). The derived line intensities were corrected for this instrumental effect. Due to strong gain shifts in the CsJ elements during the long life of the instrument we only use the NaJ elements.

RESULTS

Time profiles with high resolution [4] display a sequence of numerous individual bursts during the impuslive phases, the most exotic consists of coherent spikes on timescales of few seconds extending to energies up to 50 MeV. The individual data points in Fig. 1 (resolution of GRS-spectra limited to 16 s), starting at UT 13:56:23, are subdivided into the four impulsive burst-intervals I1-I4, a time interval of minimal intensity M and an end (decaying) phase E. The upper two panels show spectral indices and photon fluxes $\phi^B_{<0.8}$ of power law fits to the spectra between 0.3 and 0.8 MeV. The spectra, almost entirely dominated by bremsstrahlung, is well defined by a power law in this limited energy range. Above $\sim$ 0.8 MeV the spectra significantly hardened due to an increasing relative contribution from nuclear components and/or a flattening of the electronic continuum. We only use the extrapolation of the power laws as a line of reference for the analysis of the excess radiation above this line. At about 7.5 MeV the nuclear line spectrum posesses a cutoff due to the lack of strong lines above this energy [15]. Unless there is a strong production of charged pions which create by decay energetic positrons (and hence high energy bremsstrahlung), any significant excess between 7.5 and 8.5 MeV must be attributed to bremsstrahlung from primary electrons. The three panels in Fig. 1 therefore show the spectral characteristics of the bremsstrahlung. During the first burst I1, the peak excess flux $\phi^E_{7.5-8.5}$ of 1 Phot./(cm$^2 \cdot$ s) exceeds values expected from nuclear lines by two orders of magnitude which implies a strong flattening of the electronic continuum. Strong evidence was found that the short spikes (2 s) have similar spectral form indicative of a non-power law distribution of the electrons formed during the process of acceleration [12]. The indices of I1 and I2 are on the average lower than during the later bursts. Considering all phases, no sign of a significant correlation between $\phi^B_{<0.8}$ and $\phi^E_{7.5-8.5}$ could be found.

The deconvolved spectra of the time intervals I1, I3 and E are shown for the entire energy range between 0.3 MeV and 100 MeV in Fig. 2. For reasons of presentation, the spectra of I1 and I3 were multiplied by 10^4 and 10^2, respectively. The fit of the low energy power law (solid line), the extrapolation C$_1$ (dashed line) and an estimation of the electronic continuum C$_2$ (dashed and dotted) for I1 and I3 between $\sim$ 1.5 and 9 MeV are plotted

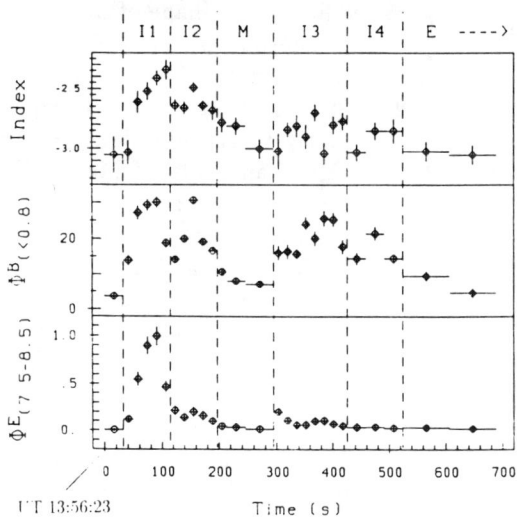

Fig. 1. *Time variation of the spectral indices, the fluxes below the fitted power laws $\phi^B_{<0.8}$ and excess fluxes $\phi^E_{7.5-8.5}$. Fluxes in units of Phot. cm^{-2} s^{-1}.*

(see also discussion of Fig.3). Although in the case of I1 and I3 the two HEM data points between 25 and 60 MeV are in good agreement with the high excess values below 10 MeV, the data point between 10-25 MeV appears to be somewhat too low for a featureless continuum of bremsstrahlung at this spectral range. However, one has to keep in mind that this maybe be due to the use of two separate response models for GRS and HEM data (see above). This also has to be considered when interpreting the last data point between 60 and 100 MeV in the spectra of I1 and I3. They indicate a very steep exponential cutoff. In contrast to the bremsstrahlung-dominated spectra of I1 and I3, the broad feature in the spectrum of E, peaking around 60 MeV, indicates contributions from π^0-decay. This is in good agreement with results of [16] who applied a more sophisticated analysis to account also for direct registered neutrons and bremsstrahlung from charged pions. The characteristic nuclear line spectrum of the time interval E is very pronounced in the excess spectrum (C_1-continuum subtracted) in the third panel of Fig. 3. Prominent de-excitation lines and a dominant neutron capture line at 2.22 MeV are visible. Note that the line flux depends on the bin width which is scaled logarithmically. Curve N shows the 'nuclear continuum' consisting of doppler-broadend and unresolved lines. The shape of this continuum corresponds to model spectra [7] used to analyse the flare of April 27, 1981. The intensity is normalized relativ to the line fluxes of Ne (1.63 MeV) and C (4.44 MeV). The fluxes for E are 0.27 ($\pm$ 0.10) and 0.22 ($\pm$ 0.10) $[Phot./(cm^2 s)]$ for the 1.64 MeV and 4.44 MeV lines, respectively. They are about a factor 2-3 lower than during I3. The ratios of the lines are relatively close to unity similar to results found for the flare on 27 April, 1981 [7]. Considering the crudeness of the applied method, the estimated continuum N agrees fairly well with the excess of E indicating that the excess is mainly produced by nuclear emission. The difference between a best fit and the continuum N is of the same

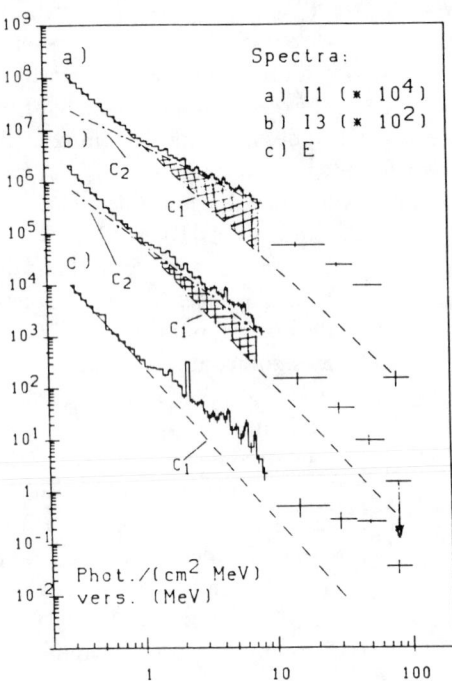

Fig. 2. Deconvolved GRS and HEM (NaJ only) data for the phases I1 (UT 13.958-13.976), I3 (14.026-14.062) and E (14.092-14.177). C_1: extrapolated power law fit to the data below 0.8 MeV, C_2: estimated bremsstrahlung continuum between $\sim$ 1.3 and 9 MeV.

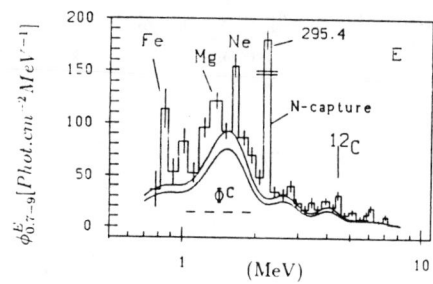

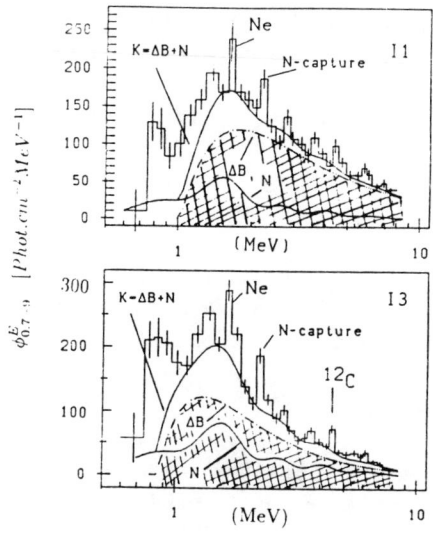

Fig. 3. *Excess (C_1 subtracted): Shown are 1) the estimated nuclear continuum N, 2) the excess due to bremsstrahlung ($\Delta B = |C_2 - C_1|$) determined from 3) the best fit under the prominent nuclear lines ($\Delta B + N$) above ~ 1.3 MeV.*

order as the estimated continuum ϕ^c from photospheric compton-scattered 2.22 MeV photons (for details see [17]). The estimated nuclear continua N in the excesses of I1 and I3 obviously cannot account for the total continuum. Note that, because of the very low line-to-continuum ratio during I1, only an approximate value for the Ne-line flux at 1.63 MeV of ~ 0.8 $[Phot./(cm^2 s)]$ could be determined. If we assume for the sake of simplicity an additional power law continuum C_2 which we attribute to bremsstrahlung, we can fit the total continuum successfully above ~ 1.3 MeV. The shaded regions in Fig. 3 and Fig. 2 give the difference $\Delta B = |C_2 - C_1|$ above 1 MeV. As can be seen from Fig. 3, the relative contribution of bremsstrahlung to the excess in the 4-7 MeV range, usually considered as a measure for nuclear emission, decreases from about 90 % during I1 to about 45 % during I3, reaching finally a value around zero during E.

CONCLUSION

We showed that the bulk of gamma-emission during the impulsive bursts of the flare of March 6, 1989 is produced by bremsstrahlung continua with spectral forms, that flatten at about 0.8 MeV and extend as the dominant radiation component to almost 100 MeV. The bremsstrahlung-dominated spectra differ from ordinary gamma-ray spectra because of this flat and intense electronic high energy continuum above 1 MeV. Since these spectra can be produced within a second or less, a very efficient and fast process is required to accelerate the electrons to highly relativistic energies. The ratio of relativistic electrons to line-producing protons is very high for these events. Assuming stochastic acceleration, the spectral form of the bremsstrahlung was explained by coulomb and synchrotron losses during the process of acceleration [18,19]. The latter loss process requires high coronal magnetic fields which are in conflict with observations. There remains other observational findings which are difficult to explain by theory: 1) all the individually analysed spectra flatten in a very limited energy range around 1 MeV ($\pm$ 200 keV), and 2) there is no direct correlation between the degree of spectral flattening and the intensity and spectral form at lower energies (< 1 MeV). Any successful model that includes acceleration, transport and interaction processes of the particles in flare loops ought to account for these detailed features in the bremsstrahlung-dominated spectra of the flare on 6 March 1989.

REFERENCES

1. D.J. Forrest and R.J. Murphy, *Solar Phys.* **118**, 123 (1988).
2. E. Rieger in Gamma-Ray Line Astrophysics, ed: P. Durouchoux und N. Prantzos, (AIP, New York, 1991), p. 421.
3. D.J. Forrest, in Positron Electron Pairs in Astrophysics, ed: M.L. Burns, A.K. Harding, and R. Ramaty (AIP, New York, 1983), p. 3.
4. E. Rieger and H. Marschhäuser, in MAX 91/SMM Solar Flares, Observations and Theory, ed: E.M. Winglee und A.L. Kiplinger, (Proc. of MAX 91 Workshop No. 3, Estes, Colorado, 1990), p. 68.
5. R. Ramaty and R.J. Murphy, *Space Sci. Rev.* **45**, 213 (1987).
6. E.L. Chupp, *Science*, **250**, 229 (1990).
7. R.J. Murphy, X.M. Hua, B. Kozlovski and R. Ramaty *Astrophys. J.* **351**, 299. (1990).
8. R.J. Murphy, R. Ramaty, B. Kozlovski and D.V. Reames *Astrophys. J.* **371**, 793 (1991).
9. J.C. Craig and I.J.D. Brown, in Inverse Problems in Astronomy, (Hilger, Bristol, (1986).
10. H. Marschhäuser and M. Boer, 21. *ICRC*, **4**, 192 (1990).
11. H. Marschhäuser, E. Rieger and G. Kanbach, 22. *ICRC*, **3**, 61 (1991).
12. H. Marschhäuser, Dissertation, Universität München (1993).
13. J.F. Cooper, et al., 19. *ICRC*, **5**, 474 (1985).
14. D.J. Forrest, private communication (1993).
15. C.J. Crannell, H. Crannell, and R. Ramaty, *Astrophys. J.* **229**, 762 (1979).
16. P.P. Dunphy and E.L. Chupp, 22. *ICRC*, **3**, 65 (1991).
17. W.T. Vestrand, *Astrophys. J.* **352**, 353 (1990).
18. F.W. Bech, J. Steinacker, and R. Schlickeiser *Solar Phys.* **129**, 195 (1990).
19. V. Petrosian, These Proceedings, (1993).

EGRET OBSERVATION OF THE JUNE 30 AND JULY 2, 1991 ENERGETIC SOLAR FLARES

B. L. Dingus, P. Sreekumar
Universities Space Research Association, NASA/GSFC, Code 662,
Greenbelt, MD 20771

D.L. Bertsch, C.E. Fichtel, R.C. Hartman, S.D. Hunter, D.J. Thompson
NASA/Goddard Space Flight Center, Code 662, Greenbelt, MD 20771

E.J. Schneid
Grumman Corporate Research Center, Bethpage, NY 11714

K.T.S. Brazier, G. Kanbach, C. von Montigny, H.A. Mayer-Hasselwander
Max-Plank Institut fur Extraterrestrische Physik, 8046, Garching bei Munich, Germany

Y.C. Lin, P.F. Michelson, P.L. Nolan
Stanford University, Stanford, CA 94305

D.A. Kniffen
Hampden-Sydney College, P.O. Box 862, Hampden Sydney, VA 23943

J.R. Mattox
COMPTON Science Support Center, Computer Sciences Corporation,
Greenbelt, MD 20771

ABSTRACT

The EGRET experiment on board the Compton Gamma Ray Observatory observed energetic gamma rays from the impulsive solar flares on June 30 and July 2, 1991. The June 20 and July 2 spectra were measured by the large NaI spectrometer in EGRET and extend up to nearly ~50 MeV and ~10 MeV respectively before appearing to cut off. These spectra can be fit by a power law model or an isothermal plasma model.

INTRODUCTION

Shortly after the launch of the Compton Gamma Ray Observatory (CGRO) in April of 1991, a period of intense solar activity occurred during the May, June, July time period. Two X-type flares occurred in the first week of June prompting the action of making the Sun a target of opportunity for CGRO and therefore all CGRO instruments were pointed at the Sun when three additional X-type flares occurred during the second week. All five X-type flares showed nuclear excitation lines in their spectra and had extended emission of high energy radiation [Schwartz et al.[1], Akimov et al.[2], Kanbach et al.[3], Ryan et al.[4], Bertsch et al.[5]]. After the active region emitting the X-type flares moved behind the limb, two subsequent M-type flares occurred on June 30 and July 2, 1991. These flares were approximately 70 degrees off the EGRET axis precluding measurement of these flares by the EGRET spark chamber

© 1994 American Institute of Physics

telescope (its FOV is approximately ±30 degrees) but allowed measurements to be made by the EGRET large NaI spectrometer.

The EGRET instrument consists of a spark chamber with interleaved tantalum plates to convert gamma rays to electron position pairs, and to image the trajectories of the pair. A NaI scintillation spectrometer measures the energy of pair as it emerges form the bottom of the spark chamber telescope. A large anticoincidence shield to reject charged particle events surrounds the spark chamber but does not extend to cover the NaI spectrometer. The NaI spectrometer (76x76x20 cm) has a special burst/flare mode for recording gamma ray bursts and solar flares. In this mode a pulse height spectrum is accumulated for all NaI events from 1 MeV to approximately 200 MeV. Spectra are routinely accumulated every 32.75 seconds but when activated by a BATSE burst trigger, EGRET accumulates 4 sequential spectra with integration times variable from 1 to 16 seconds. Details of the instrument and calibration can be found in Hughes et al.[6], Kanbach et al.[7], and Thompson et al[8].

HIGH ENERGY FLARE TIME HISTORIES

The time histories for the June 30 and July 2 solar flares were obtained with the anticoincidence dome and the NaI spectrometer and are shown in figures 1 and 2 respectively. Each figure contains three plots. The top plots show the time histories observed in the anticoincidence dome having an energy threshold of <50 keV and time resolution of 256 msec. The integrated signals in the NaI for 1 MeV (threshold) up to 10 MeV (are shown by the middle plots) and for 20 to 50 MeV (shown by the lower plots) have a time resolution of the NaI solar mode of 32.75 seconds. The June 30 flare occurred just before the earth occultation for the Observatory which can be seen by the sharp drop in the anticoincidence count rate. The BATSE experiment on the observatory observed an impulsive high energy peak than was shorter in duration than the low energy emission (Schwartz et al.[1]). For the July 2 flare, it can be clearly seen by comparing the anticoincidence count rates with the NaI spectrometer count rates that the high energy emission is impulsive and shorter in duration compared to the low energy emission detected with the adome.

HIGH ENERGY FLARE SPECTRA

Background subtracted spectra were obtained for the time integrated peak gamma ray emission for June 30 and July 2. Photon spectra for these flares were obtained using the calculated response functions for the NaI spectrometer for the direction to the flare. These calculations used the detailed CGRO mass model and the EGS4 transport code. The spectra for June 30 and July 2 are shown in figures 3 and 4 respectively. Spurious data around 12.5 MeV generated by the EGRET electronics were removed from these plots. The spectra were fit by a power law model, a thermal model for emission from an isothermal plasma (Crannell et al.[9]) and an exponential model for particles accelerated near a magnetic x line (Bruhwiler and Zweibel[10]). The fits for these models are shown in figure 3 and 4. Both the power law fits and thermal fits can explain the data whereas the exponential fit expected for particles accelerated by an electric fields parallel to the x line does not.

No strong evidence is observed for gamma ray excitation lines in the 4 to 7 MeV region or the neutron capture line at 2.223 MeV in either spectra that would indicate proton acceleration in those flares. The power law results for the June 30 flare are consistent with the BATSE results. The BATSE June 30 data discussed in Schwartz et al.[1], that indicated the high energy emission had to have a spectral index of

<3 in order for the high energy impulse peak to be observed at high energies and not at low energies. The hard impulsive spectra observed are characteristic of the events reported above 10 MeV by Rieger and Marschhauser[11] which they have designated as electron dominated flares. Even though the thermal model fits the data well, Kosugi, Dennis and Kai[12] correlated hard x-rays and microwaves from energetic electrons in impulsive flares and found that the thermal model of Crannell et al.[9] could not be used to explain the correlation between GHz microwaves and the hard x-rays but precipitating electrons with a power law dependence might better explain the microwave data.

CONCLUSIONS

The flares measured on June 30 and July 2, 1991 are impulsive in their high energy emissions and appear to be electron dominated as evidenced by their time histories gamma ray spectra and no strong evidence of gamma ray lines. The high energy emissions (> MeV) are shorter in duration than the overall emissions in the hard x-ray region observed by the anticoincidence shield counter.

The EGRET team gratefully acknowledges support from the following: Berndeministerum fur Forschung und Technologie, grant 50 QV 9095 (MPE); NASA grant NA65-1742 (HCS); NASA grant NAG 5-1605 (SU); and NASA contract NAS 5-31210 (GAC).

REFERENCES

1. Schwartz, R.A., et al., Proc of the Compton Observatory Science Workshop, 457, NASA Conference Pub. 3137 (1992)
2. Akimov, V.V., et al., 22nd International Cosmic Ray Cont. 3, 73(1991)
3. Kanbach, G., et al., Astron. Astrophys Suppl. Ser. 97, 349 (1993)
4. Ryan, J.M., et al., Proc of the Compton Observatory Science Workshop, 470, NASA Conference Pub. 3137 (1992)
5. Bertsch, D.L., to be published (1993)
6. Hughes, E.B., et al., IEEE Trans. Nucl. Sci., NS-27, 364 (1988)
7. Kanbach, G., et al., Proc. of the Gamma Ray Observatory Workshop, 2-1 (1989)
8. Thompson, D.J., et al., ApJ Suppl. Ser. 86, 629 (1993)
9. Crannell, C.J., Frost, K.J., Matzler, C., Ohki, K., and Saba, J.L., Ap. J., 223, 620 (1978)
10. Bruhwiler, D.L., and Zweibel, E.G., J. Geophys. Res. 97, 10825 (1992)
11. Rieger, E., and Marschhauser, H., Proc. of the Third Max '91 Workshop (R.M. Wingler and A.L. Kiplinger, ed.), 68 (1990)
12. Kosugi, T., Dennis, B.R., and Kai, K., Ap. J., 324, 1118 (1988)

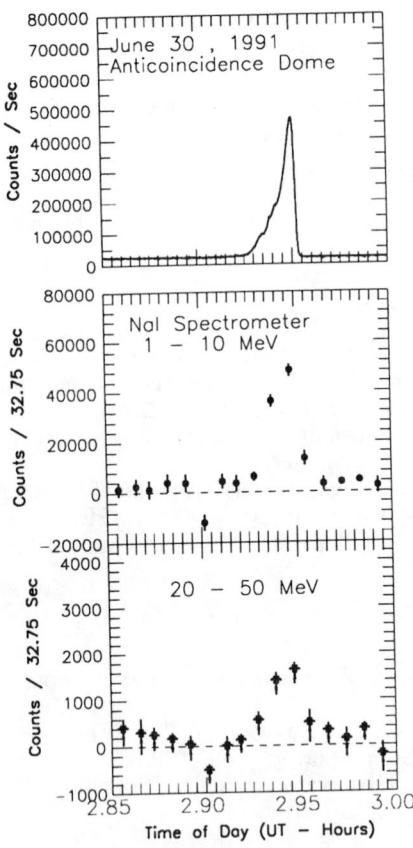
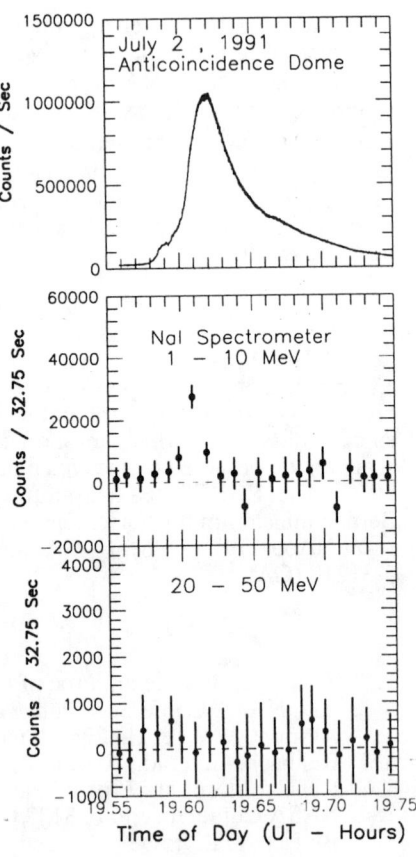

Figure 1 Time histories for the June 30, 1991 solar flare. The top curves shows a plot the count rate in the anticoincidence dome over the period of the flares. The energy threshold for the dome is <50 keV and time resolution is 0.256 seconds. The middle and bottom curves are the pulse height events in the EGRET NaI spectrometer for 1-10 MeV and 20-50 MeV respectively with a time resolution of 32.75 second. The drop in intensity at 2.96 hours corresponds to the sun being obscured by the earth.

Figure 2 Time histories for the July 2, 1991 solar flare. See figure 1 caption.

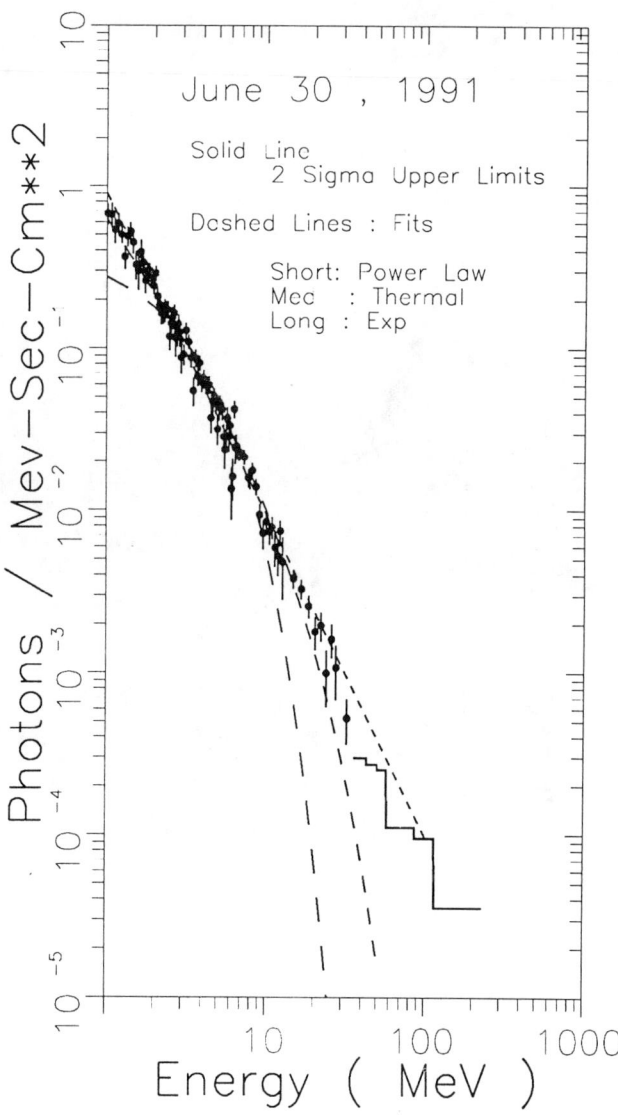

Figure 3   The energy spectrum for the June 30, 1991 solar flare. The data were fit with a power law model (short dashed line), spectral index (-1.98 ± 0.03), a thermal model (medium dashed line), kT = 9.75 ± 0.05 MeV, and an exponential model (long dashed line), kT = 2.30 ± 0.01 MeV.

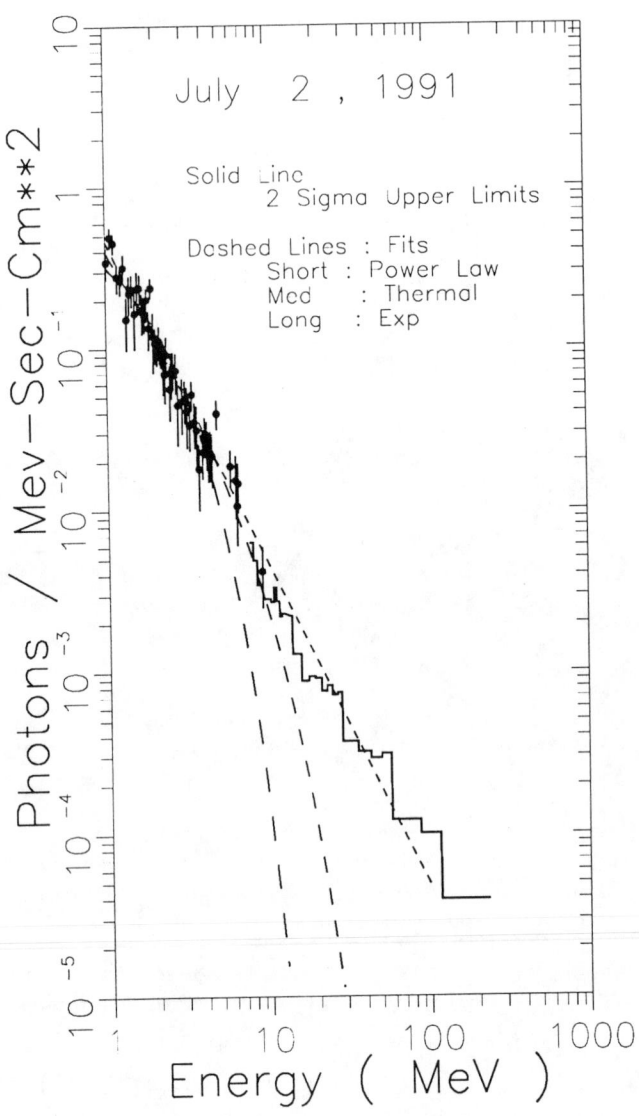

Figure 4 The energy spectrum for the July 2, 1991 solar flare. The data were fit with a power law model (short dashed line), spectral index = (-1.99 ± 0.07), a thermal model (medium dashed line), kT = 4.65 ± 0.74 MeV, and an exponential model (long dashed line), kT = 1.16 ± .83 MeV.

OBSERVATIONS OF SMALL SOLAR FLARES WITH BATSE

D. A. Biesecker, J. M. Ryan
University of New Hampshire, Space Science Center, Durham, NH 03824

G. J. Fishman
NASA / Marshall Space Flight Center, ES-62, Huntsville, AL 35812

ABSTRACT

The Burst and Transient Source Experiment on board the Compton Gamma Ray Observatory is being used to observe solar flares. The Large Area Detectors are sensitive to small solar flares. We are searching the BATSE data for solar flares with an automated algorithm that allows for independent confirmation of the event origin. With this search method, we have detected solar flares almost an order of magnitude smaller than those found in a visual search of the BATSE data. We present results that are consistent with the differential distribution of peak flare rates observed by other researchers. These results show that the rate of occurrence of the smallest flares observed by BATSE can be predicted from the rate of occurrence of larger flares.

INTRODUCTION

The study of solar flares is important in understanding energy release and the acceleration and transport of particles on the sun. One of the difficulties in studying solar flares is their complexity. By studying smaller flares, the interpretation of the data is much simpler. A *microflare* is an ideal event for investigating the basic aspects of the flare process. Microflares have been proposed as the basic component of solar flares by van Beek et al.[1] and de Jager and de Jonge.[2] They have also been suggested to be a source for heating the solar corona, based on the ideas of Gold.[3]

Earlier researchers have observed hard X-ray flares for extended periods of time.[4,5] All have found the differential distribution of peak flare rates to be well fit by a power law with a slope of ~ -1.8. The importance of this slope has been outlined by Hudson.[6] This slope implies that microflares cannot contribute significantly to the total power in all solar flares. In order to heat the corona the distribution of smaller events must be steeper than -2, assuming that flare energy scales with the peak X-ray flux.

Here the Burst and Transient Source Experiment (BATSE) on board the Compton Gamma Ray Observatory (GRO) is used to observe solar flares. The BATSE Large Area Detectors (LADs) allow for the most sensitive long term study of microflares ever conducted. The BATSE instrument also has the advantage of allowing for independent identification of source location.

INSTRUMENTATION

The BATSE instrument consists, in part, of eight uncollimated Large Area Detectors, placed on the corners of the spacecraft. The LADs are arranged so that the faces of the detectors are parallel to the faces of a regular octahedron. Every point is viewed by four detectors. The geometry of the detectors can be used to locate a source.

The LAD data are divided into four energy channels: 25-50, 50-100, 100-300, and >300 keV. The data have a time resolution of 1.024 seconds. Only the first energy channel is used in this study in order to maximize the signal-to-noise ratio. The first data used in this study begin 11 May 1991.

SOLAR FLARE SEARCH

Events are identified by a computer algorithm that automatically searches the data for count rate increases. The algorithm then uses the characteristics of the instrument to independently identify events of solar origin. The algorithm is used to accumulate a large database of microflares to study the characteristics of microflares.

The automated algorithm calculates the second difference in time of the count rate data, thereby locating peaks or bursts.[7] The four solar-facing detectors are searched concurrently. The ratio of the second difference at a local maximum to the statistical error in the second difference at that point defines the significance of the peak in units of standard deviations. A significance of at least 3σ in one BATSE detector or of at least 2.5σ in two detectors identifies peaks that undergo further analysis. There is no characteristic time scale or structure that identifies all flares, so the data are smoothed over neighboring data points on four different time scales. This allows the algorithm to pick out the weak, rapid events as well as the long, smoothly varying events.

Once significant count rate increases in the data have been detected, it is necessary to identify the sources of the increases. After background subtraction in each detector, the residual total counts in each detector are computed. Knowing the Sun's position, the counts in each detector are corrected for its angular response. A solar flare should then have equal corrected counts among the solar-facing detectors. An additional correction must be made because the BATSE detectors are not uniform in their response in the lowest energy channels. This correction is calculated by using known flares, confirmed by other spacecraft, to determine the ratio of counts between BATSE detectors expected for true solar events. The correction is applied to the unknown events, which for solar flares are then expected to have the same count rates in the solar-facing detectors.

DATA ANALYSIS

All significant count rate increases are stored in a database. The information contained in the database includes the time, peak rate, total counts in the search window, background rate, and the second difference of each peak. Data from all four solar-facing detectors are included. One can change the criteria that must be met in order for an event to be tagged as solar in origin.

Since a flare may have many local count rate maxima, the algorithm may detect a single flare several times. In addition, each day is searched with four different smoothing averages, thus resulting in the same event being detected up to four times. The data from all four searches is combined and all of the events that fall within 2 minutes of the previous event are considered to be part of the same flare, minimizing double counting of events. The largest flares are checked individually.

RESULTS

The differential distribution of peak flare rates is plotted in Figure 1 for the data processed to date. The peak rates are taken from the most solar facing detector and only corrected for angular response. The solid line is a power-law fit with spectral index of -1.79 ± 0.04. This finding is consistent with earlier results, including a visual search for solar flares in the BATSE data.[6,8] However, the automated algorithm extends the distribution plot down to peak flare rates a factor of three smaller.

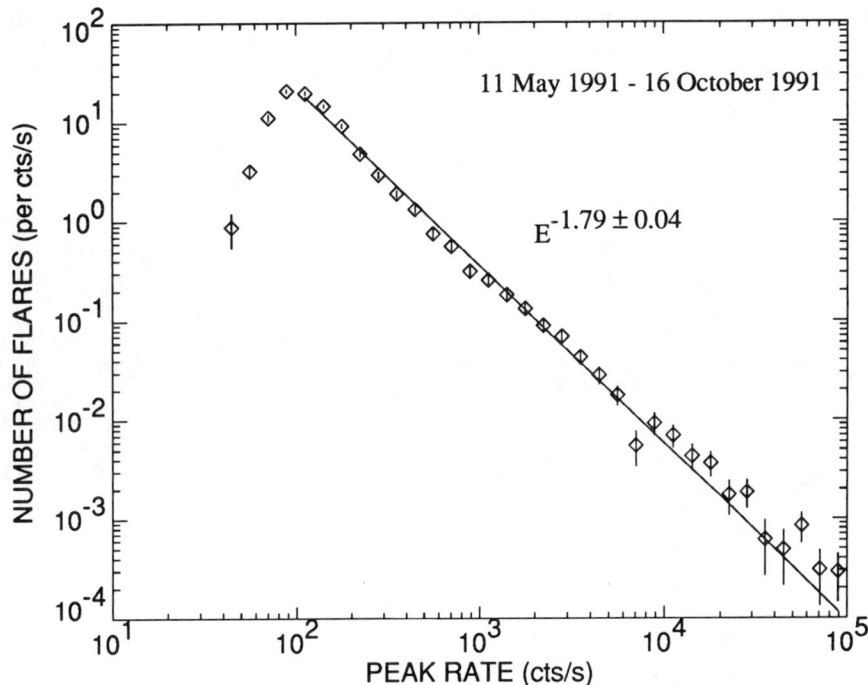

Figure 1. The differential frequency distribution of BATSE peak flare rates is plotted.

Other factors may affect the results. These are the nonuniform energy ranges of the detectors, the sensitivity of the detectors as a function of the area projected normal to the Sun's direction, and the sensitivity of the detectors as a function of the background rate.

From 11 May 1991 to 16 October 1991 the GRO spacecraft was oriented in fifteen different directions. Due to the orientation changes, four separate detectors were the most solar facing detector at some time. These detectors have a nonuniform energy response and so have different sensitivities to solar flares. The detector that can detect the lowest energy photons is then able to detect smaller flares than the other detectors. Figure 1 includes data summed over all orientations of the spacecraft. Plotted in Figure 2 are the data from the same period, but the distribution of events is broken down according to the most solar-facing detector. To illustrate this effect only two detectors are included. It is clear that the turnover at small flux values occurs at different peak rates. Thus, when the data from all four detectors is summed, as in Figure 1, the turnover in the distribution occurs at peak rates that are larger than the smallest peak rates that can be observed with BATSE.

Another factor affecting sensitivity is the projected area of a detector. As the angle between the solar direction and the detector normal increases, the sensitivity of the detector decreases. The projected area of the most solar-facing detector changed with the different orientations of the spacecraft. The differences in sensitivity due to projected area will also result in the turnover in the peak rate distribution occurring at larger peak rates.

The limiting sensitivity of the detectors is also affected by the background rate, which varies between about 800 and 2200 s^{-1}. A 3σ detection at the highest background rate corresponds to a flare with a peak rate of 140 s^{-1}. Flares above this rate can be detected independent of the background. This is the rate at the turnover as seen in Figure 1.

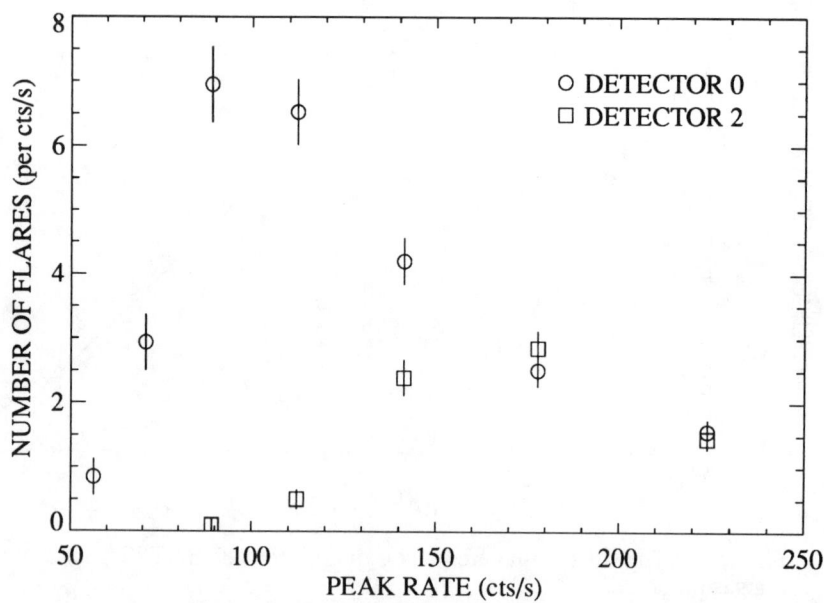

Figure 2. The differential distribution of BATSE peak flare rates for two different solar-facing detectors.

REFERENCES

1. H. F. van Beek, L. D. de Feiter, and C. de Jager, Space Research XIV, 30 (1974).
2. C. de Jager and G. de Jonge, Sol. Phys. 58, 127 (1978).
3. T. Gold, in The Physics of Solar Flares (NASA SP-50), edited by Wilmot N. Hess, 389 (1964).
4. D. W. Datlowe, M. J. Elcan, and H. S. Hudson, Sol. Phys. 39, 155 (1974).
5. B. R. Dennis, Sol. Phys. 100, 465 (1985).
6. H. S. Hudson, Sol. Phys. 133, 357 (1991).
7. M. A. Mariscotti, Nucl. Instrum. Methods 50, 309 (1967).
8. R. A. Schwartz, B. R. Dennis, G. J. Fishman, C. A. Meegan, R. B. Wilson, and W. S. Paciesas, in Proceedings of The Compton Observatory Science Workshop (NASA CP-3137), edited by Chris R. Shrader, Neil Gehrels, and Brian Dennis, 457 (1992).

ENERGETIC ELECTRON INJECTION INTO THE HIGH CORONA DURING THE GRADUAL PHASE OF FLARES: EVIDENCE AGAINST ACCELERATION BY A LARGE SCALE SHOCK

Karl-Ludwig Klein, Gérard Trottet
Observatoire de Paris - Section de Meudon, DASOP
F-92195 MEUDON PRINCIPAL CEDEX

ABSTRACT

Large scale coronal shock waves are observed by their radio signature associated with many solar flares. In this contribution it is demonstrated that such coronal shock waves are inefficient accelerators of electrons observed both in the downstream region (<1 R_0 above the photosphere) and the upstream region (>10 R_0), as compared to other, though unspecified, mechanisms which act in the lower solar atmosphere.

INTRODUCTION

Radio bursts of narrow bandwidth which slowly drift from high to low frequencies in the decimetre-to-decametre wave band are generally considered as tracers of large scale, MHD-like shock waves propagating from the middle to the high corona. Such "type II bursts" are most often observed in association with flares, after the impulsive phase (e.g. /1/). The electromagnetic signature arises at once or twice the local plasma frequency, but it is not yet clear whether the Langmuir waves are generated by electron streams, by a cloud of accelerated electrons which moves with the shock front, or whether it is due to a mixture of different particle populations. The occasional detection of fast drift bursts emanating from the type II band indicates nevertheless that electron streams are produced in the vicinity of the shock front.

Although the electrons which ultimately generate the radio emission are not very energetic, the coronal shock wave was in the early times supposed to be the accelerator of relativistic electrons and protons which accompanied large flares (e.g. /2,3/). More recently these shocks have been claimed to accelerate electron streams which emit bright hectometre radio emission at about 10 R_0 above the photosphere /4/. These studies were based on the statistical association of the energetic phenomena with type II radio bursts.

In the following we summarise and extend our former work on the comparison of the time histories of electromagnetic emissions over a large spectral range in order to investigate relevant associations between energetic electrons injected into various regions of the solar corona and the type II burst, which is used as a tracer of the coronal shock. It is recalled in the next section that the available observations of gradual hard X-ray/radio bursts point to a different site of particle acceleration than the vicinity of the shock wave. This is well accepted for electrons in the downstream region, i.e. the low corona. Then the time profiles of hectometre radio emission and radio emission at shorter wavelengths as well as X-ray emission are compared. This observational evidence suggests that the electrons radiating hectometre emission at altitudes above 10 R_0 above the photosphere were injected in the low or middle corona, in a process closely related to or identical with the one providing the electrons which emit hard X-rays and broad band radio waves during gradual flares.

ON THE ASSOCIATION BETWEEN GRADUAL HARD X-RAY/RADIO BURSTS AND TYPE II BURSTS

The early idea that the large scale coronal shock revealed by type II radio emission might be the accelerating agent of mildly relativistic electrons came from the statistical association of type II bursts with energetic particles as revealed by hard X-ray or microwave emission from the corona, downstream of the shock, and by detection in the interplanetary space, upstream. When imaging observations became possible at X-ray energies above 20 keV, it was realised that the gradual hard X-ray emission does not come from the middle or high corona where the shock is observed, but from more compact sources in the low corona. The relevance of these observations to the idea of shock acceleration has been discussed by /5/. The majority of metric type II bursts is not associated with the signature of energetic electrons in the low corona /6/, and this shows that the acceleration mechanism producing electrons in the impulsive phase is more efficient than the acceleration in the vicinity of the large scale coronal shock during the gradual phase.

A key observation as to the location of the accelerator during the gradual flare phase comes from a comparison of hard X-ray and broad band radio continuum emission (type IV bursts). It was shown in /7-9/ that the time histories of gradual events in these ranges are similar, although the emissions are generated by different mechanisms at widely separated altitudes in the corona (ranging from some 10^3 km for the hard X-rays to several 10^5 km for the metric radio waves). This demonstrates that also during the gradual phase electrons are injected into several magnetic structures which cover a wide range of altitudes. Given that the lifetime of a 100 keV electron in the source of the hard X-ray emission is of the order of some tens of seconds at most, the injection must continue over the whole duration of the gradual phase, during which the coronal shock ascends to greater and greater heights. Although we are presently unable to identify the acceleration process during gradual flares, we can conclude that the large scale shock wave in the corona does not play an important part in it, as far as the downstream region is concerned /10/.

THE X-RAY/RADIO COUNTERPART OF BRIGHT HECTOMETRE ("SHOCK ASSOCIATED") EVENTS

The discovery of bright hectometre emission (frequency below 2 MHz) by space-borne radio detectors has recently lead to a revival of the discussion on electron acceleration in the large scale shock wave /4,11,12/. The hectometre emission comes from altitudes greater than 10 R_0 above the photosphere, i.e. from upstream of the coronal shock wave identified in metre/decametre radio waves. The frequency drift of the elements of such events resembles type III bursts. The emission is consequently attributed to electron streams which are formed somewhere in the middle or high corona. Because of their association with metric type II bursts these electrons were claimed to be accelerated by the shock /4/. However, type II bursts often occur together with metric continua, which reveal electron injection over extended time periods in the low/middle corona. We therefore undertook a detailed comparison of the time histories of 12 hectometre "shock associated events" selected by /11/ with radio and X-ray emission from the middle to the low corona /13/. As a rule, it was found that the hectometre signature of energetic electrons in the high corona is indeed accompanied by injection into a wide range of structures in the low and middle corona. Two

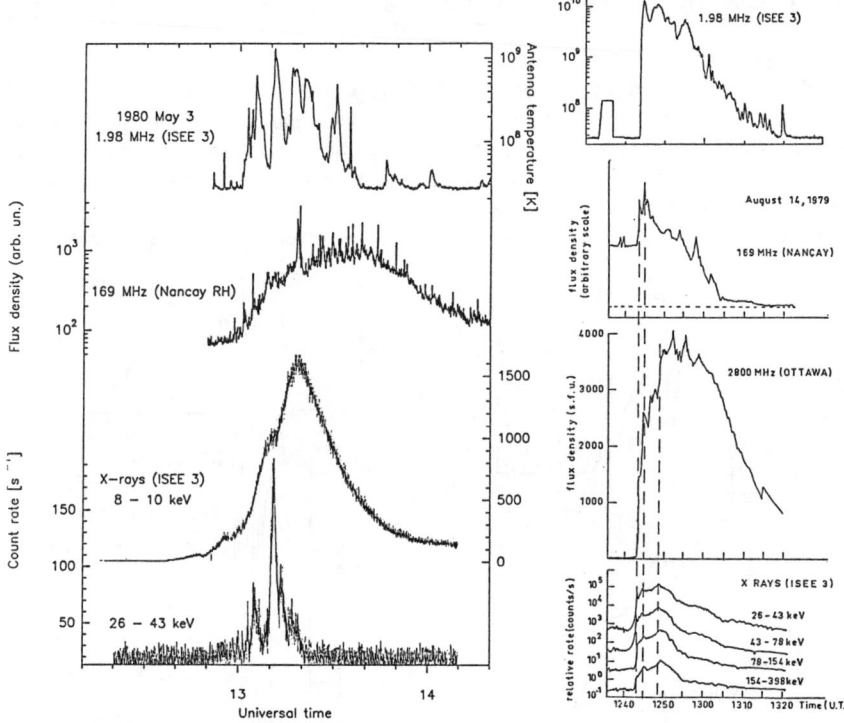

Fig. 1: Hard X-ray and radio emission during two intense hectometre bursts ("shock-associated" events; from /13/).

kinds of simultaneous signatures of energetic electrons in the low corona are illustrated in fig. 1:
1. a long-lasting decimetre/metre continuum ("stationary" type IV or noise storm) where only the onset is accompanied by hard X-ray emission (e.g. left panel of fig. 1);
2. a gradual hard X-ray/radio burst where during the whole duration of the metric radio event hard X ray and microwave emission is detected, revealing the acceleration of electrons up to several hundreds of keV (right panel).

The type II burst which occurred in the course of the hectometre emission was observed at metre/decametre wavelengths, so that the electrons radiating the radio and hard X-ray emissions in fig. 1 are downstream of the shock wave. The fact that the hectometre emission is accompanied by a long-lasting metric continuum demonstrates time-extended injection of energetic electrons into the middle corona. The rather close correspondence of the duration during the 1979 aug 14 event (fig. 1) shows that the hectometre emission ceases as soon as no electrons are injected any more in the low/middle corona. As the shock wave cannot provide the energetic electrons in the hard X-ray source in the low corona, the similar time histories over a wide spectral range argue against a causal relationship between the type II shock and the bright hectometre emission. The observations rather suggest that some of the electrons injected in the low and middle corona attain also great heights where they emit the hectometre radio waves, which thus trace the electrons escaping into interplanetary space.

THE HECTOMETRE COUNTERPART OF SOME GRADUAL HARD X-RAY/RADIO BURSTS

If this interpretation is correct, it will be expected that hectometre radio emission similar to the "shock associated" events is a natural counterpart of gradual hard X-ray/radio bursts. We checked this hypothesis by investigating the hectometre emission detected by ISEE 3 during five selected gradual events: the four events whose hard X-ray and radio emission was studied in /7/ and the case of 1981 apr 26 which had been analysed by /14-16,9/. Among the four gradual events studied in /7/ two were selected as possibly "shock

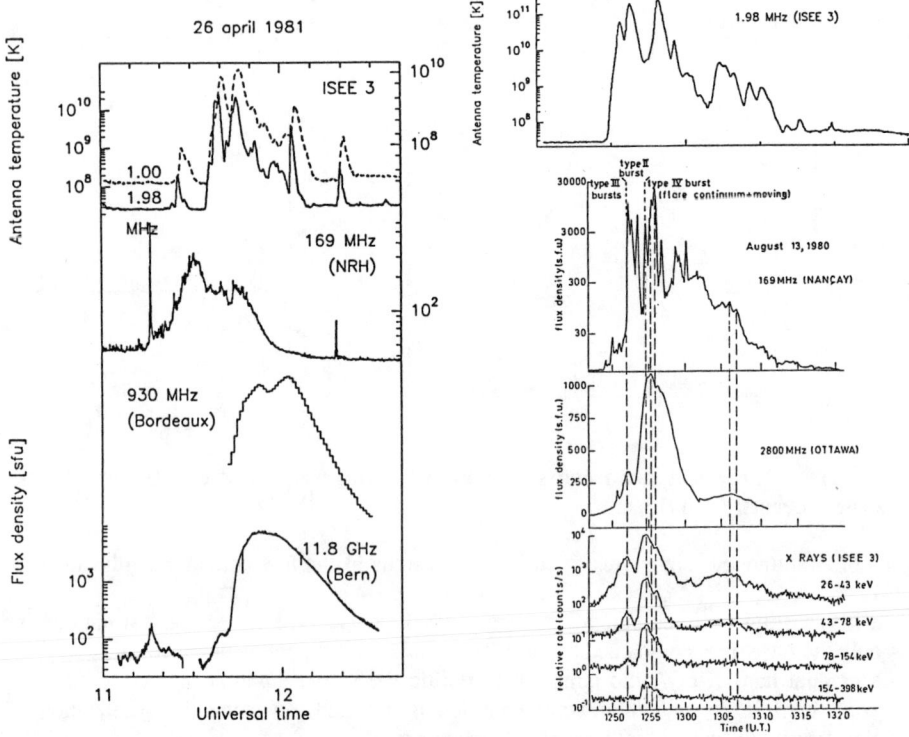

Fig. 2: Hard X-ray and radio emission during gradual events, and the associated intense hectometre burst (1.98 MHz: solid, 1.00 MHz: dashed for 1981 apr 26).

associated" by /11/. Among them is the 1979 aug 14 event of fig. 1. The time profiles of two of the three remaining events are plotted in Fig. 2. The flare of 1980 aug 13 (right panel) is very similar to that of 1979 aug 14: the electromagnetic emission has a similar duration and displays a globally similar evolution at all wavelengths from hard X-rays to metric radio waves, as discussed in /7/. Fig. 2 shows that this similarity extends to hectometre waves. The 1981 apr 26 event is somewhat different: microwave emission is delayed with respect to the hard X-rays by roughly one minute /14-16/ and lasts much longer than the radio emission at 169 MHz, while the hectometre emission continues during the bulk of the gradual phase seen at microwaves. Observations with the Nançay Radioheliograph /16/ show that the 169 MHz source becomes destabilised around 11:48 UT. The source moves upward in the corona, and its emission fades within about 5 minutes. During that time the injection of energetic electrons into the low and middle corona continues, as evidenced by the centimetre and decimetre radio emission in fig. 2. The hectometre emission persists as long as the emission from energetic electrons in the low corona. This argues again in favour of a common injection of electrons into different structures. The destabilisation of the structure where the 169 MHz emission was generated does not affect the escape of energetic electrons from the low to the high corona.

This cursory check hence shows that gradual hard X-ray/radio events are accompanied by bright hectometre emission which by its duration, flux density and time profile is close to satisfying the selection criteria of candidate "shock associated" events of /11/. The events may be accompanied by a metre/decametre type II burst (1980 aug 13) or not (1979 aug 14, 1981 apr 26). Their basic characteristic is the continuum radio and hard X-ray emission from the low and middle corona with similar duration which cannot be explained by electrons accelerated by the large scale shock wave in the overlying corona.

SUMMARY AND INTERPRETATION

These observations are summarised as follows:
1. Those "shock associated" hectometre events identified by /11/ which were also observed at the Nançay Radio Observatory are accompanied by long-lasting continuum emission in the low/middle corona which requires the time-extended injection of energetic electrons into closed magnetic structures.
2. In a few gradual hard X-ray/radio events which were not comprised in the list of "shock associated" events /11/ a hectometre counterpart is found. Its properties (flux density, duration, complexity of the time profile) are similar to those of the "shock associated" events, but they are not necessarily associated with a type II burst. The global temporal evolution of hectometre, centimetre to metre and hard X-ray emissions is most often similar. An exception from this similarity can be ascribed to the instability of a coronal structure which does not affect either the particle acceleration/injection into lower lying structures or the propagation towards the high corona.

We interpret these results on the gradual phase signatures both upstream and downstream of the coronal shock wave as follows:
1. During the gradual flare phase electrons are injected (by an unspecified process) over an extended time period into the middle and low corona ($<1\ R_0$), as witnessed by broad band radio continua and hard X-ray emission. Some electrons get access to the high corona ($>10\ R_0$) and eventually to interplanetary space. Upon travelling through the corona they form electron beams which become detectable in the high corona through bright hectometre radio waves.
2. The detailed correspondence of radiative signatures at different levels seems to depend on the large scale magnetic structure and its stability.
3. The type II associated coronal shock is not an efficient accelerator of electrons of subrelativistic and mildly relativistic energies. During the gradual phase of flares it accelerates electrons which are revealed by the type II emission. The accelerated electrons are either trapped in the vicinity of the shock or, when they escape, lose their energy rapidly. For example the electron streams revealed by the occasional fast drift bursts which emanate from the slowly drifting lane of the type II emission are not

energetic enough to generate detectable signatures elsewhere than in the immediate neighbourhood of the shock.

The difference between our conclusions and those of /4,11,12/ is due to the different approach: while those authors used a statistical study of the common occurrence of different types of emission, we undertook a comparison of time profiles in different wavelength ranges. It has long been known that the common occurrence of type II bursts and type IV bursts gives misleading results on the causal relationship between interplanetary protons and coronal shock waves /17/. This seems also the case of the association between electron streams in the corona above 10 R_0 and coronal shocks. Furthermore, the use of dynamic spectra recorded on film may be misleading because their limited dynamic range tends to privilege the detection of the bright and structured type II bursts, while the more smoothly evolving continuum emission appears only on discrete frequency records.

Acknowledgements: The authors are grateful to J.L. Steinberg and J.M. Robillot for providing ISEE3 and Bordeaux radio data.

REFERENCES

1. G.J. Nelson, D.B. Melrose, in Radio Physics of the Sun, ed. D.J. McLean, N.R. Labrum, Cambridge: Cambridge University Press, 333 (1985).
2. K.J. Frost, B.R. Dennis, Astrophys. J. 165, 655 (1971).
3. H.S. Hudson, R.P. Lin, R.T. Stewart, Solar Phys. 75, 245 (1982).
4. H.V. Cane, R.G. Stone, J. Fainberg, J.L. Steinberg, S. Hoang, Geophys. Res. Letters 8, 1285 (1981).
5. S. Kahler, 1984, Solar Phys. 90, 133 (1984).
6. K.-L. Klein, G. Trottet, A.O. Benz, S.R. Kane, in *Plasma Astrophysics*, ESA-SP 285 Vol. 1, 157 (1988).
7. L. Klein, K.A. Anderson, M. Pick, G. Trottet, N. Vilmer, S.R. Kane, Solar Phys. 84, 295 (1983).
8. K. Kai, H. Nakajima, T. Kosugi, R.T. Stewart, G.J. Nelson, S.R. Kane, Solar Phys. 105, 383 (1986).
9. E.W. Cliver, B.R. Dennis, A.L. Kiplinger, S.R. Kane, D.F. Neidig, N.R. Sheeley, M.J. Koomen, Astrophys. J. 305, 920 (1986).
10. G. Trottet, Solar Phys. 104, 145 (1986).
11. R.J. MacDowall, R.G. Stone, M.R. Kundu, Solar Phys. 111, 397 (1987).
12. S.W. Kahler, E.W. Cliver, H.V. Cane, Solar Phys. 120, 393 (1989).
13. K.-L. Klein, in *Solar Wind Seven*, ed. E. Marsch, R. Schwenn, Oxford: Pergamon Press, 635 (1992).
14. B.R. Dennis, Solar Phys. 100, 465 (1985)
15. T. Bai, B. Dennis, Astrophys. J. 292, 699 (1985).
16. G. Bruggmann, N. Vilmer, K.-L. Klein, S.R. Kane, Solar Phys., submitted (1993).
17. S. Kahler, 1982, Astrophys. J. 261, 710 (1982).

NEW TECHNIQUE FOR DYNAMIC BURST POSITION DETERMINATION WITH HIGH SPATIAL DEFINITION/TIME RESOLUTION AT A MM-WAVELENGTH

J.E.R. Costa, E. Correia, P. Kaufmann
Center of Radio Astronomy and Space Applications, CRAAE, S. Paulo, Brasil.

R. Herrmann, A. Magun
Institute of Applied Physics, University of Bern, Bern, Switzerland.

ABSTRACT

The position of solar burst maximum emission can be monitored with spatial resolution of few arcseconds and millisecond time resolution, simultaneously, using a multifeed/multiradiometer front-end system placed at the focus of the 13.7-m Itapetinga radio-telescope. This new concept was developed in a cooperative program between the Institute of Applied Physics of the University of Bern, and the Center of Radio Astronomy and Space Applications, São Paulo. The five independent radiometer front-end operates at 48 GHz and produces five beams in space, partially overlapping to each other. Dynamic position images of the centroid of burst emission can be derived. Various events have been investigated, and the first results suggest that the fast time structures superimposed to the bursts' time profiles, may originate from distinct spatial positions in certain events, or close to a main generating stable source for other events. The dynamic images suggest apparent burst source displacements, or distinct and independent burst sites (for time resolved fast structures).

Solar burst time profiles at cm-mm wavelengths are believed to describe the rate the energy release mechanisms actuate in flares. High sensitivity and high time resolution observations made at short microwaves have suggested that bursts are a response to processes producing multiple, discrete and fast sub-second spikes which may or may not be convoluted in time and in space [1]. These features were confirmed at other short microwaves [2,3], and seem to occur also at hard X-rays [4] with good association to cm-mm wave burst emissions [5,6]. Recent high sensitivity and high time resolution hard X-ray observations of solar bursts by BATSE experiment on board of GRO satellite suggest that fast sub-second time structures are common to most events analysed so far [7].

These evidences at cm-mm waves have raised new questions necessary for a complete diagnostics: (a) Where these simple and fast burst structures are located in the active regions? (b) Are the multiple fast structures produced in single flaring sites, or not? (c) Are the multiple fast spikes representative of actual moving sources or would they indicate discrete and independent flaring sites? In order to investigate these questions a new observational concept was developed within a cooperation between the solar groups of the University of Bern, Switzerland and of Itapetinga (operated by CRAAE), Brazil. It consists in a multi-receiver, multi-feed 48GHz front-end, built to be placed at the Cassegrain focus of the radome-enclosed 13.7-m Itapetinga antenna, producing five 2 arcmin beams in space, partially overlapping to each other [8,9,10].

Figure 1 shows the Itapetinga 13.7-m antenna, and the upper right box showing the five 48GHz feed-horns, connected to five mixer-receivers. The antenna tracking accuracy is of 2 arcseconds. The position uncertainty of the burst centroid of emission is limited by tracking, the atmospheric "seeing" and the burst source intensity and position relative to the beams. It is usually better than few arcseconds. More technical details on the system are found in Herrmann et al. [9], and Costa et al. [10].

The 48GHz five-beam system is able to define instantaneously the position of burst emission and permits the derivation of "dynamic" images over a certain interval of time. The basic observational requirements are (a) burst sources must be smaller than the beams' half power widths (< 2 arcmin); (b) beam shapes must be well known in order to correlate the outputs and restore burst positions.

Figure 2 shows the example of a complex solar burst, which is being studied separately by Correia et al. [11]. At the botton the 48GHz time profile is shown, restored from the outputs of the various beams/radiometers. At the top it is shown a time development of the power spectra. Faster repetition rates of sub-second time structures are found, being more pronounced for the burst structure labeled H.

Figure 3 shows dynamic images in frames of 20x20 arcsec, for several sections of the event, indicated by the corresponding labels. The various time structures are clearly not coincident in space, being separated from each other by several arcseconds.

In conclusion, the 48GHz solar burst dynamic images obtained at Itapetinga suggest that time resolved fast structures are often resolved spatially into distinct spatial locations. These results may be interpreted on two alterate ways:
(a) Emitting particles are magnetically trapped, moving, mirroring, acquiring and losing energy. In this case one would be dealing with a single burst, which complexity is due to the magnetic topology and the dynamics of the event. (b) Discrete bursts, independent energy conversion sites within an active region who suddenly became unstable. In this case one would be dealing with multiple discrete accelerations, and multiple bursts, in space and time.

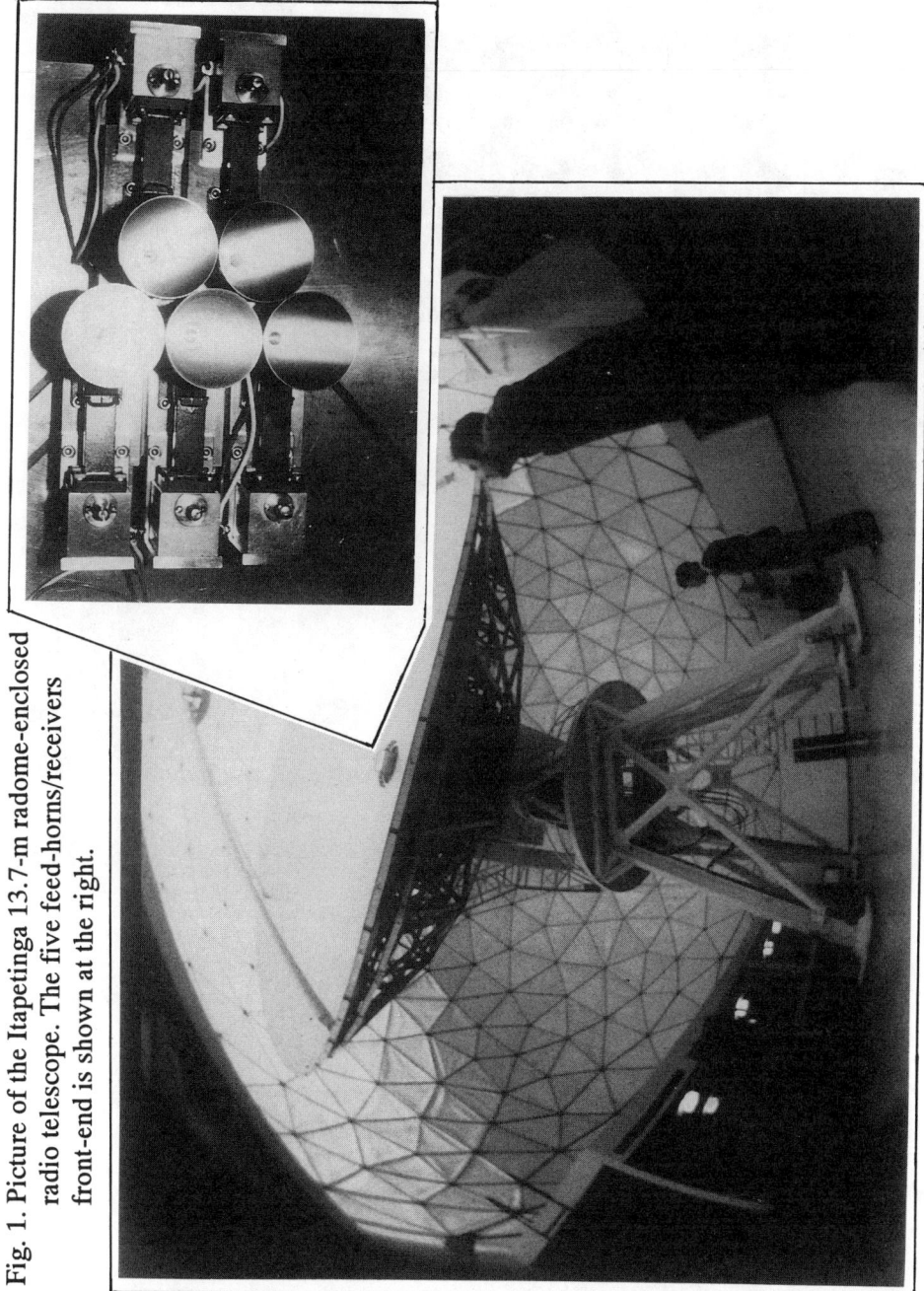

Fig. 1. Picture of the Itapetinga 13.7-m radome-enclosed radio telescope. The five feed-horns/receivers front-end is shown at the right.

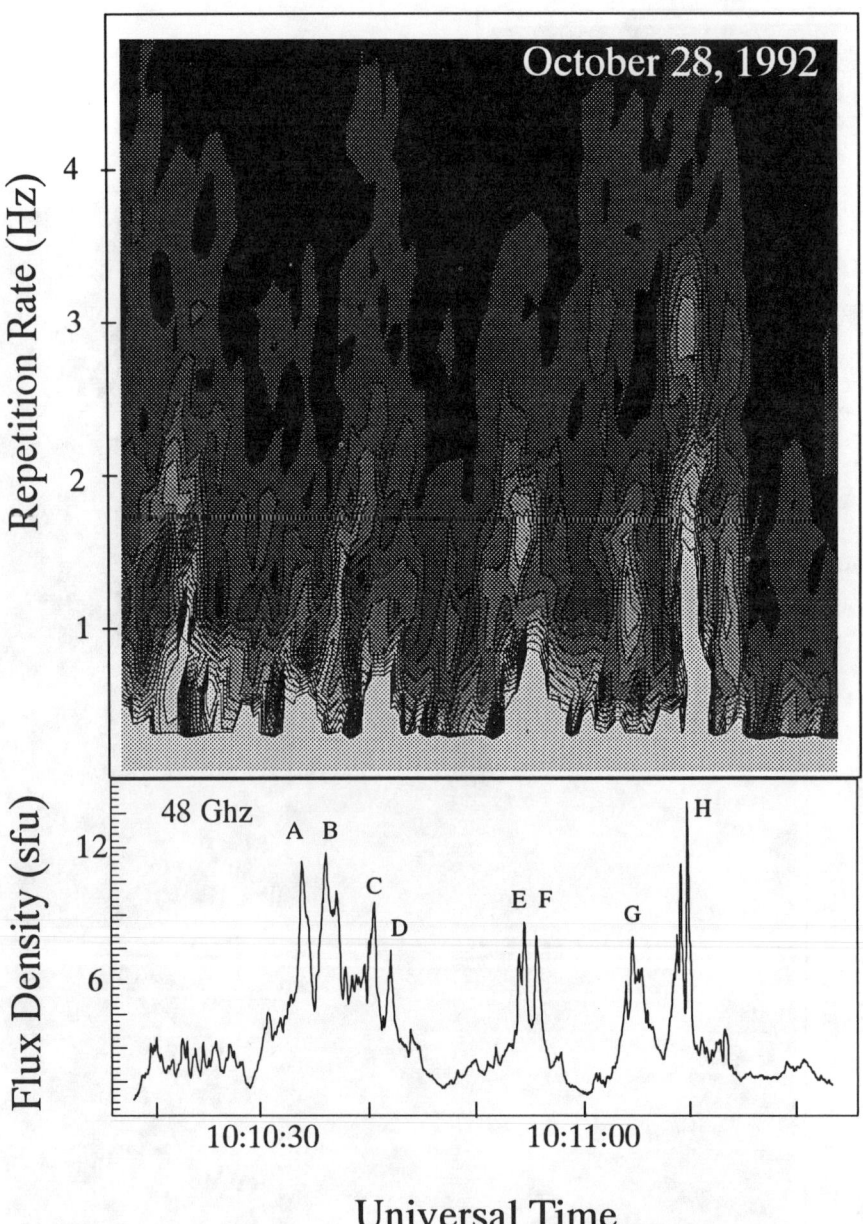

Fig. 2. The 28 October 1992 solar burst observed at 48GHz. Restored time profile is shown at the bottom and power spectra at the top indicate the repetition rates at different sequential time intervals of the event (after Correia et al.[11]).

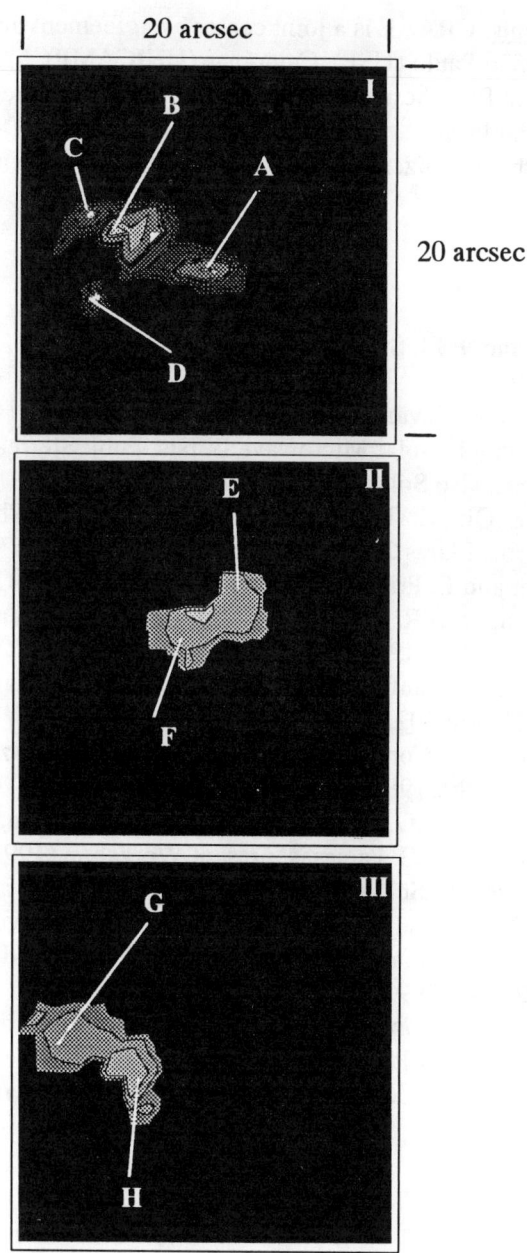

Fig. 3. Dynamic images for the 28 October 1992 event, in 20x20 arcsec frames, for different time intervals and labels shown in Fig. 2 (after Correia et al.[11]).

Acknowledgments: CRAAE is a joint center by agreement between the Universities of São Paulo (USP), Campinas (UNICAMP), Mackenzie, and the Space Institute INPE. These researches are partially supported by agencies FAPESP (São Paulo State, Brazil) and NSF (Switzerland). One author (PK) acknowledges the organizers for providing travel support to attend the Workshop.

REFERENCES

1. P. Kaufmann, F.M. Strauss, R. Opher, and C. Laporte, Astron. Astrophys. 87, 58 (1980).
2. M.A.F. Allaart, J. van Nieuwkoop, C. Slottje, and L.H. Sondaar, Atlas of Fine Structure in Solar Microwave Bursts, Publ. Sterrekundig Instituut, Utrecht, NL, also Solar Phys. (1980).
3. F. Qijin, Z. Qin, Z. Gao, Z. Li, G. Huang, S. Jin, X. Cheng, C. Huo, and S. Wei, Proc. Flares 22 Workshop: Dynamics of Solar Flares (ed. by B. Schmieder and E. Priest), Observatoire de Paris, DASOP, p.77 (1991).
4. A.L. Kiplinger, B.R. Dennis, A.G. Emslie, K.J. Frost, and L.E. Orwig, Astrophys. J. 265, 199 (1983).
5. T. Takakura, P. Kaufmann, J.E.R. Costa, K. Degaonkar, K. Ohki, and N. Nitta, Nature 302, 317 (1983).
6. P. Kaufmann, E. Correia, J.E.R. Costa, A.M. Zodi Vaz, and B.R. Dennis, Nature 313, 380 (1985).
7. M.E. Machado, A.G. Emslie, K.K. Ong, G.J. Fishman, C. Meegan, R. Wilson, and W.S. Paciesas, present Workshop (1993).
8. C.B. Georges, R. Schaal, J.E.R. Costa, P. Kaufmann, and A. Magun, Proc. SBMO International Microwave Symposium/Brazil, vol. II, p.447 (1989).
9. R. Herrmann, A, Magun, J.E.R. Costa, E. Correia, and P. Kaufmann, Solar Phys. 142, 157 (1992).
10. J.E.R. Costa, E. Correia, P. Kaufmann, A. Magun, and R. Herrmann, to be published in Solar Phys. (1993).
11. E. Correia, J.E.R. Costa, P. Kaufmann, and P. Herrmann, in preparation (1993).

ENERGETIC ELECTRON POPULATIONS IN SOLAR FLARES

Stephen M. White
University of Maryland, College Park MD 20742

ABSTRACT

Millimeter–interferometer observations of flares are used to study the MeV–energy electrons accelerated in solar flares. The focus of this study is a remarkable similarity found in the time profiles of emission associated with the impulsive onset of a flare. In a large fraction of flares, the impulsive phase emission at millimeter wavelengths consists of a rapid rise ($\sim$ 5 seconds) linear in time to a sharp peak, followed by an exponential decay with a decay constant of order 15 seconds. The onset of millimeter emission may be delayed by several seconds with respect to the onset of hard X–rays. The implications of this homologous property are discussed briefly.

INTRODUCTION

Millimeter–wavelength observations are one of the few means presently available for studying the production of electrons with energies of order 1 MeV and higher in "ordinary" solar flares. The bremsstrahlung γ–ray continuum of these electrons can be detected in very large solar flares, but present detectors are generally not sensitive to the small numbers of γ–rays produced by even moderate–sized flares. On the other hand, the gyrosynchrotron emission of relatively small numbers of nonthermal MeV–energy electrons can be detected with the millimeter–wavelength interferometers developed in recent years. The typical range of magnetic field strengths in the corona requires that millimeter emission be at high harmonics of the electron gyrofrequency, and since nonrelativistic electrons have low emissivity at such high harmonics the observed millimeter emission should be dominated by the contribution of MeV–energy electrons [1]. Detectable millimeter–wavelength emission is also produced by bremsstrahlung in the hot dense thermal plasma present in the corona in the decay phase of flares, but the temporal signature of this thermal bremsstrahlung is generally quite different from the emission in the impulsive phase.

Since 1989 the solar group at Maryland has been studying the millimeter emission of flares using the 3–element Berkeley–Illinois–Maryland millimeter interferometer (BIMA) during campaigns. These observations have established that even small flares can produce significant amounts of millimeter emission in the impulsive phase, implying that electrons can be accelerated to MeV energies in flares of all sizes [2]. Not all flares do produce detectable impulsive–phase emission at millimeter wavelengths, but certainly most do. The millimeter emission can be delayed with respect to the microwave emission attributed to lower–energy electrons [3].

As more flares have been observed at millimeter wavelengths, a striking result has been noticed: this is the recurrence of a particularly simple morphology in the time profiles of the impulsive phase, which is the topic of this paper. The first event of this type to be studied was discussed by White *et al.* (1992) [4]: it showed a linear rise to a sharp peak in 6 seconds, followed by an immediate decay of exponential form with a decay constant of 18 seconds. This particular event was peculiar in a number of respects not shared by the rest of the events to be discussed here, but the time profile has subsequently recurred frequently. The motivation for this work is the possibility of finding a property of electron acceleration common to many flares which can then be used to constrain models for acceleration of energetic electrons. Such shared properties are notoriously rare, and this is a limit to progress in flare research.

MILLIMETER TIME PROFILES IN THE IMPULSIVE PHASE

Figure 1 presents the time profiles of the impulsive phase emission of 8 events observed at BIMA. The time profiles have been overlaid on one another so that their peaks are coincident; the amplitudes have all been scaled to unity, but there is no scaling on the time axis. The events are of varying signal-to-noise levels, but the apparent linearity of the rise to the peak, the sharpness of the peak and the apparently exponential nature of the decay are all striking. We note that a rapid rise may appear linear without actually being so; perhaps the more important observation is that the rise starts abruptly. Rise times are generally between 4 and 8 seconds, and the e-folding decay times are 10 – 20 seconds. The time resolution is 0.4 seconds for all these events, so the rise and decay are both well resolved in all cases. Two of the events show a small and more gentle rise immediately preceding the linear part of the rise, but this is actually rare: it is more usual for these profiles to be linear from onset.

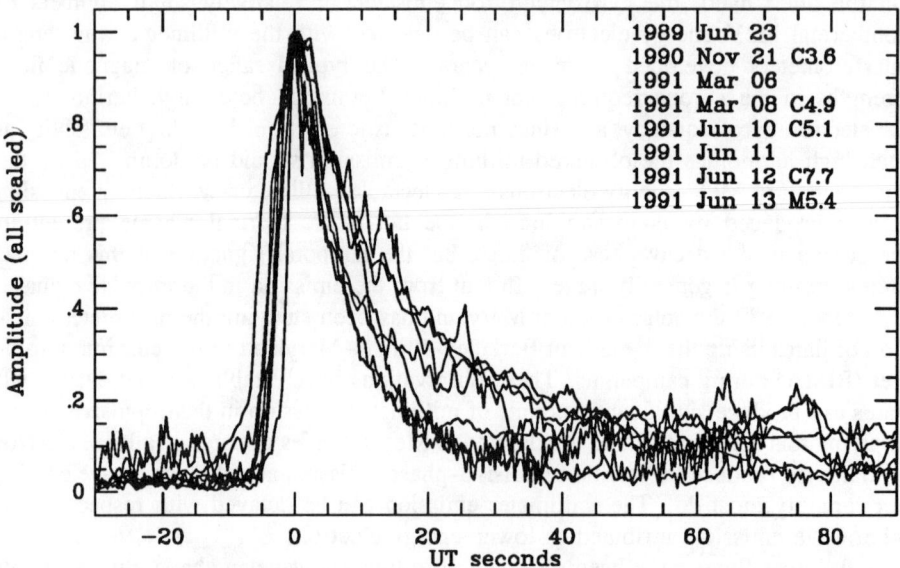

Figure 1. The time profiles of eight events observed by BIMA, scaled to the same height and placed so that their peaks are coincident in time. The dates and GOES soft X-ray classes of the events are given.

One of the difficulties in drawing conclusions about the properties of solar flares is the well–known axiom that "no two flares are alike". Here we seem to have a common property shared by a large range of flares. However, in order to be able to draw general conclusions from this observation we must ensure that it is not the result of selection effects, and that it is a common occurrence. As of early 1993, 59 events have been observed at 86 GHz by BIMA operating as a non-imaging interferometer. In 23 of these 59 events (40%), a simple impulsive spike of the shape described here is the only emission associated with the impulsive rise. In 10–20% of events there was no nonthermal impulsive–phase emission: rather, the data are consistent with thermal bremsstrahlung from dense material in the corona (e.g., some examples are given in Kundu et al. 1993 [5]). In a number of cases the millimeter emission shows both an impulsive spike in the rise phase and a smoother time profile in the decay phase consistent with thermal emission. In the remaining cases the time profiles are consistent with nonthermal emission but with more complicated time profiles (e.g., Lim et al. 1992 [3]).

From the flare sizes listed in Figure 1, it is clear that this time profile is not restricted to small flares alone: it is seen in events ranging in GOES class from less than C1.0 up to M5. The corresponding range in millimeter flux at the peak is about a factor of over 50, from less than .3 sfu up to 15 sfu. One obvious question that must be asked is whether these time profiles differ from the microwave time profiles for the same events: it is well known that the most common morphology for the time profile of a microwave burst is "simple", and the templates for "simple" bursts in many cases resemble the time profile seen here [6]. We have microwave data for approximately half of these events (mostly from the RSTN patrol telescopes and from the Owens Valley frequency–agile interferometer), and while the microwave profiles at high frequencies often have a simple shape, they generally have a gentler, less linear rise, the peak is less sharp, microwave onset often precedes the onset at millimeter wavelengths [3], and the decay can show quite different behaviour from the millimeter data. The most extreme example of this seen so far is shown in Figure 2, where the hard X–ray data from the *Gamma Ray Observatory*, the BIMA 86 GHz time profile and the microwave time profiles for one event are superimposed [7]. In this event the microwave and hard X–ray profiles generally match over a period of several minutes, but the millimeter emission (which is assumed to be associated with electrons more energetic than those associated with the other wavelengths) shows the linear rise–exponential decay discussed here. This figure also shows the microwave profiles having a more gradual initial rise and a more rounded peak than does the millimeter profile, which is typical of the microwave profiles of simple events in general.

Thus we can draw several conclusions: this linear rise–exponential decay time profile is not shared by the microwave emission or the hard X–ray emission in the same events. Since we interpret the nonthermal impulsive–phase millimeter emission as a diagnostic of the MeV–energy electrons, this means that they have a property not shared by the lower–energy electrons responsible for the hard X–rays (typically 25 – 100 keV) and microwaves (probably 100 – 500 keV). There is also some evidence that the spectral energy distribution of the millimeter–emitting electrons differs from that of the hard X–ray–emitting electrons [7].

DISCUSSION

It seems that for these events the production of MeV–energy electrons follows a very similar pattern. The simplest interpretation of the morphology seems to be that the electrons are accelerated, or somehow injected into a coronal loop, on a timescale of several seconds (the linear rise), but acceleration then ceases abruptly and the number of electrons in the corona decreases exponentially. The similarity in the rise time and decay time among a large number of events would then impose strong constraints on the mechanism of acceleration and the conditions in which it occurs.

Generally it is assumed that a trap model is appropriate for describing the evolution of the number of radio–emitting electrons in a coronal loop, and the observed time profile must be "deconvolved" for the effects of the different physical timescales relevant to the evolution of the particle distribution in the trap (e.g., see reviews by Petrosian [8,9]). In a trap the relevant timescales are: the injection/acceleration time; the time required to propagate along the trapping loop; and the energy loss or precipitation time. As is well known, the propagation time is of the order of 10 milliseconds for relativistic electrons in the compact loops believed to be involved in these bursts (several arcsec in dimension [3]), and thus is shorter than either of the observed timescales: propagation effects should not be observable at this level. The observed decay time can be treated as the particle lifetime as long as acceleration/injection of energetic electrons does not continue throughout the event but stops at or shortly after the peak, which seems to be the simplest assumption (that does not make it a correct assumption, of course). The expected decay time due to collisions in the corona alone encompasses two effects: actual energy loss, and pitch–angle scattering into the loss

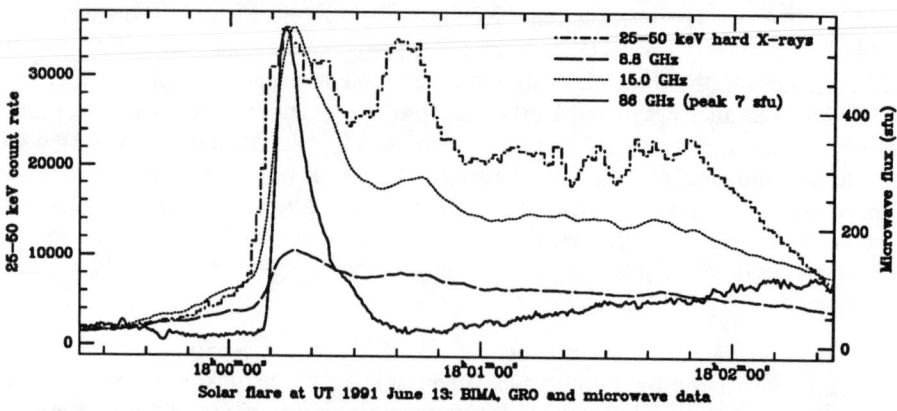

Figure 2. The hard X–ray (GRO/BATSE data, dash–dotted histogram), millimeter (BIMA data, solid line) and microwave (RSTN data; 8.8 GHz, long–dashed line; 15.0 GHz, dotted line) time profiles for a GOES class M5 flare on 1991 June 13.

cone where particles are then lost within a propagation time by precipitation into the chromosphere. The energy loss time is of order 300 seconds for a 1 MeV electron in a loop of density 10^{10} cm^{-3}, and thus can probably be ignored. The precipitation loss time depends on the pitch–angle scattering mechanism, and has a minimum value of V/Ac in the strong–diffusion limit, where V is the loop volume, A the area of the footpoint in the chromosphere and c is an estimate of the particle velocities at the footpoint. This reduces to the loop propagation time ($\ll 1$ second) for a loop of constant cross–section area, but is larger for a tapered loop. If scattering is not strong enough to maintain a filled loss cone throughout the loop, which is the case for collisions of relativistic electrons, the lifetime will be longer. For a 1 MeV electron in a plasma of density 10^{10} cm^{-3}, collisions alone will produce a precipitation time of order 500 seconds (note that collisional energy diffusion becomes more effective relative to pitch–angle diffusion as particle energy increases), so it seems unlikely that collisions play much of a role in the observed loss time. Note that It seems that wave–particle are required to increase the scattering and thus reduce the precipitation loss time, but then the narrow range in the observed loss times must be explained.

In principle instrumental effects can also affect the appearance of time profiles, but in this case the intrinsic time constant of the telescopes is of order 0.5 seconds, and this is much shorter than either the rise or decay times.

CONCLUSIONS

The millimeter emission in the impulsive phase of solar flares frequently shows the same time profile: a rise approximately linear in time which ends abruptly and is followed by a decay which is nearly exponential in time. Both the rise times and the decay times show a surprisingly narrow range of values: 3–6 seconds for the (total) rise time, 10–20 seconds for the (e–folding) decay time. This is seen in roughly half of all flares which show nonthermal millimeter emission in the impulsive phase. The millimeter emission should be due to gyrosynchrotron by MeV–energy electrons, and the observed time profiles should represent the time behaviour of the number of MeV–energy electrons present in the corona. The simplest interpretation of this profile, ignoring for the moment any theoretical models, is that MeV energy electrons are accelerated or otherwise injected into a coronal loop during the rise; acceleration/production then ceases abruptly and the number of electrons decays exponentially as wave–particle processes scatter the electrons into the loss cone and they precipitate into the corona. For plausible conditions in the corona, collisional processes appear to be too slow to play any role in the observed evolution. The relative smoothness of the time profiles at our resolution of 0.4 seconds seems to imply simple mechanisms, but it should also be mentioned that there are suggestions from observations with millisecond time resolution that millimeter emission often shows subsecond time structures [11].

In the same events, emissions traditionally associated with lower–energy electrons can have quite different time profiles. The millimeter emission generally decays before the microwave or hard X–ray emission. One can argue that the γ–ray emission from electrons in the same energy range ought to show the same time profile. All of the events discussed here were too weak to be detected above 300 keV by the BATSE detectors on the *Compton Gamma Ray Observatory*, so we cannot carry out a direct comparison. We have looked at the time profiles of the 300 – 1000 keV emission detected by BATSE from a large number of large solar flares, and as is well known they are generally more complex than the simple template shown in Figure 1. The time profiles of impulsive features within these time profiles do consistently show similar profiles, but these are almost always symmetric in their rise and fall, with both appearing to be exponential in shape, not linear (the time profile of photons due to precipitation should be the time derivative of the trapped population once acceleration/injection has ceased). Thus they provide no confirmation that this time profile represents a "universal" form for MeV–energy electrons.

Clearly more work needs to be done: firstly a greater sample of events is required to firmly establish the frequency of this time profile, and secondly the implications need to be explored further. With the increasing availability of millimeter interferometers, and with the ability to make images of these events which was not available for those discussed here, both these topics will be addressed further in the next few years.

ACKNOWLEDGMENTS

Solar radio research at the University of Maryland is supported by NASA under the GRO GI program (NAG–W–1540) and grants NAG–4–1541 and NAG–W–2172, and by the NSF under grant ATM 90–19893. The use of BIMA for scientific research at the University of Maryland is supported by NSF grant AST 91–00306.

REFERENCES

1. S. M. White and M. R. Kundu, Solar Phys. <u>141</u>, 347 (1992).
2. M. R. Kundu, S. M. White, N. Gopalswamy, J. H. Bieging, and G. J. Hurford, Ap. J. <u>358</u>, L69 (1990).
3. J. Lim, S. M. White, M. R. Kundu and D. E. Gary, Solar Phys. <u>140</u>, 343 (1992).
4. S. M. White, M. R. Kundu, T. S. Bastian, D. E. Gary, G. J. Hurford, T. Kucera and J. H. Bieging, Ap. J. <u>384</u>, 656 (1992).
5. M. R. Kundu, S. M. White, N. Gopalswamy and J. Lim, Ap. J., submitted (1993).
6. *Solar Geophysical Data* <u>499</u> (Supplement): Explanation of Data Reports (1986).
7. S. M. White, R. Murphy, R. A. Schwartz, M. R. Kundu, J. Lim and N. Gopalswamy, in preparation (1993).
8. V. Petrosian, in Basic Plasma Processes on the Sun, eds. E. R. Priest and V. Krishan, p. 391 (1990).
9. V. Petrosian, these proceedings (1993).
10. P. Kaufmann, these proceedings (1993).

NEUTRON MONITOR
AND GLE EVENTS

NEUTRON MONITOR MEASUREMENTS AS A COMPLEMENT TO SPACE MEASUREMENTS OF ENERGETIC SOLAR PARTICLE FLUXES

H. Debrunner
Physikalisches Institut, University of Bern, CH-3012 Bern, Switzerland

ABSTRACT

Neutron monitors remain indispensable cosmic ray detectors for primary energies from about 500 MeV to 30 GeV. They respond to variations of the cosmic ray intensity in near-Earth space that are not measured by space experiments. The records of the world wide network of standardized neutron monitors, therefore, complement cosmic ray spacecraft measurements. In this review paper we first describe the design of a neutron monitor and its characteristic parameters as a continuous ground-based cosmic ray instrument. We then discuss how the data from the world wide network of neutron monitors are evaluated to determine the energetic solar proton fluxes in near-Earth space during solar cosmic ray events. We present, as an example, the solar proton fluxes near Earth during the May 7, 1978 solar cosmic ray event. These fluxes were derived from neutron monitor data and measurements of the cosmic ray telescopes on board IMP-7 and cover the energy range from 50 MeV up to 10 GeV. Finally, we describe the method of analysing solar neutron events. As an example, we summarize the observations made during the solar flare on June 3, 1982 by the Gamma-Ray-Spectrometer on the Solar Maximum Mission Satellite and the Jungfraujoch neutron monitor. These measurements were used to determine the directional solar neutron emissivity spectrum of the June 3, 1982 solar neutron event over the energy range from 100 MeV to about 3 GeV.

INTRODUCTION

The basic ideas for the development and construction of the neutron monitor (NM) as a continuous recorder of the cosmic ray intensity originated from measurements by Simpson [1,2]. He found that the latitude dependences of the intensities of the energetic nucleonic component and of evaporation neutrons from the secondary cosmic radiation in the atmosphere are equal and that these latitude variations are several times larger than those of the ionizing and the hard components. The anticipation that the secondary nucleonic component and the production of evaporation neutrons could be used to study the time variations of the primary cosmic rays at energies lower than previously possible with ionization chambers or muon counters led a number of cosmic ray physicists to develop new detectors so as to investigate the local neutron production induced by the secondary cosmic radiation in the atmosphere (Tongiorgi [3], Simpson [2,4], Cocconi et al. [5], Adams and Braddick [6] and Simpson et al. [7,8,9]). The NM designed by Simpson [10] was adopted as the standard detector to study the time variations of the primary cosmic ray intensity at GeV-energies in the near-Earth space during the International Geophysical Year 1957/1958 and was

© 1994 American Institute of Physics

designated as an IGY-NM. The period following the IGY before the spacecraft era began was the heyday of NMs. The records of the world wide network of NMs substantially increased our knowledge about the interactions of the galactic cosmic radiation with the interplanetary magnetic field and about the production of energetic cosmic rays in solar flares. Later, the IGY network was supplemented by the larger NM64 neutron monitor designed by Carmichael [11]. This NM which had an increased count rate was recommended as the standard ground-based cosmic ray detector for the International Quiet Sun Year 1964.

The present world wide network of NMs consists of about 50 detectors, the majority of which conform to the specifications of the NM64 or the IGY neutron monitor. In this paper we will show that this network is still a powerful tool for cosmic ray research and complements space experiments. After a short description of the design of a NM64 neutron monitor and its characteristic parameters as a primary cosmic ray detector, we will discuss how measurements by NMs are evaluated to determine the fluxes of energetic solar protons in near-Earth space during solar cosmic ray events. We present, as an example of the results of such an analysis, the deduced solar proton fluxes near Earth for the solar cosmic ray event on May 7, 1978. These fluxes were derived by Debrunner et al. [12] from measurements of the cosmic ray telescopes on board IMP-7 for energies from 50 MeV to about 500 MeV and from the data of the world wide network of NMs for energies $\geq$ 500 MeV . Since the May 7, 1978 solar cosmic ray event was characterized by an essentially scatter-free propagation of the solar protons between the Sun and the Earth, Lockwood et al. [13, 14] used the solar proton spectra observed at Earth to construct the proton emission spectra at the Sun. These results covering more than two decades in energy were then compared with the predictions of different solar flare particle acceleration models and interpreted in terms of the diffusive coronal shock acceleration model of Ellison and Ramaty [15]. This analysis enabled us to study the acceleration and release of the solar protons on May 7, 1978 in considerable detail.

Next we present, as an example of the response of NMs to solar neutrons, the observations made during the June 3, 1982 solar cosmic ray event with the Gamma-Ray-Spectrometer on the Solar Maximum Mission Satellite and with the Jungfraujoch neutron monitor. We then describe the method used by Chupp et al. [16] to derive the directional solar neutron emissivity spectrum of the June 3, 1982 solar flare over the energy range from 100 MeV to about 3 GeV. We also show that only the combination of the spacecraft measurements and the NM data has enabled us to determine that the neutron emission at the Sun was time-extended. We, therefore, conclude that the world wide network of NMs does actually complement the space measurements of solar proton and solar neutron fluxes. The detailed knowledge of the solar particle emission in the non-relativistic and the relativistic energy regime combined with solar X- and γ-ray measurements should extend our understanding of the physical processes in solar flares.

DESIGN OF THE NM64 NEUTRON MONITOR

Neutron monitors are relatively simple instruments to build and to maintain in stable operation over long periods of time. They are ideally designed to measure the intensity of the secondary nucleonic component of the cosmic radiation in the atmosphere and respond to lower energy primary particles than ionisation chambers and muon telescopes. A detailed review of the design of NMs, their mode of operation and their response to the ambient secondary cosmic radiation has been given by Hatton [17]. The core of the NM64 neutron monitor consists, as illustrated in Figure 1, of a layer of lead with an average depth of 13.8 cm (156 g/cm^2) corresponding to about 75% of the inelastic mean-free path of nucleons in lead. The lead target is interspersed with BF$_3$ proportional counters. Each counter is surrounded by 2.0 cm of polyethylene which acts as a neutron moderator. The counters detect evaporation neutrons which are produced by the nucleonic component of the secondary cosmic radiation interacting with lead nuclei and slowed down by the polyethylene. The reaction used is:

$$^{10}B_5 + n \rightarrow {}^7Li_3 + {}^4He_2$$

where the α-particle and the Li-nucleus generate the electrical discharge in the counter. The whole assembly of counters, inner polyethylene moderator and lead is enclosed by polyethylene of an average thickness of 7.5 cm which moderates and reflects the evaporation neutrons from the lead. This polyethylene has also the function of reflecting and absorbing the low energy neutrons produced in surrounding materials outside the NM.

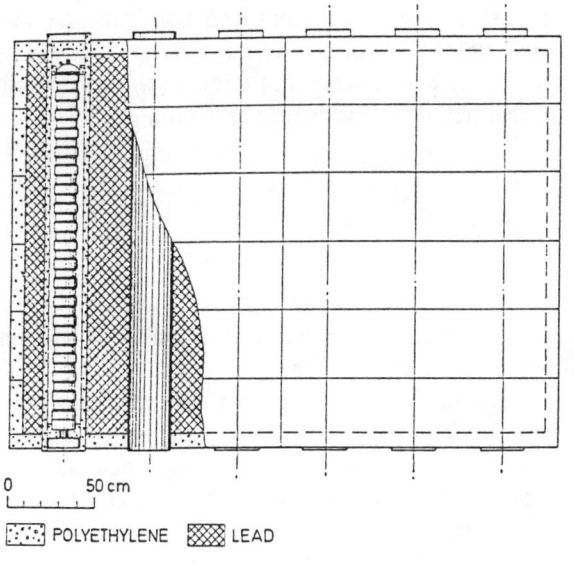

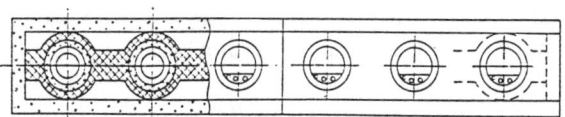

Fig. 1. A 6-NM64 neutron monitor [Carmichael [11]]

The probability for a cosmic ray neutron or proton which hits the neutron monitor to interact with a nucleus of the lead target is ~50%. The average number of evaporation neutrons produced per nuclear reaction in the lead amounts to ~15 and the detection probability for evaporation neutrons to ~5.7%. With these parameters, the average count rate of a high latitude sea level NM64 neutron monitor with 6 BF$_3$ counters and a geometrical area of 1.91 x 2.07 m^2 is ~2.5·10^5 counts per hour or ~70 counts per second which is ~1.4 times that of an equatorial sea level NM64 neutron monitor. This latitude variation of the count rate shows that NMs respond to the primary cosmic radiation at GeV-energies.

THE NEUTRON MONITOR AS A TOOL TO STUDY ENERGETIC SOLAR PROTON EVENTS

Primary cosmic rays are for the most part high energy protons. They also include nuclei of the heavier elements as well as electrons and positrons. As these charged particles approach the Earth, they are deflected by the geomagnetic field and the effects of this magnetic deflection are best described in terms of the particle rigidity. The rigidity of an electrically charged particle is defined as:

$$P\,[\text{Volts}] = \frac{p \cdot c}{Z \cdot e} \qquad (1)$$

where p is the momentum and $Z \cdot e$ the electric charge of the particle, and c is the velocity of light. From the Lorentz force equation follows that in a static magnetic field, particles of the same rigidity exhibit the same orbits.

The first effect of the geomagnetic field on cosmic ray measurements is that cosmic ray particles can reach an observer at the Earth from a given direction only at given rigidities. Figure 2 shows, as an example, the vertical viewing conditions of the Jungfraujoch NM with the geographic coordinates 46.5° N and 8.0° E. These viewing conditions were determined by evaluating trajectory calculations of cosmic ray particles in a model of the quiescent geomagnetic field [Flückiger [18]]. The black regions indicate the rigidities of the particles which can reach the top of the atmosphere above Jungfraujoch from the vertical direction. No cosmic rays with $P < 3.9$ GV are observed over Jungfraujoch from the vertical. On the other hand, all rigidities with $P > 4.8$ GV are allowed. The effect of the complex penumbral region, which for the vertical direction at Jungfraujoch extends from $P_s = 3.9$ GV to $P_m = 4.8$ GV, is usually described by introducing the effective geomagnetic cutoff rigidity, P_c, defined as:

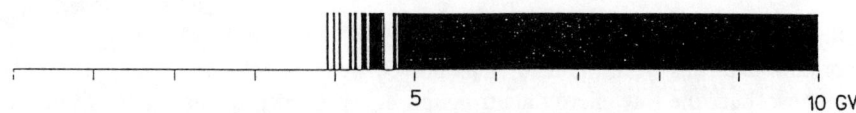

Fig 2. Allowed and forbidden rigidities for cosmic rays hitting the atmosphere above Jungfraujoch from the vertical [Flückiger [18]]

$$P_c = P_m - \left[\int_{P_s}^{P_m} dP \right]_{\text{allowed}} \qquad (2)$$

The effective geomagnetic cutoff rigidity depends on the location of the observer and the viewing direction. The vertical cutoff rigidities for ground-based cosmic ray detectors vary from $P_c \approx 0$ near the magnetic poles to $P_c \approx 15$ GV in the equatorial region. The geomagnetic field, therefore, acts as a rigidity filter for the world wide network of NMs. From a comparison of the count rates of standardized NMs at different locations at sea level, the spectral variations of the primary cosmic rays in space from ~1 GV to ~15 GV can be deduced.

The second effect of the deflection of cosmic rays in the Earth's magnetic field is that the angle of incidence of a particle at the top of the atmosphere differs from the direction of the velocity vector of the particle outside the magnetosphere. Figure 3, taken from Flückiger [18], shows the direction of origin outside the geomagnetic field of the particles hitting vertically the atmosphere above Jungfraujoch. These so-called asymptotic directions depend on rigidity. We see from Figure 3 that the bulk of the cosmic rays observed at the Jungfraujoch originates from equatorial regions. The deflection in longitude is ~45° for a 20 GV particle and increases with decreasing rigidity to ~310° for $P = 4.84$ GV.

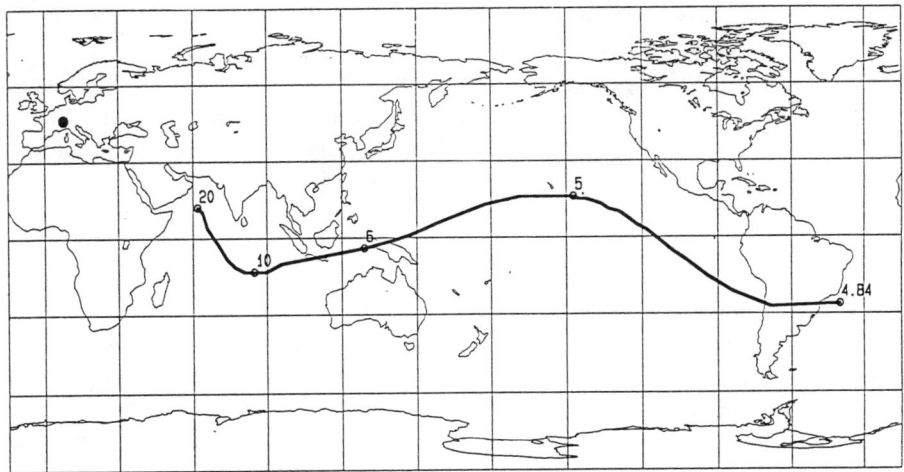

Fig. 3. The asymptotic directions of the Jungfraujoch NM for cosmic rays entering the atmosphere from the vertical [Flückiger [18]]

When the primary cosmic rays enter the Earth's atmosphere, they collide with atoms and nuclei of the atmospheric gases. They rapidly loose energy and give rise to the different components of the secondary cosmic radiation in the atmosphere [Rossi [19]]. The response of NMs to the primary cosmic radiation is, therefore, determined by the effects of the geomagnetic field, the interactions of the primary cosmic rays with the atmosphere and the detector characteristics for measuring the secondary cosmic radiation in the atmosphere.

In analyses of energetic solar proton events, the response of a NM at geographic longitude Φ and latitude Λ to the anisotropic flux of solar protons can be written as [Smart et al. [20], Debrunner and Lockwood [21]]:

$$\delta N_p(\Phi, \Lambda, t) = \int_{P_c}^{\infty} \Psi_p(P, t) \, F(\chi, t) \, S_p(P) \, dP \qquad (3)$$

where:

$\delta N_p(\Phi, \Lambda, t)$ is the increase in the count rate of a NM at location Φ, Λ and time t due to solar protons, corrected to sea level and measured in units of the average count rate of an equivalent equatorial NM at sea level.

P_c is the effective geomagnetic cutoff rigidity of the NM for vertically incident cosmic rays.

$\Psi_p(P, t)$ is the solar proton flux (protons/m^2 sr s GV) in interplanetary space near the Earth averaged over the whole solid angle of 4π.

$F(\chi, t)$ is the pitch angle distribution of the solar protons near Earth taken to be independent of rigidity. χ represents the angular distance between the asymptotic direction of the vertically incident protons at the NM and the direction of the apparent source.

$S_p(P)$ is the normalized specific yield function of NMs at sea level for priprimary protons.

In applying equation (3) the measured count rate increases are first corrected to sea level using the two-attenuation length method proposed by McCracken [22] and then normalized to the average count rate of the equivalent equatorial NM at sea level using the latitude survey by Carmichael and Bercovitch [23] corrected for the solar modulation present at the time of the observation. Ψ_p and F can be found by first assuming a source location which is shifted from the actual location of the Sun by the effect of the interplanetary magnetic field upon the transport of the solar protons from the Sun to the Earth. Next, the calculated data, $\delta N_p(\Phi, \Lambda, t)$, are fitted as function of Ψ_p and F to the normalized observed increases, $\delta N_p(\Phi, \Lambda, t)$, by a trial and error method. This procedure is discussed in detail in Debrunner and Lockwood [21].

We present an example of such an analysis to determine the differntial energy spectrum of the solar protons in interplanetary space near Earth. In Figure 4 we show

$\Psi_p(E)$ derived by Debrunner et al.[12] for the first 10 minutes of the May 7, 1978 solar cosmic ray event. The open circle points were deduced from proton measurements with the cosmic ray telescopes on board the IMP-7 spacecraft which cover an energy range from ~50 MeV to ~500 MeV. The solid circle points were obtained from the data of the world wide network of NMs. At the overlapping energy region near $E \approx 500$ MeV, it is seen that the agreement between the two different data sets is excellent. This shows that the NM measurements when used with proper specific yield functions can yield the absolute primary proton flux. Figure 5 illustrates the time dependence of the solar proton flux near Earth on May 7, 1978. The complete energy spectra cannot be described by a simple exponential or power law either in kinetic energy or rigidity. A pronounced maximum is observed at lower energies during the earlier phase of the event. This turnover of the spectra and the shift of the maximum with time to lower energies can be partly understood by the energy dependence of the transit time of the protons for propagating from the Sun to the Earth.

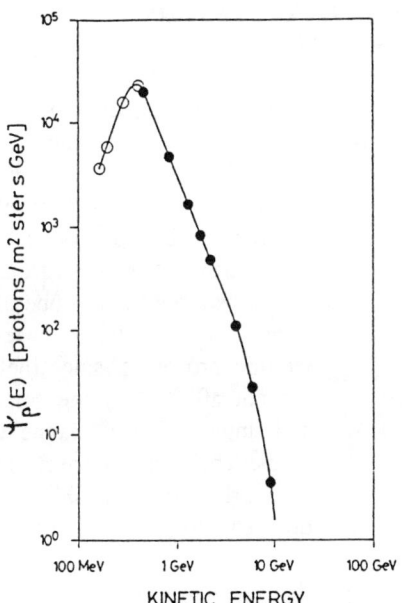

Fig. 4. Differential energy spectrum of the solar protons during the first 10 minutes of the May 7, 1978 solar cosmic ray event [Debrunner et. al.[12]]

On May 7, 1978, the interplanetary magnetic field between the Sun and the Earth was practically undisturbed. The solar protons reaching the Earth propagated nearly scatter-free along the the spiral magnetic field lines connecting the Earth to the Sun. This direct transport led Lockwood et al.[13,14] to construct the relative proton emission spectra at the Sun from the solar proton spectra observed at Earth. The relative proton energy spectra at the Sun varied with time and do not agree with those predicted for stochastic acceleration processes [Ramaty[24]]. They agree, however, with the predictions of the diffusive coronal shock acceleration model of Ellison and Ramaty[15] for a shock size of $\geq 10^9$ cm and a shock compression ratio decreasing from ~2.1 at onset of the event to ~1.8 at the end of the high energy proton emission.

We conclude from the analysis of the May 7, 1978 solar cosmic ray event that the combined evaluation of the data from the world wide network of NMs and from cosmic ray space experiments is still a powerful tool to study energetic solar proton events. The NM measurements complement the space measurements and expand our knowledge about the primary solar proton spectra near Earth up to energies of ~15 GeV. They facilitate a detailed comparison of the experimental data with the predictions of the various acceleration models and deepen the insight into the physical processes of large solar flares.

THE NEUTRON MONITOR AS A TOOL TO STUDY SOLAR NEUTRON EVENTS

Shortly after the first reports on solar cosmic ray proton events, Biermann et al.[25] pointed out that the GeV-ions accelerated at the Sun could interact with nuclei of the solar atmosphere and produce a sufficient flux of neutrons to be observable at the Earth. These neutrons would be a powerful probe of solar flare particle acceleration processes, since they are not affected by the solar and interplanetary magnetic fields. Their energy spectrum and arrival time would give detailed information about the spectral characteristics and the time history of the charged particles which are accelerated in the flare and deposit their energy in the solar atmosphere. However, attempts to detect solar neutrons at or near the Earth were unsucessful for about 30 years. The first direct observations of solar neutrons were made by the Gamma-Ray-Spectrometer (GRS) on the Solar Maximum Mission (SMM) satellite during the impulsive solar flares on June 21, 1980 [Chupp et al.[26]] and June 3, 1982 [Chupp et al.[16]]. During the second event the solar neutrons were also detected by NMs [Debrunner et al.[27], Efimov et al.[28], Iucci et al[29]].

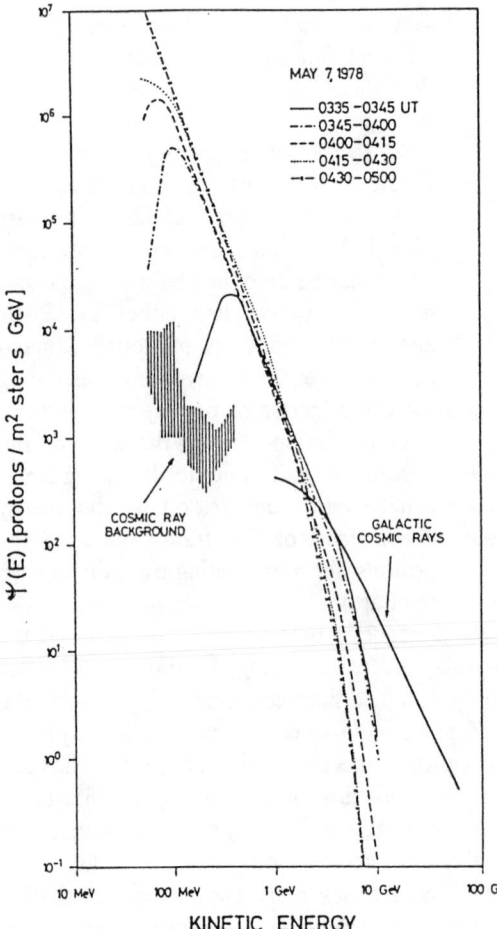

Fig. 5, Energy spectrum of solar protons of the May 7, 1978 solar cosmic ray event at various time intervals [Debrunner et. al.[12]]

The response of a NM to solar neutrons can be written [Debrunner et al.[30]]:

$$\Delta N_n^x (h,\vartheta,t) = \int \Psi_n (E_n, t)\, S_n^x (h, \vartheta, E_n)\, dE_n \qquad (4)$$

where:

ΔN_n^x is the absolute count rate increase of a NM of type x (size; type: IGY or NM64) due to the solar neutrons.

h   is the atmospheric depth of the NM.

ϑ is the angular distance between the location of the NM and the subsolar point.

t   is the time.

$\Psi_n (E_n, t)$ is the solar neutron flux at Earth, measured in neutrons/m² s MeV, as a function of energy E_n.

$S_n^x (h, \vartheta, E_n)$ is the yield of a NM of type x at the atmospheric depth h and angular distance ϑ from the subsolar point for solar neutrons of energy E_n.

The unambiguous determination of whether a count rate increase observed by the world wide network of NMs was due to solar protons or solar neutrons is fraught with difficulties. This fact has been demonstrated by the various papers evaluating the first peak in the count rate of NMs during the solar particle event on May 24, 1990 [Shea et al.[31], Debrunner et al.[32,33], Kovaltsov et al.[34]]. For solar cosmic ray ground level events due to solar protons the NM count rate increases corrected to sea level and measured in units of the average counting rate of an equivalent equatorial NM, $\delta N_p(\Phi, \Lambda, t)$, should be well ordered in respect to the effective geomagnetic cutoff rigidity of the stations, P_c, and the angular distance between the mean asymptotic viewing direction of the NMs and the apparent source direction. In addition, the absolute increases measured in counts per unit time which may be observed by two neighbouring NMs at different atmospheric depths should in most cases lead to an attenuation length of the event in the order of $L_p \approx 100$ g/cm². For solar cosmic ray ground level events due to solar neutrons a count rate increase is observed only by NMs on the day side hemisphere. As a further criterion to differentiate between solar cosmic ray ground level events produced by solar neutrons and those produced by solar protons, the dependence of the response of the NMs on the atmospheric depth of the stations and their angular distance from the subpolar point can be used. In applying equation (4) it has, however, to be recognized that $\Delta N_n^x (h,\vartheta,t)$ and $S_n^x (h, \vartheta, E_n)$ depend on the type and the size of the NMs. The observed absolute increases in the NM count rate must, therefore, be normalized to the increases which would have been measured under the same conditions by a given NM type. In reducing the observed increases to the normalized increases, the different response of IGY and NM64 neutron monitors to the cosmic radiation has to be taken into account. This difference in response between the IGY and the NM64 neutron monitor has been discussed in detail by Hatton [17]. For cosmic ray events due to solar neutrons the normalized absolute increases should then depend on the atmospheric depth, h, and the angular distance from the subsolar point, ϑ, as determined by Debrunner et al. [30,35] and Kovaltsov et al. [34].

If the occurence of a solar neutron event has been established, the normalized measured NM increases, $\Delta N_n (h,\vartheta,t)$, can be evaluated in terms of the neutron emission at the Sun. Equation (4) can be rewritten as:

$$\Delta N_n (h,\vartheta,t) = R^{-2} \int_{t_{min}}^{\infty} \mu (t - t_s) \, Q (E_n) \, dE_n / dt_s \, P (E_n) \, S_n (h, \vartheta, E_n) \, dt_s \qquad (5)$$

with:

$$\int_0^{\infty} \mu (t) \, dt = 1 \qquad (6)$$

where:

R	is the Sun-Earth distance.
μ	is the normalized time profile of the neutron emission at the Sun.
t_s	is the transit time of the neutrons for propagating from the Sun to the Earth.
$E_n = E_n (t_s)$	is the neutron energy defined by t_s.
$Q (E_n)$	is the time integrated directional solar neutron emissivity measured in neutrons/sr MeV.
dE_n / dt_s	is the neutron energy-time dispersion relation given by Lingenfelter and Ramaty [36].
$P (E_n)$	is the survival probability for a neutron in reaching the Earth.
t_{min}	corresponds to the high energy cutoff of the neutron emissivity spectrum $Q (E_n)$ at the Sun.

The time profile of the neutron emission at the Sun can be assumed to fit the time profile of the γ-emission at $E_\gamma \approx 60$ MeV which is most probably due to π° decays where the π°'s result from the neutron production by the high energy protons in the solar atmosphere [Murphy et. al. [37]]. It is then possible to determine the time integrated directional neutron emissivity at the Sun, $Q (E_n)$, by fitting $\Delta N_n (h,\vartheta,t)$ calculated according to equation (5) to the observed count rate increase of a NM by a trial and error method.

We present an example of such an analysis to determine the time integrated neutron emissivity spectrum of a solar neutron event. In Figure 6 we show the count rates of several energy channels of SMM GRS and the relative excess count rate of the Jungfraujoch IGY neutron monitor during the June 3, 1982 solar cosmic ray event

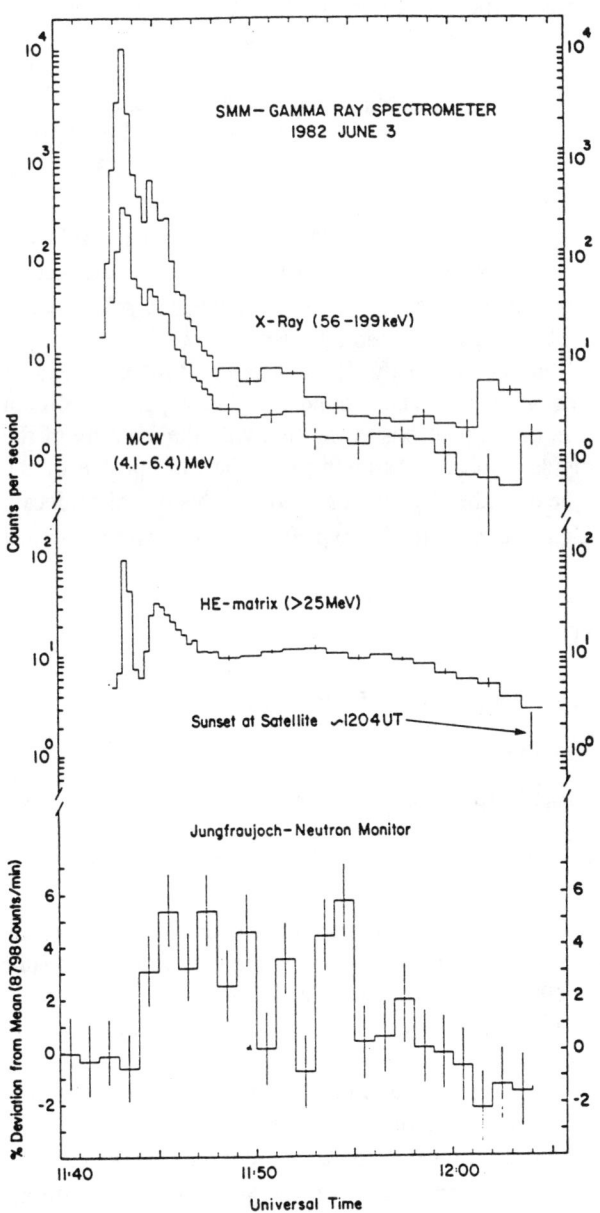

Fig. 6. Time history for several data channels from the SMM GRS and for the Jungfraujoch NM count rate for the June 3, 1982 solar neutron event.

[Chupp et al.[16]]. The 56-199 keV channel responds predominantly to electron bremsstrahlung and the 4.1-6.4 MeV γ-ray channel to high energy nuclear line emission indicative of ion interactions at the Sun. The count rate of the high energy matrix HE of SMM GRS results from interactions with energy losses in the detector of ≥25 MeV and is caused by hard γ-rays and neutrons. The intensity-time profile of this channel shows a first peak with a maximum in the time interval 1143.0-1143.5 UT and a second peak with a maximum at ~1145.0 UT, followed by a high excess count rate until satellite sunset. The contributions of γ-rays from electron bremsstrahlung, nuclear line emission and π° decay and of solar neutrons to this complex time history of the HE matrix count rate have been analyzed and discussed by Forrest et al.[38]. The relative excess count rate of the Jungfraujoch NM is expressed in percent of the average count rate of two reference intervals extending on June 3, 1982 from 1110 to 1140 UT and from 1200 to 1230 UT.

Evidence that the increase in the count rate of the Jungfraujoch NM was produced by energetic solar neutrons and not by solar protons came from the world wide network of NMs. Similar increases were also measured by the neighbouring NMs at Lomnicky Stit, Czechoslovakia, [Efimov et al. [28]] and in Rome [Iucci et al. [29]]. The high altitude NM at Lomnicky Stit and the sea level NM in Rome recorded in the time interval 1145-1150 UT relative enhancements in the count rate of $(2.80\pm0.33)\%$ and $(2.04\pm0.71)\%$, respectively. All the other NMs of the world wide network showed no increase. It is further important to note that the angular distance from the subsolar point of the three stations which measured an enhancement in the counting rate was $\vartheta \leq 31°$. All these facts lead to the generally accepted conclusion that the solar cosmic ray ground level event on June 3, 1982 was produced by solar neutrons.

Chupp et al. [16] evaluated the records of the HE matrix of SMM GRS and of the Jungraujoch NM applying equation (5). They found that the experimental data required a time extended emission of the neutrons at the Sun with the majority of the neutrons produced after the impulsive phase. Assuming that the time profile of the neutron emission at the Sun can be described by the observed time history of $\pi°$ decay γ-rays reasonable fits of the calculated ΔN_n to the experimental data were obtained for:

$$Q(E_n) \propto E_n^{-s} \qquad E_n < E_c \qquad (7)$$
$$Q(E_n) = 0 \qquad E_n > E_c$$

with $\qquad 2.2 < s < 2.8 \quad$ and $\quad 2 \text{ GeV} < E_c < 4 \text{ GeV}$

and

$$Q(E_n) \propto E_n^{3/8} \exp\{-(E_n/(3.26\,\varphi^2))^{1/4}\} \qquad (8)$$

with $\qquad 0.05 < \varphi < 0.09$

All these fits are consistent with a time integrated neutron emissivity at the Sun for energies $E_n > 100$ MeV of $\sim 8 \times 10^{28}$ neutrons / sr and agree very well with results of Evenson et al. [39] for $E_n < 100$ MeV which were deduced from ISEE 3 measurements of protons resulting from the decay of solar neutrons in interplanetary space.

Murphy et al. [37] suggested that the first and second peak in the count rate of the HE matrix of SMM GRS were produced by two different proton populations. Early in the time history of the flare, the observed γ-rays due to nuclear deexcitation and bremsstrahlung from electrons were dominant compared with the γ-rays due to $\pi°$ decay. The solar proton spectrum which produced the observed ratio between γ-rays from nuclear lines and γ-rays from $\pi°$ decay was found to be much steeper than the spectrum of the solar flare protons observed in interplanetary space by the Goddard cosmic ray detector on board the Helios I spacecraft [McDonald and Van Hollebeke [40]]. On the other hand, later in the event the ratio between γ-rays from nuclear lines and γ-rays from π^0 was consistent with a proton spectrum at the Sun described by a power law in rigidity with spectral index γ = -2.4. In the analysis of Chupp et al.[16] it was also possible to find different spectral forms during the impulsive and extended emissions of hard γ-rays which give reasonable fits to the combined SMM GRS and Jungfraujoch NM measurements. It is clear, however, that the experimental neutron data are not adequate to determine the large number of parameters necessary for such emission models in an unambiguous way.

In summary, we learned from the analysis of the June 3, 1982 solar cosmic ray event that only by combining the neutron observations by the SMM GRS and the Jungfraujoch NM was it possible to determine that the neutron emission at the Sun was time-extended. Using both data sets Chupp et al.[16] were then able to find the neutron emissivity and the neutron spectrum.

CONCLUSIONS

NMs are relatively simple detectors and can be maintained in stable operation for long periods of time. Since they respond to variations of the primary cosmic ray intensity in near-Earth space for energies from about 500 MeV to 30 GeV, they complement satellite measurements in an indispensable way to expand our knowledge about cosmic rays. From the two examples of the solar cosmic ray events on May 7, 1978 and June 3, 1982 we see that NMs provide invaluable information about solar proton and neutron emission in the transition region of nonrelativistic to relativistic energies and help to deepen our understanding about the particle acceleration mechanisms in large solar flares.

ACKNOWLEDGMENTS

The author expresses his profound appreciation to J.A. Lockwood who helped to prepare this review paper by stimulating discussions and helpful criticism. He also thanks E.O. Flückiger for providing results of an analysis of the geomagnetic cutoff and the asymptotic directions of the Jungfraujoch NM. This work was supported by the Swiss National Science Foundation, grant 20-31130.91.

REFERENCES

1. J.A. Simpson, Phys. Rev. 73, 1389 (1948)
2. J.A. Simpson, Echo Lake Conf. on Cosmic Rays, 175 (1949)
3. V.C. Tongiorgi, Phys. Rev. 75, 1532 (1949)
4. J.A. Simpson, Echo Lake Conf. on Cosmic Rays, 252 (1949)
5. G. Cocconi, V. Cocconi-Tongiorgi and M. Widgoff, Phys. Rev. 79, 768 (1950)
6. N. Adams and H.J. Braddick, Z. Naturforsch. 6a, 592 (1951)
7. J.A Simpson and R.D. Uretz, Phys. Rev. 90, 44, (1953)
8. J.A. Simpson, W. Fonger and S.B. Treiman, Phys. Rev. 90, 934, (1953)

9. J.A. Simpson and W.C. Fagot, Phys. Rev. 90, 1068, (1953)
10. J.A. Simpson, Ann. I.G.Y. (London, Pergamon Press, 1957) part VII
11. H. Carmichael, Cosmic Rays, IQSY Instruction Manual No. 7 (IQSY Secretariat, London, 1964)
12. H. Debrunner, E. Flückiger, J.A. Lockwood and R.E. McGuire, J. Geophys. Res. 89, A2, 769 (1984)
13. J.A. Lockwood, H. Debrunner and E.O. Flückiger, J. Geophys. Res. 95, A4, 4187 (1990)
14. J.A. Lockwood, H. Debrunner, E.O. Flückiger and H. Grädel, Ap. J. 355, 287 (1990)
15. D.C. Ellison and R. Ramaty, Ap. J. 298, 400 (1985)
16. E.L. Chupp, H. Debrunner, E. Flückiger, D.J. Forrest, F. Golliez, G. Kanbach, W.T. Vestrand, J. Cooper and G. Share, Ap. J. 318, 913, (1987)
17. C.J. Hatton, Progress in Elementary Particle and Cosmic Ray Physics, (North-Holland Publishing Company, Amsterdam and London, 1971), Vol.X, p. 1
18. E.O. Flückiger, private communication (1993)
19. B. Rossi, Cosmic Rays (McGraw-Hill Book Company, New York, 1964)
20. D.F. Smart, M.A. Shea and P.J. Tanskanen, Proc. 12th Int. Cosmic Ray Conf. 2, 483 (1971)
21. H. Debrunner and J.A. Lockwood, J. Geophys. Res. 85, 6853 (1980)
22. K.G. McCracken, Geophys. Res. 67, 423 (1962)
23. H. Carmichael and M. Bercovitch, Canad.. J. Phys. 47, 2037 (1969)
24. R. Ramaty, Particle Acceleration Mechanisms in Astrophysics (American Institute of Physics, New York, 1979), p. 135
25. L. Biermann, O. Haxel and A. Schluter, Z. Naturforsch. 6a, 47 (1951)
26. E.L. Chupp, D.J. Forrest, J.M. Ryan, J. Heslin, C. Reppin, K. Pinkau, G. Kanbach, E. Rieger and G.H. Share, Ap.J. 263, L95 (1982)
27. H. Debrunner, E. Flückiger, E.L. Chupp and D.J. Forrest, 18th Int. Cosmic Ray Conf. 4, 75 (1983)
28. Yu.E. Efimov, G.E. Kocharov and K. Kudela, 18th Int. Cosmic Ray Conf. 10 276 (1983)
29. N. Iucci, M. Parisi, C. Signorini, M. Storini and G. Villoresi, 19th Int. Cosmic Ray Conf. 4 134, (1985)
30. H. Debrunner, E.O. Flückiger and P. Stein, Nuclear Instr. and Methods, A278, 573 (1989)
31. M.A. Shea, D.F. Smart and K.R. Pyle, Geophys. Res. Letts., 18, 1655 (1991)
32. H. Debrunner, J.A. Lockwood and J.M. Ryan, Ap.J. Letters 387, L51 (1992)
33. H. Debrunner, J.A. Lockwood and J.M. Ryan, Ap.J., in press (1993)
34. G.A. Kovaltsov, Yu.E. Efimov and I.G. Kocharov, Solar Physics Letters, in press (1993)
35. H. Debrunner, E.O. Flückiger and P. Stein, 21st Int. Cosmic Ray Conf. 5 129, (1990)
36. R.E. Lingenfelter and R. Ramaty, High Energy Nuclear Reactions in Astrophysics (Benjamin, New York, 1967)

37. R.J. Murphy, C.D. Dermer and R. Ramaty, Ap. J. Suppl. 63, 721, (1987)
38. D.J. Forrest, W.T. Vestrand, E.L. Chupp, E.L. Rieger, J. Cooper and G. Share, Adv. Space Res. (COSPAR) 6, No.6, 115 (1986)
39. P. Evenson, P. Meyer and K.R. Pyle, Ap. J. 274, 875, (1983)
40. F.B. McDonald and M.A.I. Van Hollebeke, Ap. J. Letters 290, L67, (1985)

THE RELATIVISTIC SOLAR PROTON GROUND-LEVEL ENHANCEMENTS ASSOCIATED WITH THE SOLAR NEUTRON EVENTS OF 11 JUNE AND 15 JUNE 1991.

D. F. Smart and M. A. Shea
Space Physics Division, Geophysics Directorate
Phillips Laboratory, Hanscom AFB, Bedford, MA 01731-3010

L. C. Gentile
Boston College Institute for Space Research
885 Centre Street, Newton, MA 02159

ABSTRACT

The Solar Cosmic Ray Ground-Level Enhancements (GLEs) observed on 11 and 15 June 1991 were distinctly different in character. The small GLE on 11 June was mildly anisotropic with an approximately 2-to-1 ratio in the relativistic proton flux observed by "forward viewing" high latitude neutron monitors as compared with the flux observed by "reverse viewing" high latitude neutron monitors. In contrast the 15 June GLE was almost isotropic in spite of the fact that the source solar flare position was at heliolongitudes that were presumably "well-connected" to the earth via the average interplanetary magnetic field topology. A differential power law in rigidity seems to fit the data in the region between 1 and 6 GV for both events. For the 11 June GLE maximum our derived slope is -5.5. For the 15 June GLE maximum our derived slope is -6. It is our opinion that the lack of observed flux anisotropy during the 15 June GLE is probably due to the very disturbed interplanetary propagation conditions rather than solar source characteristics.

I. INTRODUCTION

The episode of solar activity that occurred during June 1991 generated a number of energetic phenomena including intense X-ray and gamma ray emission, solar neutron emission, and the acceleration of ions to relativistic energies detectable at the earth. The details of these solar flares and their energetic X-ray, gamma ray and solar neutron emission are described elsewhere in this volume and will not be repeated here. The powerful solar flares in this activity episode generated interplanetary shocks that propagated through the heliosphere. Six sudden commencement geomagnetic storm onsets were recorded at the earth between 4 and 12 June. The intense solar activity contributed to the historic cosmic ray intensity minimum observed during this month. While the effects of this activity episode on the propagation conditions in the heliosphere have not been fully ascertained, all of these effects strongly suggest that propagation conditions were not quiescent. Also suggestive of the non-quiescent propagation conditions were the variations in the pre-event cosmic ray background which exceeded the variations expected from Poisson statistics.

The energy source for the solar cosmic ray ground-level enhancements (GLEs) studied in this paper were energetic solar flares in NOAA region 6659. The 11 June 1991 GLE is time-associated with the X12/3B solar flare at heliographic coordinates N31, W17 having an H-alpha onset of 0156 UT. The 15 June 1991 GLE is time-associated with the X12/3B solar flare at heliographic coordinates N33, W69 having an H-alpha onset of 0810 UT. The historic cosmic ray intensity low was recorded on 13 June 1991, in the interval between these two GLEs.

II. METHOD OF DETERMINING HIGH ENERGY SOLAR PARTICLE SPECTRA FROM THE ANALYSIS OF NEUTRON MONITOR DATA

An introduction to the general concept of using cosmic ray neutron monitor data for the analysis of high energy solar proton events is given in this volume by Debrunner[1] and that material will not be repeated in this paper. We have used our standard technique for the analysis of GLEs[2] to determine the spectral characteristics and flux anisotropy of the June GLEs. The method is designed to reproduce the increase observed by the individual neutron monitors around the world. This is done from a numerical analysis of the solar particle spectrum, the flux anisotropy, the asymptotic cone of acceptance for each station and the neutron monitor yield function[3]. For this analysis we have used the Debrunner[4] et al. neutron monitor yield functions to reproduce the observed increases at each neutron monitor. We have limited the form of the spectral parameters used to a differential power law in rigidity which seems to produce a satisfactory fit to the neutron monitor data in the rigidity range between 1 and 6 GV.

We model the increase utilizing the functional form,

$$I = \sum_{R_c}^{\infty} J_\alpha(\alpha,R) \, S(R) \, G(\alpha) \, \Delta R \tag{1}$$

where I is the increase at the neutron monitor, R_c is the cutoff rigidity, $J_\alpha(\alpha,R)$ is the differential flux in the interplanetary medium at pitch angle α and rigidity R that is allowed through the asymptotic cone of acceptance, S(R) is the neutron monitor specific yield as a function of rigidity, and $G(\alpha)$ is the anisotropic pitch angle distribution. In our modeling approach we sum the spectrum-yield response for each station from 0.7 (or the cutoff rigidity) at 0.1 GV intervals to 25 GV.

In describing our method it is necessary to explain the concept of asymptotic directions of approach. Charged particles of a specified energy arriving at a detector from a specific direction can be "mapped" through the geomagnetic field to a specific direction in space[5,6,7]. The asymptotic direction of approach defines an allowed particle's direction in space prior to its interaction with the earth's magnetic field. From the "geomagnetic optics" of high latitude neutron monitors, we can determine the orientation of the asymptotic cone of acceptance to the interplanetary magnetic field direction and estimate the flux arriving at each station. In our model calculations we define pitch angle zero as the direction of the maximum particle flux which generally corresponds to the direction of the interplanetary magnetic field.

III. THE GLE OF 11 JUNE 1992

The world-wide network of neutron monitors on the earth recorded a small, mildly anisotropic GLE on 11 June. For this event there was an impulsive onset at the forward viewing stations (those having asymptotic directions of approach viewing into the solar particle flux propagating along the interplanetary magnetic field direction away from the sun) in the five-minute interval 0235-0240 UT[8]. The onset for reverse viewing stations (those having asymptotic directions of approach viewing into the particle flux propagating along the interplanetary magnetic field back toward the sun) was after 0300 UT. At the time of the GLE maximum at about 0330 UT stations viewing in the probable forward direction such as Apatity, Russia and Mawson, Antarctica recorded an increase of ~7 percent while stations viewing in the probable reverse direction such as Tixie Bay, Russia and Inuvik, Canada recorded an increase of ~3 percent. An overall conceptual view of this small GLE can be obtained from Figure 1. In this composite figure we show the asymptotic viewing directions for selected high latitude neutron monitors on the right. The top left shows the increase observed by a station viewing into the forward propagating flux and a station viewing into the reverse particle flux. The bottom left illustrates the pitch angle distribution required to generate the observed particle anisotropy at the GLE maximum[8]. This form is similar to the exponential form derived by Beeck and Wibberenz[9].

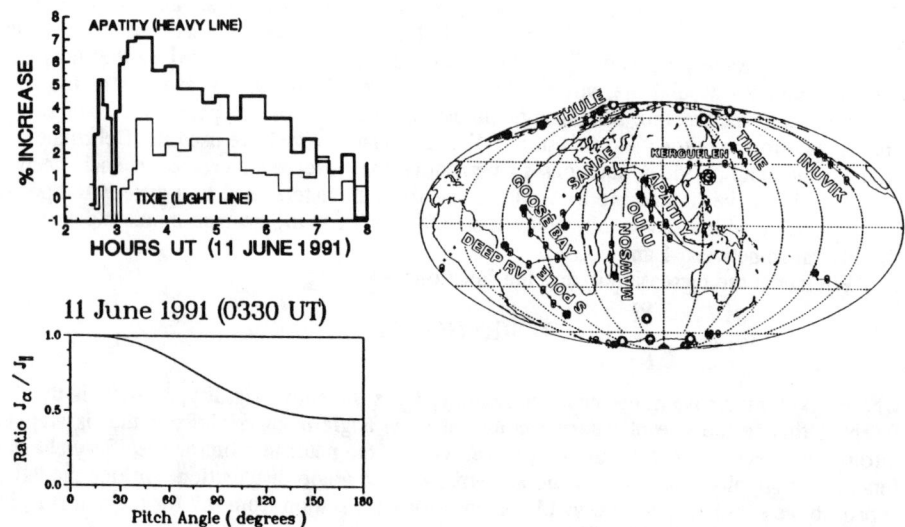

FIG. 1. *The 11 June 1991 GLE. Left top: relative increase observed by Apatity, Russia (forward viewing) and Tixie Bay, Russia (reverse viewing). Right: Display of the asymptotic viewing directions responsible for 10 to 90 percent of the response of selected high latitude neutron monitors. (Five minute data are displayed until maximum; 15 minute averages thereafter.) At the time of the GLE maximum (0330 UT), the subsolar point was at 23° N, 127° E. Bottom left: Solar particle flux pitch angle distribution necessary to produce the observed variation in the high latitude neutron monitors at the GLE maximum.*

Unfortunately, there are no interplanetary magnetic field (IMF) measurements by earth-orbiting spacecraft for 11 June 1991. However, we can use the anisotropy of the observed increase and the onset time to approximate the probable IMF direction, at least to the proper octant. The geomagnetic field was severely disturbed, and a geomagnetic storm was in progress. At the GLE onset the Dst was -96 nT and increasing toward the maximum of -140 nT which was observed at 06 hours UT. In our analysis of this event, the IMF direction did not appear to be stable during the GLE, but in our opinion, in this case this is not a serious impediment to making a useful spectral determination.

A power law in rigidity having a slope of -5.5 yielded a satisfactory fit between the increases of the various neutron monitors as a function of latitude. For this event there were no significant increases reported for stations whose geomagnetic cutoff exceeded 4 GV. A harder spectrum would predict increases at stations having a cutoff rigidity > 4 GV. We determined the magnitude of the particle flux parallel to the IMF direction and the flux averaged over all directions. For this event the differential power law in rigidity that fits the neutron monitor data in the range of 1 to 4 GV at the 0330 UT GLE maximum is:

$$J_{||} = 4.93 \, P^{-5.5}; \quad \text{and} \quad J_{(avg)} = 3.49 \, P^{-5.5}. \tag{2}$$

$J_{||}$ is the flux in units of $(cm^2\text{-s-ster-GV})^{-1}$ parallel to the interplanetary magnetic field (i.e. pitch angle of zero). $J_{(avg)}$ is obtained by summing the anisotropic flux over 4π steradians.

A. Comparison of the 11 June 1991 GLE Spectra with Spacecraft Data.

We can compare the spectrum of the more rigid particles (>1 GV or >433 MeV) derived from the analysis of neutron monitor data with the GOES 6 and 7 five minute data. We have integrated the GLE spectrum derived from the analysis of the neutron monitor data and extended it to 30 MeV as illustrated in Figure 2. The heavy line in this figure indicates the spectrum that is derived from the high energy flux observed by the neutron monitors at the 0330 UT GLE maximum. The light line is an extension of this spectrum to the spacecraft measurement energies. The + symbol identifies the spacecraft observed fluxes at the time of the GLE maximum.

In addition to data at 30, 50, 60 and 100 MeV, there is a higher energy particle detector on the GOES-6 spacecraft from which integral flux above 355, 433 and 505 MeV can be obtained[10]. We have plotted these data in Figure 2 for comparison. The empirical correction for side penetration of the sensor by high energy particles provided by Sauer[10] has been applied to these data. The spacecraft data show a velocity dispersive time-of-maxima in the initial part of the event. The velocity dispersive flux maximum for energies >300 MeV to >30 MeV occurred after the 0330 UT GLE flux maximum, between 0430 and 0500 UT. These data (indicated by the • symbol in Figure 2) can be used to construct a time-of-maximum spectrum.

The GOES spacecraft particle flux data[10] for this event are shown in Figure 3. To our biased observations, there are two maxima displayed in these data. A velocity dispersive flux maximum occurs between 0330 UT and 0500 UT. There is a second, larger and non-velocity-dispersive flux maximum that occurs at about 14 hours UT during an extended period of increased magnetic activity when Dst exceeds -100 nT. It is our opinion that this strongly indicates an interplanetary source of the particles contributing to the second maximum. In view of this we are reluctant to take the integrated fluence observed by earth-orbiting satellites for this event and extrapolate it back to the sun to estimate the number of protons released from the acceleration site.

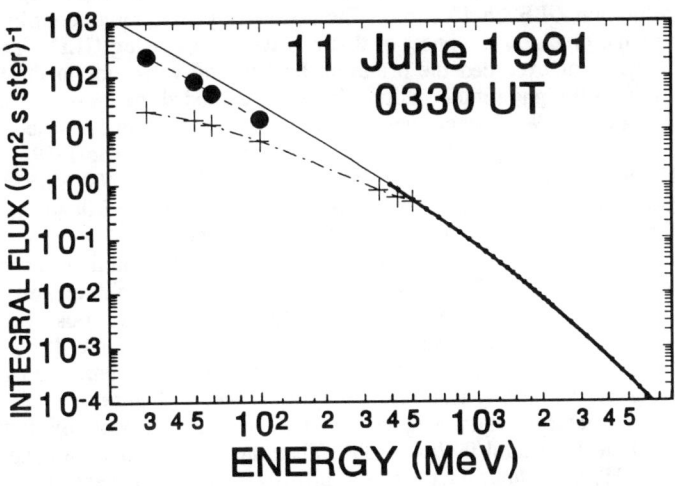

FIG. 2. Integral energy spectra for the GLE of 11 June 1991. The heavy dark line indicates the spectrum derived from neutron monitor data. The light line is the extension to satellite measurement energies. The + symbol indicates the spacecraft measured integral flux at the time of the GLE maximum (0330 UT). The • symbol indicates the spacecraft measured integral flux during the 0430-0500 UT maximum.

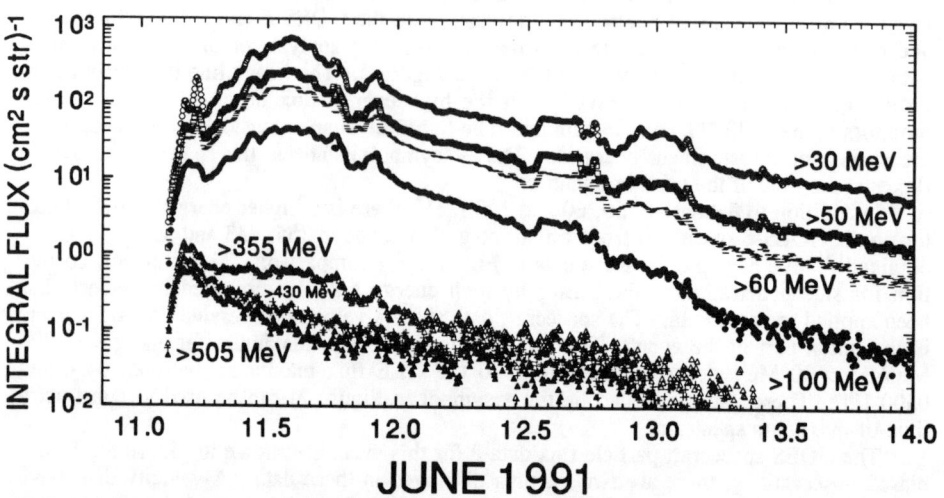

FIG. 3. *The solar particle flux observed by the GOES spacecraft for the 11 June 1991 event. Note the non-velocity-dispersive maximum at about 14 hours UT which corresponds in time to the maximum of the geomagnetic storm.*

IV. THE GLE OF 15 JUNE 1991

The world-wide network of neutron monitors recorded a small, approximately isotropic, long-duration GLE on 15 June. The disturbed propagation conditions make it difficult to determine precisely the onset of this relatively slow-rising GLE. The increase systematically equaled or exceeded the pre-event background variations in the 0835-0840 UT time interval[12]. All high latitude stations definitely exceeded the pre-event background variations after 0840 UT. We cannot identify a definite anisotropy in the onsets of the forward viewing stations as compared to reverse viewing stations. The IMP-8 spacecraft recorded a velocity-dispersive onset in its measurement range of 8 to 400 MeV[13]. The onset for the highest energy measurement (190-400 MeV) is in the 0835-0840 time interval, essentially in time coincidence with the neutron monitor onsets.

At the time of the GLE maximum at about 0930 UT all high latitude neutron monitors recorded an increase of $\sim 20 \pm 4\%$. There was a very small flux amplitude anisotropy with stations viewing in the forward flux propagation direction such as Goose Bay, Canada recording an ~ 22 percent increase while stations viewing in the reverse flux propagation direction such as Tixie Bay, Russia observed an increase of ~ 17 percent. An overall view of this long lasting, approximately isotropic GLE is presented in Figure 4.

There are direct interplanetary magnetic field (IMF) measurements by earth-orbiting spacecraft for 15 June 1991 until 09 hours UT. Then there is a 4 hour data gap, one hour of IMF data at 14 UT, a one hour data gap, and then IMF data are present for UT hours 16 through 20. Fortunately there did not appear to be large variations in the hourly averaged IMF direction and we assumed that the IMF direction does not have large deviations during the data gaps. At 09 hours UT the observed IMF was at GSE latitude -23°, and GSE longitude 145°. Therefore the probable viewing direction into the solar proton flux was at GSE longitude of -35°.

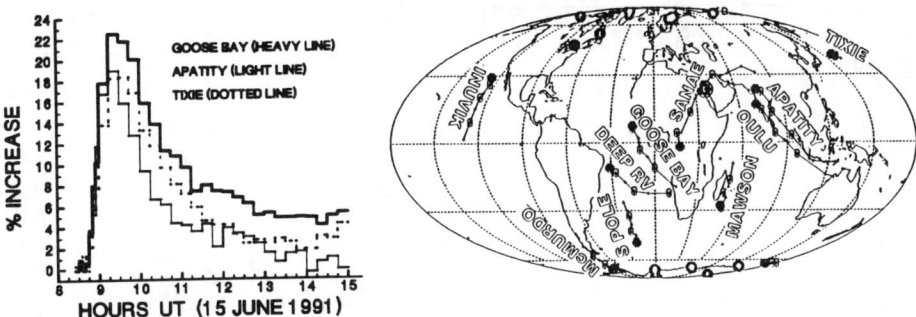

FIG. 4. *Illustration of the essentially isotropic GLE of 15 June 1991. Left: Relative increase observed by selected neutron monitors. Right: Display of the asymptotic viewing directions responsible for 10 to 90 percent of the response of selected high latitude neutron monitors. At the GLE maximum (0930 UT), the subsolar point was at 23° N, 37.5° E.*

The geomagnetic field had been severely disturbed, and was undergoing a slow recovery from a major geomagnetic storm. At the GLE onset the Dst was -41 nT and was slowly recovering during the remainder of the day.

A power law in rigidity having a slope of -6.0 yields a satisfactory fit to the increases observed by the various neutron monitors as a function of latitude. For this event there were measurable increases for stations at a quiescent geomagnetic cutoff of ~6 GV. In our analysis method, if we assume an omnidirectional flux, a differential power law in rigidity with a slope of -6.0 generates the observed 0.7% increase at Rome (R_c = 6.3 GV) and an equivalent increase when corrected to sea level at the 18-NM-64 neutron monitor (3340 meters altitude) at Alma-Ata, Kazakhstan (R_c = 6.6 GV). When we include the slight anisotropy we find that the spectrum cannot be harder than -5.5 or we would predict a larger increase than was observed at these stations. Since this was a long duration GLE we can also determine the spectrum at other times. We derive a differential rigidity spectrum which gives the flux, J, in units of $(cm^2$-s-ster-$GV)^{-1}$. We find the high energy solar cosmic ray differential rigidity spectrum to be

$$J = 19.7\ P^{-6.0} \text{ at 0930 UT, and } J = 12.5\ P^{-6.0} \text{ at 1030 UT.}$$

A. Comparison of the 15 June 1991 GLE Spectra with Spacecraft Data.

In Figure 5 we have integrated the spectrum derived from the analysis of the neutron monitor data and extended this spectrum to the lower energies for comparison with data from the GOES spacecraft. The heavy line in each panel indicates the spectrum derived from the high energy flux observed by the neutron monitors; the left panel shows the spectrum at the 0930 UT GLE maximum, the right panel one hour later. The light line is an extension of this spectrum to the spacecraft measurement energies. The + symbol identifies the spacecraft observed fluxes at each time for energies of >505 MeV, >433 MeV >355 MeV, >100 MeV, >60 MeV, >50 MeV and >30 MeV. The cosmic ray intensity was rapidly recovering from its historic intensity low on 13 June and overwhelmed the remnant of the high energy solar cosmic ray flux after 17 June.

Inspection of this figure shows that the power law in rigidity with a slope of -6.0 does not smoothly extrapolate to the spacecraft energies below 100 MeV. Further inspection of this figure suggests that a "broken power law" type of spectrum may be a better representation of the solar particle flux which evolves extremely slowly for a "well connected" solar particle event.

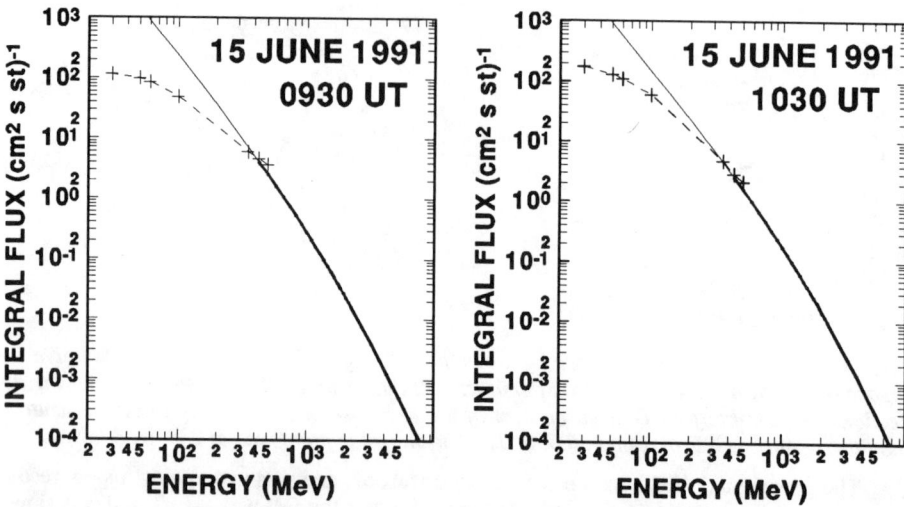

FIG. 5. *Spectrum derived for the GLE of 15 June 1991 at 0930 UT (left) and 1030 UT (right). The differential rigidity spectrum has been integrated and converted to an integral energy spectrum for comparison purposes. The heavy dark line indicates the spectrum derived from neutron monitor data. The light line is this spectrum extended to the satellite measurement energies. The + symbol indicates the measured satellite integral flux at each indicated time.*

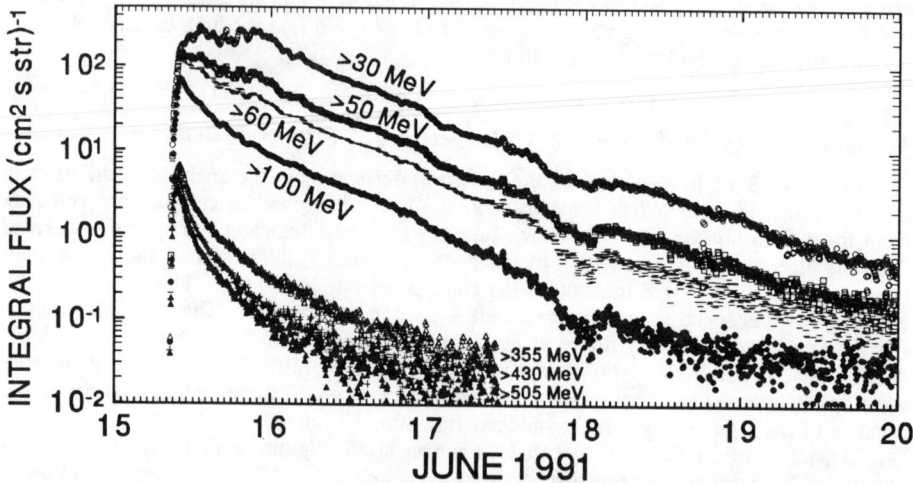

FIG. 6. *The solar particle flux observed by the GOES spacecraft for the 15 June 1991 GLE and associated solar proton event. The empirical correction for side penetration of the sensor by high energy particles provided by Sauer[10] has been applied to these data.*

The GOES spacecraft particle flux data for this event are shown in Figure 6. In spite of the velocity dispersive onset observed by the IMP-8 spacecraft[13], the flux maximum has only a weak velocity dispersion. The GLE maximum (particles with energies $> \sim 1$ GeV) occurred at about 0930 UT, and the maximum for energies between >50 MeV to >100 MeV occurred at about 1000 UT. It is our opinion that the intensity-time profile indicates that interplanetary diffusion controls both the spectral evolution and the intensity-time profile. Under these circumstances, these observations at 1 AU may not be representative of the particle source release profile.

Acknowledgment

We thank all the principal investigators who have contributed to the GLE data base. This data base is accessible to those who have the capability for remote network connections. Contact Gentile@PLH.AF.MIL, or AFGLSC::Gentile for access instruction. We wish to express special thanks to E. Eroshenko who provided data from the Russian neutron monitor network.

The research at Boston College was supported by U.S. Air Force Materiel Command contract No. F19628-90-K-0006.

References

1. Debrunner, H. This volume (1993).
2. Shea, M.A., and D.F. Smart, Space Sci. Rev., 32, 251 (1982).
3. Lockwood, J.A., W.R. Webber, and L. Hsieh, J. Geophys. Res., 79, 4149 (1974).
4. Debrunner, H., J.A. Lockwood, and E.O. Flückiger, Preprint, 8th ECRC (1982).
5. McCracken, K.G., J. Geophys. Res., 67, 423 (1962).
6. McCracken, K.G., U.R. Rao, B.C. Fowler, M.A. Shea, and D.F. Smart, in Annals of the IQSY, 1, (The MIT Press, Cambridge, 1968) Chapter 14, p. 198.
7. Gall, R., A. Orozco, C. Marin, A. Hurtado, and G. Vidargas, Tech. Rep., Instituto de Geofisica, Universidad Nacional Autonoma de Mexico (1982).
8. Smart, D.F., M.A. Shea and L.C. Gentile, 23rd Int. Cosmic Ray Conf., (Calgary), in press (1993a).
9. Beeck, J. and G. Wibberenz, Astrophys. J., 311, 437 (1986).
10. H. Sauer (private communication, 1993).
11. GOES Space Environment Monitor, National Geophysical Data Center, Boulder, Colorado (1992).
12. Smart, D.F., M.A. Shea and L.C. Gentile, 23rd Int. Cosmic Ray Conf., (Calgary), in press (1993b).
13. T. Armstrong (private communication, 1993).

The NASA High-Energy Solar Physics Mission (HESP)

B. R. Dennis
Laboratory for Astonomy and Solar Physics
Goddard Space Flight Center, Greenbelt, MD 20771

A. G. Emslie
University of Alabama, Huntsville, AL 35899

R. Canfield
Institute for Astronomy, University of Hawaii, Honolulu, HI 96822

G. Doschek
Naval Research Laboratory, Washington, DC, 20375

R. P. Lin
Physics Department and Space Sciences Laboratory
University of California, Berkeley, CA 94720

R. Ramaty
Laboratory for High Energy Astrophysics
Goddard Space Flight Center, Greenbelt, MD 20771

ABSTRACT

The NASA High Energy Solar Physics (HESP) mission offers the opportunity for major breakthroughs in our understanding of the fundamental energy release and particle acceleration processes at the core of the solar flare problem. HESP's primary strawman instrument, the High Energy Imaging Spectrometer (HEISPEC), will provide X-ray and gamma-ray imaging spectroscopy, i.e., high-resolution spectroscopy at each spatial point in the image. It has the following unique capabilities: (1) high-resolution ($\sim$ keV) spectroscopy from 2 keV - 20 MeV to resolve flare gamma-ray lines and sharp features in the continuum; (2) hard X-ray imaging with 2" angular resolution and tens of millisecond temporal resolution, commensurate with the travel times and stopping distances for the accelerated electrons; (3) gamma-ray imaging with 4 - 8" resolution with the capability of imaging in specific lines or continuum regions; (4) moderate resolution measurements of energetic (20 MeV to $\sim$1 GeV) gamma-rays and neutrons. Additional strawman instruments include a Bragg crystal spectrometer for diagnostic information and a soft X-ray/XUV/UV imager to map the flare coronal magnetic field and plasma structure. The HESP mission also includes extensive ground-based observational and supporting theory programs. Recently, the HESP mission has been adapted to "lightsats" - lighter, smaller, cheaper spacecraft that can be built faster - and the baseline plan now includes two Taurus-class and one Pegasus-class spacecraft. A launch by the end of the year 2000 is desirable to be in time for the next solar activity maximum.

INTRODUCTION

The overarching scientific objective of the High Energy Solar Physics (HESP) mission is to explore the processes of impulsive energy release and particle acceleration in the magnetized plasmas of the solar atmosphere. The fundamental importance of these high-energy processes transcends their significance in solar physics since they are found to play a major role throughout the universe at sites ranging from planetary and neutron star magnetospheres to active galaxies. The detailed understanding of these processes is one of the major goals of space physics and astrophysics, but in essentially all cases, we are only just beginning to perceive the relevant basic physics. Nowhere can one pursue the study of this basic physics better than in the active Sun, where solar flares are the direct result of impulsive energy release and particle acceleration. The accelerated particles, notably the electrons with energies of tens of keV, appear to contain a major fraction of the total flare energy, thus indicating the fundamental role of the high-energy processes. The acceleration of electrons is revealed by hard X-ray and gamma-ray bremsstrahlung; the acceleration of protons and nuclei is revealed by nuclear gamma-rays, pion-decay radiation, and neutrons. The proximity of the Sun means that these high-energy emissions appear orders of magnitude more intense than from any other cosmic source, plus they can be better resolved, both spatially and temporally. Consequently, the Sun is the only astrophysical object where the phenomena can be studied with the detail necessary to understand the fundamental processes.

The high-energy processes in solar flares that HESP is designed to study involve the rapid release of energy stored in unstable magnetic configurations, the equally rapid conversion of this energy into kinetic energy of accelerated particles and hot plasma, the transport of these particles, and the subsequent heating of the ambient solar atmosphere. Observations of hard X-rays, gamma rays, and neutrons serve as the best diagnostic of these processes by providing direct evidence for the interaction of accelerated particles in solar flares. The necessary spatial and temporal resolving powers must match the spatial and temporal scales that characterize the processes of energy release, acceleration, and transport. The sensitivity should be high enough to detect the initial energy release and particle acceleration, and also to provide observations over a wide range of intensities from microflares to large flares. Equally important, the spectral resolving power must be high enough to allow the deciphering of the rich information encoded in both the gamma-ray lines and the highly-structured photon continuum. The observations should provide imaging with spectroscopy and should cover the entire photon energy range from soft thermal X-rays to high-energy gamma rays, as well as energetic neutrons. It is the primary goal of HESP to provide, for the first time, such comprehensive observations. Additional objectives of HESP include the study of the composition of the solar atmosphere, using gamma-ray spectroscopy, and the investigation of microflares and their contribution to the heating of the corona.

The importance of HESP was recognized by the Astronomy and Astrophysics Survey Committee[1] and by the NASA Space Physics Strategy Implementation Study of the Space Physics Subcommittee for the Space Science and Applications Advisory Committee.[2] In both cases, HESP was the highest priority intermediate space mission for solar physics. The NASA Space Sciences and Applications Advisory Committee (SSAAC) placed HESP second among all new NASA intermediate, moderate, and flagship missions in their Strategic Plan 1992. A pre-phase-A study was initiated in January 1991 and an initial report was issued in July 1991.[3] Recently, studies have been initiated to accommodate the HESP mission on Lightsats: smaller, lighter, cheaper spacecraft that can be built in a shorter time. These studies have resulted in a multi-spacecraft HESP program with more schedule and programmatic flexibility. To provide significant coverage of the next solar maximum, a HESP launch by the end of the year 2000 and three years of observations are desirable.

SCIENTIFIC OBJECTIVES

SUB-RELATIVISTIC ELECTRONS

One of the most outstanding signatures of solar flares is the hard X-ray emission, commonly observed during the impulsive phase. Hard X-rays are generally attributed to bremsstrahlung produced in collisions between suprathermal electrons and the constituents of the ambient solar atmosphere. When the electron energy E_e is much larger than the kT of the ambient gas, the energy lost by the electrons to bremsstrahlung is only a small fraction ($\sim 10^{-5}$ in the deka-keV range) of the energy lost to Coulomb collisions. Thus, the fast electrons must contain many orders of magnitude more energy than the X-rays that they produce. Since observed hard X-ray spectra are typically steep power-laws produced by steep electron spectra, the total energy content of the electrons is critically dependent on the low-energy cutoff of the accelerated electron spectrum. Even for assumed cutoffs in the 20 to 30 keV range, the inferred electron energy can amount to a substantial fraction, $\sim 10 - 50\%$, of the total energy released in the flare. Thus, particle acceleration must be a significant part of the flare energy release process.

The bulk of the radiative output of flares appears at lower energies, in the visible, UV, and soft X-ray bands, with a substantial fraction of the total output radiated in soft X-rays.[4] Because so much energy is contained in the hard X-ray producing electrons, it is reasonable to assume that it is the energy deposited by these electrons into the ambient solar atmosphere which produces the observed lower-energy emissions. The thick-target model[5] assumes that the primary result of the energy release process is the impulsive acceleration of electrons, probably in the coronal part of a loop or arcade of loops. These electrons propagate along the magnetic field lines and deposit their energy predominantly in the lower corona and the chromospheric portions of the loops, heating the ambient gas. This model has been shown to be consistent with much of the observational data.[6] However, the very large number of accelerated electrons required in this model and the substantial currents they carry pose difficult theoretical challenges.

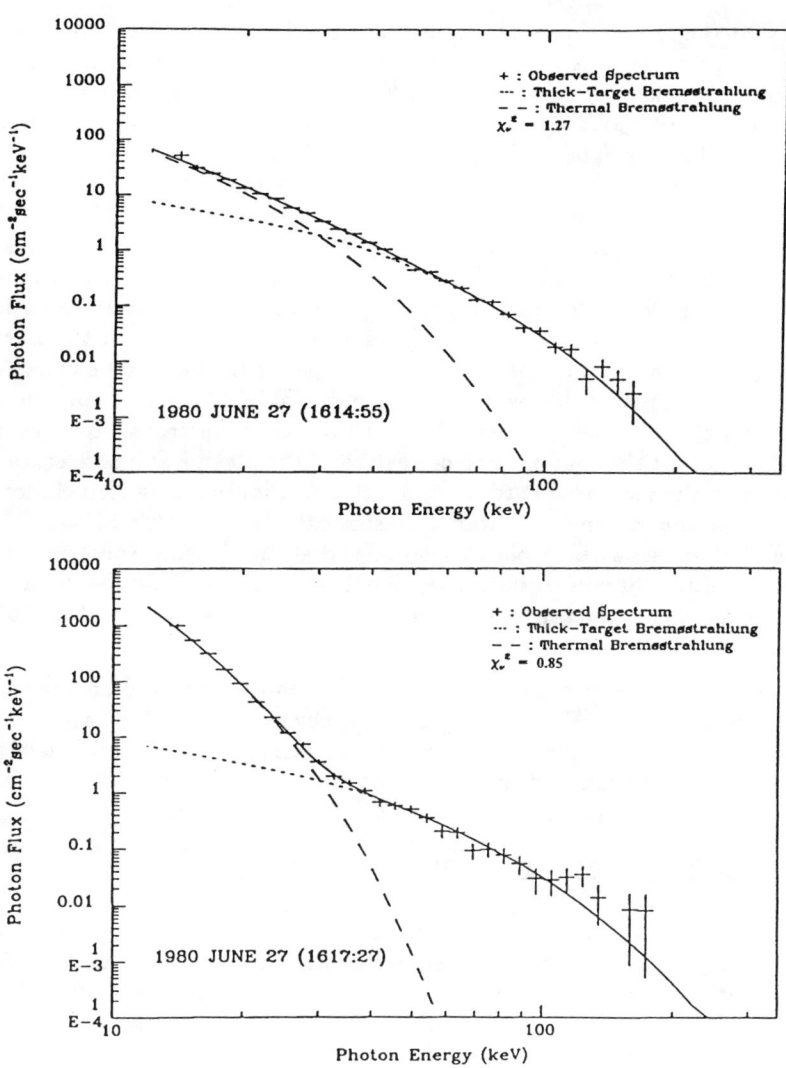

Figure 1 Measured high-resolution spectra during the hard X-ray burst observed on 27 June 1980[7,8] and the predicted spectra from a model of an electric field applied parallel to the magnetic field.[9] The electric field results in both Joule heating to produce the hot thermal plasma and runaway electron acceleration to produce the thick-target bremsstrahlung.

The fact that the impulsive energy release and particle acceleration are assumed as a starting point in these models illustrates how little is known about these fundamental processes. A clue to their nature may be contained in high spectral resolution (~ 1 keV FWHM) observations of hard X-rays (Figure 1), presently available only for a single medium-sized flare observed with a balloon-borne instrument.[7,8] These measurements show that the hard X-ray spectrum has a characteristic double power-law shape with a sharp downward break at an energy that varied during the flare from ~ 30 keV to greater than 100 keV. This sharp break may result from the electrons being accelerated by quasi-static electric fields parallel to the magnetic field; the break energy would then be a measure of the potential drop. This type of acceleration occurs in the Earth's aurora but with potential drops of ~ 10 kV compared to the ~ 100 kV required in flares. The electric fields will also produce Joule heating of the thermal electrons as well as runaway acceleration of the faster electrons. Holman and Benka[9] showed that the combination of heating of the plasma and acceleration is consistent with the measured hard X-ray spectrum (Figure 1), which includes the "superhot" thermal component from a plasma at a temperature of $\sim 30 \times 10^6$ K.[7] Note that it is not possible to resolve the steep thermal spectrum or to resolve the spectral breaks using conventional scintillation X-ray detectors but these important spectral features are clearly resolved with the spectral resolution of germanium detectors.

In spite of the success of a simple electric field model, the fundamental questions relating to energy release and particle acceleration in solar flares remain unanswered. Nevertheless, it appears at present that a hybrid model involving both particle acceleration and direct heating is required to account adequately for all of the observed complexities and subtleties of flare emission.

IONS AND RELATIVISTIC ELECTRONS

Gamma-ray and neutron emissions from solar flares are signatures of ion and relativistic electron interactions.[10] Gamma-ray bremsstrahlung continuum from primary relativistic electrons has been observed up to at least 100 MeV.[11] Gamma-ray lines are produced in nuclear reactions of accelerated protons and heavier nuclei interacting with the ambient solar atmosphere. A rich spectrum of lines, resulting from de-excitations of all the abundant constituents of the solar atmosphere up to Fe, has been observed from many flares.[12,13] Such a spectrum is shown in Figure 2a, with the predicted spectrum shown in Figure 2b.[14] The nuclear reactions also lead to the production of neutrons and positrons. Capture of thermalized neutrons by protons in the photosphere produces the very narrow deuterium deexcitation line at 2.223-MeV. Annihilation of positrons produces the 0.511-MeV line, whose shape and accompanying positronium annihilation continuum are sensitive probes of the temperature, density, and state of ionization of the ambient solar atmosphere.[15] Neutrons from flares have been detected directly,[16] as have the protons resulting from the decay of the neutrons in interplanetary space.[17] The nuclear reactions also produce pions[18,19] and these have been detected through their decay products. The neutral pions decay into

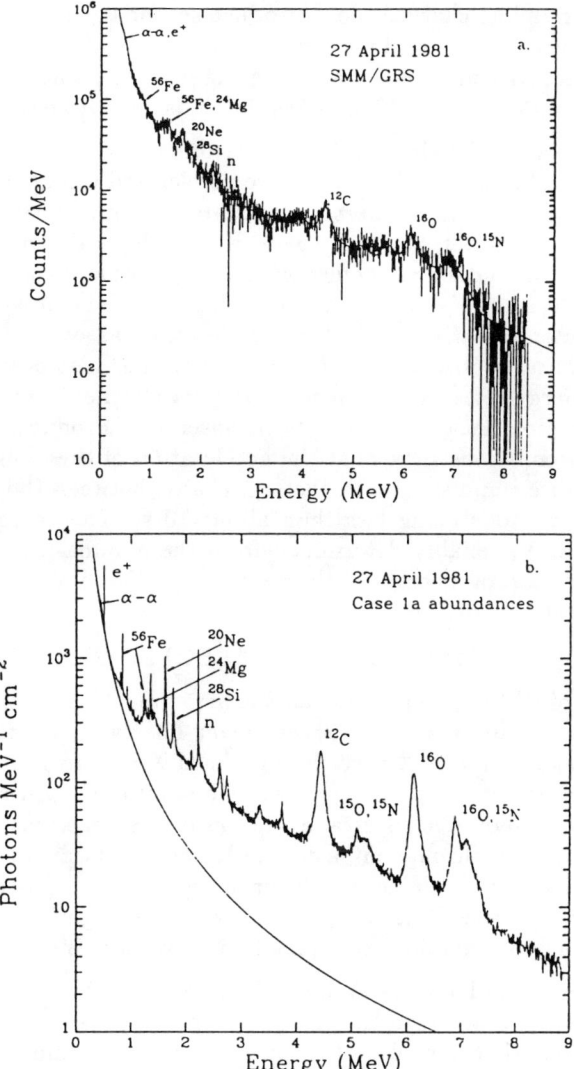

Figure 2 (a) Count spectrum of the 27 April 1981 flare measured with the NaI(Tl) Gamma Ray Spectrometer (GRS) on SMM, together with a best-fit calculated spectrum.[14] (b) The corresponding calculated incident photon spectrum showing the ions responsible for many of the lines. The smooth curve represents the bremsstrahlung continuum spectrum. Note that the widths of all the lines in the measured spectrum are attributable to the GRS instrument resolution whereas all the lines in the calculated spectrum, except the neutron-capture deuterium line at 2.223 MeV with a predicted width of <0.1 keV, can be resolved with a cooled HPGe detector.

gamma rays directly. The charged pions produce secondary positrons and electrons, which produce bremsstrahlung gamma rays. In addition, the positrons also produce high-energy photons by annihilating in flight. The resultant high-energy gamma-ray emission (> 10 MeV) has been observed from several flares.[20]

Before observations with the Gamma Ray Spectrometer (GRS) on SMM, it was thought that only subrelativistic electrons are accelerated impulsively in flares. While the acceleration of ions in flares was established much earlier by direct particle observations in interplanetary space, it was thought that this acceleration was merely a secondary phenomenon that occasionally accompanied the much more frequent impulsive acceleration of electrons. This now appears not to be the case. Ion and relativistic electron acceleration, as evidenced by impulsive gamma-ray emission observed from many flares, must also be closely linked to the primary energy-release mechanisms. Relativistic neutrons and prompt gamma rays from pion decay have also been observed. Produced in GeV ion interactions, these emissions show that the acceleration of these highest-energy particles is also quite impulsive, with the lag, if any, between the acceleration of MeV and GeV protons being less than about 10 s. The energy contained in ions above 10 MeV, reliably determined from the gamma-ray observations, amounts to at least several percent of the flare energy. Even more energy could be contained in lower-energy ions.

OBSERVATIONAL APPROACH AND EXPECTED PERFORMANCE

The primary goal of HESP is to provide, for the first time, high spatial resolution and high spectral resolution *imaging spectroscopy* observations over the entire photon energy range from soft X-rays through hard X-rays to gamma-rays. By this we mean high-resolution spectroscopy at each point of the X-ray or gamma-ray image. This will allow the spectral evolution of the emissions to be traced in both space and time throughout a flare. It represents an important new capability not previously available in this wavelength range. Furthermore, the Sun is the only astrophysical X-ray or gamma-ray source bright enough and close enough to allow such observations to be made with present instrumentation.

In order to achieve a full understanding of the acceleration of particles and ions, and their transport through the solar atmosphere, it is essential to obtain support observations that can place the hard X-ray and gamma-ray events in the context of the important lower-energy (thermal) processes in flares. Such observations include measurements that must be made from space and those that can be made from the ground. Observations that require instruments in space include high-resolution spectroscopy and/or imaging at soft X-ray, XUV, and/or UV wavelengths. Sufficiently high spectral resolution is required to study Doppler shifts, line profiles, and line ratios. Such spectroscopy will provide detailed information on the plasma in which particle acceleration and transport occur, including electron temperatures and densities; turbulence, anisotropic mass motions, and atomic excitation processes; element abundances; and departures from ionization balance and Maxwellian velocity distributions. An imaging resolution commensurate with that achieved in hard X-rays would be sufficient to locate the positions of the hard X-ray and gamma-ray sources relative to magnetic field structures delineated by trapped plasma emitting in the X-ray, XUV, or UV spectral regions. Complementary direct spacecraft measurements of the energetic solar flare particles that escape to the interplanetary medium are also highly desirable.

Crucial ground-based observations include vector magnetic field measurements, and radio and optical imaging and spectroscopy. These observations will provide context information on the lower-energy components of the flaring region, including the morphology and dynamics of the thermal plasma and the magnetic field strengths and morphology in both the photosphere and the corona. In addition, they will provide complementary information on the high-energy components of flares such as the energetic electrons and ions, shocks and other plasma phenomena, and other indications of non-thermal processes. A more complete description is given in reference 21.

The primary observational objectives of HESP, then, are to obtain the following:

- Hard X-ray images with an angular resolution of as fine as 2 arcseconds and a temporal resolution of tens of milliseconds, commensurate with the known size scales of the flaring magnetic structures and the travel and stopping distances and times of the accelerated electrons. The images will be obtained with sufficient sensitivity to detect and locate the initial flare energy release and to study microflares.

- High-resolution X-ray spectra with ~1-keV resolution down to energies as low as 2 keV. Since the bremsstrahlung cross-section is well known, the hard X-ray spectrum can provide detailed quantitative information on the distribution of X-ray producing electrons at the Sun. Recently, a numerical inversion technique has been developed which derives the parent electron spectrum from the bremsstrahlung X-ray spectrum for optically thin sources.[22] The measurement of the precise shape of the X-ray continuum made possible with such fine energy resolution thus will provide unique, detailed information on the spectrum of the accelerated electrons and of the heated plasma, thus allowing the thermal and non-thermal aspects of individual flares to be clearly distinguished.

- Spectrally resolved hard X-ray images. Apart from the separate scientific objectives of the imaging and spectroscopy alone, the combination of the two with sub-second time resolution will allow spectral changes to be measured as a function of space and time as the electrons propagate along the magnetic field in the flaring loop or loops, providing for the first time powerful constraints on the mechanisms of energy gain and loss.

- Gamma-ray images with an angular resolution as fine as 4 arcseconds. The high-resolution gamma-ray imaging spectroscopy will allow images to be obtained in specific gamma-ray lines or energy ranges such that, for example, the proton- and alpha-induced lines could be imaged separately, as could the 511-keV positron annihilation line and the 4-to-7-MeV photons, which are primarily from accelerated heavy nuclei. The intercomparison of these images from different types of particles and their comparison with the images in the electron-produced X-rays will allow the effects of differences in charge and mass on the acceleration and propagation processes to be explored for the first time.

- High-resolution gamma-ray spectra with a few keV resolution to energies as high as 20 MeV. This resolution is sufficient to resolve the gamma-ray lines and to measure their shapes, thus allowing the full potential of gamma-ray line spectroscopy to be realized for the first time. Such high-resolution spectra would provide unique information on the directionality of the interacting particles, the composition of both the ambient gas and the accelerated ions, and the temperature, density, and state of ionization of the ambient gas.

Each of the above observational capabilities are new and unique; no previous solar mission has provided such capabilities. Other observational objectives, which are highly desirable but secondary to those above, are to obtain the following:

- High-energy gamma-ray and neutron spectra and images for large flares at energies from 20 MeV to $\gtrsim 1$ GeV. These measurements should provide information on the acceleration of ions to the highest energies.
- Soft X-ray, XUV, and/or UV spectra or images with spectroscopic resolution of about 7000 (corresponding to a velocity of 40 km/s) and/or spatial resolution as fine as 2 arcseconds, both with a temporal resolution of a few seconds. These observations will provide detailed context information on the flaring region in which the high-energy (nonthermal) processes take place.

It is important to realize that HESP will provide hard X-ray imaging spectroscopy, not just for a few flares, but for many thousands of flares in its $\geq$ 3-year lifetime. The smallest bursts detectable by the SMM HXRBS instrument would give $\sim 10^3$ counts s^{-1} above 20 keV in HESP. In those events, rapid spatial changes in the X-ray sources could be followed on time scales of 0.1 s. Imaging spectroscopy, i.e., obtaining the spectrum as a function of spatial location, could be done with images in each of ten energy intervals with $\sim$2-s resolution. For the once-per-day or larger flare, (i.e., 10^4 HESP counts s^{-1} above 20 keV), imaging information could be obtained in tens of ms and imaging spectroscopy every $\sim$100 ms.

HESP will provide, for the first time, images and high-resolution spectroscopy of gamma-ray lines and continuum for solar flares. Assuming that every flare accelerates ions with fluxes proportional to the hard X-ray flux, the 2.223-MeV line should be detected, on average, once every $\sim$2 to 3 days, the positron annihilation line every $\sim$8 days, and prompt nuclear lines every $\sim$20 days.

The narrow lines (Figure 2) are produced by protons or alpha particles colliding with the solar atmosphere. The primary background is the continuum emission from the flare itself. This consists of two components – bremsstrahlung emission produced by relativistic electrons, and broad lines produced by the inverse process of accelerated heavy ions colliding with hydrogen and helium in the solar atmosphere. The line-to-continuum ratio is ∼3 to 10 for the narrow, prompt, inelastic-scatter lines in this flare, but these ratios appear to vary from flare to flare. The line widths are typically 5 to 10 times narrower than the energy resolution of scintillation detectors. The high spectral resolution of Ge detectors, thus, is essential to obtain unambiguous images of the accelerated ions.

HESP STRAWMAN INSTRUMENTS

The original HESP strawman instrument payload consists of a primary high energy instrument, HEISPEC, and two context instruments to provide thermal plasma diagnostic measurements. In addition, an extensive ground-based observational program (described in reference 21) and supporting theory program are integral parts of the HESP mission.

The High Energy Imaging Spectrometer (HEISPEC), the main instrument of the HESP payload (Figure 3), is designed to image the entire range of flare energetic photons from soft X-rays (∼2 keV), to π^0-decay gamma-rays (∼200 MeV), as well as energetic neutrons. Furthermore, HEISPEC has the capability to perform spatially resolved spectroscopy with high spectral resolution, to allow the full diagnostic power of hard X-rays and gamma-rays to be applied on a spatial point-by-point basis within solar flare.

The imaging is based on a Fourier transform technique using a set of rotating modulation collimators (RMCs), each of which is similar to those used on previous missions such as the US SAS-C and the Japanese Hinotori spacecraft. Each RMC consists of two widely-spaced, fine-scale linear grids, which temporally modulate the photon signal from sources in the field of view as the RMC rotates about its long axis. The modulation can be measured with a detector having no spatial resolution placed behind the RMC. The modulation pattern over half a spin for a single RMC provides the amplitude and phase of many spatial Fourier components over a full range of angular orientations but for a small range of spatial source dimensions. Multiple RMCs, each with a different slit width, can provide coverage over a full spectrum of flare source sizes. An image is constructed from the set of measured Fourier components in exact mathematical analogy to multi-baseline radio interferometry.

The grid diameters and thicknesses are chosen to give full-Sun fields of view. Thus, upper and lower grids have diameters of 12.5 cm and 7.5 cm, respectively, and grid thicknesses range from ∼2.8 mm for the finest (100-micron pitch) grids up to a maximum chosen thickness of 4 cm. With ∼5-m separation between the grids, HESP will provide spatial resolution of ∼2 arcseconds at hard X-ray energies (below ∼400 keV), 4-8 arcseconds for gamma-ray lines and continuum above ∼0.5 MeV, and ∼40 arc seconds for neutrons. The chosen rotation rate of ∼15 rpm provides a complete image with the maximum number of Fourier components ($\sim 5 \times 10^3$) in 2 s, but spatial information is still available on timescales down to tens of ms, provided the count rates are sufficiently high. This high time resolution capability is important in following the propagation of electrons along typical magnetic loops.

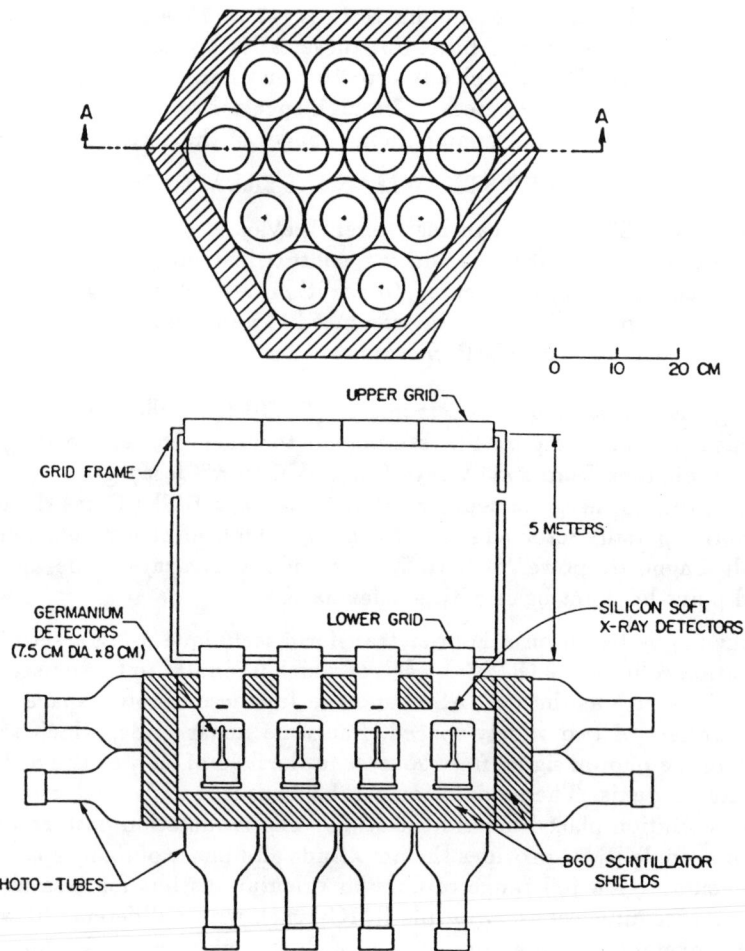

Figure 3 Schematic cross sections of the High Energy Imaging Spectrometer (HEISPEC). The upper and lower tungsten grids, separated by 5 m, form the rotating modulation collimators (RMCs); the two-segment germanium detectors provide high spectral resolution measurements from ~10 keV to 20 MeV; the combination of the Ge detectors and the bismuth germanate (BGO) shield extends the gamma-ray range to > 200 MeV and provides neutron coverage from ~20 MeV to $\gtrsim$ 1 GeV; the silicon detectors cover the energy range from ~2 keV to $\gtrsim$ 20 keV; the BGO forms an active anti-coincidence shield and collimator to reduce the background and protect the Ge detectors from radiation damage in the polar orbit.

This high time resolution capability is important in following the propagation of electrons along typical magnetic loops.

Behind the RMCs, dual-segment coaxial germanium detectors provide high spectral resolution from ~10 keV to ~20 MeV. The front ~2-cm thick segment measures hard X-rays up to ~200 keV, while the rear 8-cm thick segment provides undistorted high-resolution gamma-ray line measurements in the presence of very intense hard X-ray fluxes in large flares. The spectral resolution of Ge detectors is sufficient to resolve all of the solar gamma-ray lines with the exception of the neutron-capture deuterium line, which has an expected FWHM of about 0.1 keV. High spectral resolution is also required to resolve sharp breaks in the non-thermal continuum and the steep super-hot thermal component of solar flares.

The bismuth germanate (BGO) scintillator acts as an active anticoincidence shield and collimator to reduce the background, provide excellent Compton rejection, and extend the energy range to $\gtrsim$ 200 MeV for moderate spectral and spatial resolution gamma-ray imaging. The combination of BGO and Ge detectors also provides coarse spectral and spatial resolution imaging of energetic neutrons up to ~1 GeV. A third important role of the BGO is to protect the Ge detectors from radiation damage in the polar orbit.

Silicon semiconductor detectors (or alternatively, proportional counters) placed in front of the germanium detectors serve to extend the imaging spectroscopy down to ~2 keV to cover the transition from non-thermal to thermal emission and to relate the high energy measurements to the thermal soft X-ray flare.

THE HESP MISSION

The initial HESP studies[3,21,23,24] resulted in a single 4800-lb spacecraft accommodating the HEISPEC and context instruments, to be launched by a Delta 7920 rocket into a 600-km Sun-synchronous, 98-degree orbit. In order to accommodate HESP on Lightsats, the HEISPEC instrument's weight was reduced from 800 kg to $\lesssim$ 400 kg by removing the BGO shield and associated electronics. This modified HEISPEC, together with a Bragg crystal spectrometer, is accommodated on a spinning lightsat spacecraft that could be launched on a Taurus-class rocket. With the BGO removed, it was necessary to choose an equatorial orbit to minimize the background and the radiation damage to the Ge detectors. Furthermore, to fit inside a Taurus fairing, the telescope length must be reduced from 5 m to 2.5 m. Thus, grids with a pitch as fine as 50 microns must be produced in order to maintain the 2-arcsecond angular resolution.

A second Taurus-class spacecraft in equatorial orbit carries a compact, non-imaging germanium spectrometer with active BGO shield for high resolution gamma-ray line spectroscopy and for moderate resolution gamma-ray and neutron measurements up to ~1 GeV. A third, Pegasus-class (~500 lb), non-spinning spacecraft carries two context instruments in a Sun-synchronous polar orbit. This three-spacecraft HESP mission provides comparable imaging spectroscopy up to 20 MeV, better context measurements, and improved gamma-ray/neutron spectroscopy, compared to the original HESP, but it does not provide $\gtrsim$ 20 MeV imaging of gamma-rays and neutrons.

Recently, we have shown that a basic HEISPEC instrument could be launched on a Pegasus-class rocket provided that the telescope is shortened still further to $\sim$1.8 m. New techniques are now being developed for making grids with even finer pitch than previously thought possible and these hold promise that the 2-arcsecond resolution could still be attained with such a short telescope. However, the need to reduce weight and moment of inertia would require that all grids be thinner than $\sim$1 cm, thus limiting the imaging capability to less than a few hundred keV. Full high-resolution spectroscopy with 12 cooled Ge detectors would still be possible to 20 MeV.

Based on the flare hard X-ray burst rates measured by the SMM HXRBS in the last maximum, and current best estimates for the period of the solar activity cycle, the average flare rate should rise significantly at the beginning of 1999 and decrease substantially by the middle of 2004, with an uncertainty of one year. Thus, to be reasonably certain that the three-year mission life lies within the active portion of the next solar cycle, a HESP launch by the end of 2000 is desired.

Acknowledgments

We are grateful for the efforts of the other members of the HESP Science Study group, and the support of NASA Headquarter's Solar Physics discipline office. J. Phenix in the Advanced Missions Analysis Office at Goddard Space Flight Center is providing much of the technical support for the spacecraft and instrument studies.

REFERENCES

1. Bahcall, J., et al., The Decade of Discovery in Astronomy and Astrophysics, National Research Council, 1991.
2. Space Physics Strategy - Implementation Study, The NASA Space Physics Program for 1995 to 2010, report to the Space Physics Subcommittee of the Space Science and Applications Advisory Committee, 1991.
3. Lin, R. P., et al., The High Energy Solar Physics Mission (HESP), Scientific Objectives and Technical Description, NASA, 1991.
4. Canfield, R. C., in Solar Flares, ed. P.A. Sturrock, Colorado Assoc. Univ. Press, Boulder, 1980, p. 451.
5. Brown, J. C., Solar Phys., 18, 489, 1971.
6. Emslie, A. G., Solar Phys., 86, 133, 1986.
7. Lin, R. P., R. A. Schwartz, R. M. Pelling, and K. C. Hurley, Astrophys. J., 251, L109, 1981.

8. Lin R. P., and R. A. Schwartz, Astrophys. J., 312, 462, 1987.
9. Holman G. P., and S.G. Benka, Astrophys. J. Lett., 400, L79, 1992.
10. Ramaty, R., and R. J. Murphy, Space Sci. Rev., 45, 213, 1987.
11. Forrest, D. J., 19th Internat. Cosmic Ray Conf. Papers, 4, 146, 1985.
12. Chupp, E. L., in Gamma Ray Transients and Related Astrophysical Phenomena, eds. R.E. Lingenfelter, H.S. Hudson, and D.M. Worrall, New York Assoc. Univ. Press, 1982, p. 363.
13. Murphy, R. J., G. H. Share, J. R. Letaw, and D. J. Forrest, Astrophys. J., 358, 290, 1990.
14. Murphy, R. J., R. Ramaty, B. Kozlovsky, and D. V. Reames, Astrophys. J., 371, 793, 1991.
15. Crannell, C. J., G. Joyce, R. Ramaty, and C. Werntz, Astrophys. J., 210, 52, 1976.
16. Chupp, E. L., Astrophys. J., 318, 913, 1987.
17. Evenson, P., P. Meyer, and K. R. Pyle, Astrophys. J., 274, 875, 1983.
18. Murphy, R. J., C. D. Dermer, and R. Ramaty, Astrophys. J. Suppl., 63, 721, 1987.
19. Mandzhavidze, N. Z., Thesis, Physico-Technical Inst., Leningrad, USSR, 1987.
20. Rieger, E., Solar Phys., 121, 323, 1989.
21. Lin, R. P., B. R. Dennis, R. Ramaty, A. G. Emslie, R. Canfield, G. Doschek, The NASA High Energy Solar Physics (HESP) Mission for the Next Solar Maximum, in proc. of Yosemite meeting on Solar System Plasma Physics, Resolution of Processes in Space and Time, Feb. 1993, Geophysical Monograph Series, 1993.
22. Johns C. M., and R. P. Lin, Solar Phys., 137, 121, 1992.
23. "Pre-Phase-A Study Report on the High Energy Solar Physics Mission (HESP) - Spacecraft Study Report" prepared by the HESP Spacecraft Study Team (J. Phenix, Study Manager), GSFC, 1991.
24. "Preliminary Mission Study for NASA Space Physics Division (Code SS) High Energy Solar Physics (HESP) Addendum" by the Ball Aerospace Systems Group, 1991.

AUTHOR INDEX

A

Akimov, V. V., 106, 130

B

Belov, A. V., 106
Bennett, E., 118
Bennett, K., 51, 55, 89, 100
Bertsch, D. L., 94, 177
Biesecker, D. A., 183
Bloeman, H., 51
Boyer, R., 124
Brazier, K. T. S., 94, 177

C

Canfield, R., 59, 230
Chertok, I. M., 106, 130
Chuikin, E. I., 45
Chupp, E. L., 112
Cliver, E. W., 65
Correia, E., 193
Costa, J. E. R., 193
Crosby, N. B., 65

D

Debrunner, H., 89, 207
Dennis, B. R., 65, 230
Diehl, R., 51, 55, 100
Dingus, B. L., 94, 177
Doschek, G., 230
Dunphy, P. P., 112

E

Emslie, A. G., 230

F

Fichtel, C. E., 94, 177
Fishman, G. J., 183
Forrest, D. J., 15, 51, 55, 89, 100, 143

G

Gentile, L. C., 222
Grabelsky, D. A., 15
Grove, J. E., 15

H

Hanlon, L., 51, 55, 89
Hartman, R. C., 94, 177
Heristchi, D., 124
Hermsen, W., 51
Herrmann, R., 193
Hudson, H. S., 151
Hunter, S. D., 94, 177

J

Jensen, C. M., 15
Johnson, W. N., 15
Jung, G. V., 15

K

Kanbach, G., 94, 171, 177
Kaufman, P., 193
Kinzer, R. L., 15
Klein, K.-L., 187
Kniffen, D. A., 94, 177
Kocharov, G. E., 45
Kocharov, L. G., 45
Kovaltsov, G. A., 45
Kroeger, R. A., 15
Kurfess, J. D., 15
Kurt, V. G., 106, 130

L

Lau, Y.-T., 71
Lee, M. A., 118, 134
Leikov, N. G., 106, 130
Lichti, G., 51, 55, 100
Lin, R. P., 230
Lin, Y. C., 94, 177
Lingenfelter, R. E., 77
Lockwood, J., 89
Loomis, M., 89, 100

M

Macri, J., 100
Magun, A., 106, 193
Mandzhavidze, N., 26
Marschhäuser, H., 171
Mattox, J. R., 94, 177
Matz, S. M., 15
Mayer-Hasselwander, H. A., 94, 177
McConnell, M., 21, 51, 55, 89, 100
Melnikov, V. F., 106
Metcalf, T., 59
Michelson, P. F., 94, 177
Mickey, D., 59
Morris, D., 89
Murphy, R. J., 15

N

Nolan, P. L., 94, 177

P

Petrosian, V., 162
Purcell, W. R., 15

R

Ramaty, R., 26, 71, 230

Rank, G., 51, 55, 89, 100
Rieger, E., 171
Ryan, J., 51, 55, 89, 100, 118, 183

S

Schneid, E. J., 94, 177
Schönfelder, V., 51, 55, 89, 100
Share, G. H., 15
Shea, M. A., 222
Smart, D. F., 222
Sreekumar, P., 94, 177
Strickman, M. S., 15
Strong, A., 51
Suleiman, R., 51, 55
Swanenburg, B. N., 51, 55, 89, 100

T

Thompson, D. J., 94, 177
Trottet, G., 3, 187

U

Ulmer, M. P., 15
Usoskin, I. G., 45

V

van Dijk, R., 100
Varendorff, M., 51, 55, 100
Vestrand, W. T., 15, 143
von Montigny, C., 94, 177

W

Webber, W., 89
White, S. M., 199
Winkler, C., 51, 55, 89, 100
Wülser, J.-P., 59